# Advances in
# Mercury
# Toxicology

# ROCHESTER SERIES ON ENVIRONMENTAL TOXICITY

**Series Editors:** Thomas W. Clarkson and Morton W. Miller

A Continuation Order Plan is available for this series. A continuation order will bring delivery of each new volume immediately upon publication. Volumes are billed only upon actual shipment. For further information please contact the publisher.

# Advances in Mercury Toxicology

**Edited by**

**Tsuguyoshi Suzuki**
*University of Tokyo*
*Tokyo, Japan*

**Nobumasa Imura**
*Kitasoto University*
*Tokyo, Japan*

**and**

**Thomas W. Clarkson**
*University of Rochester School of Medicine*
*Rochester, New York*

**Plenum Press** • **New York and London**

Library of Congress Cataloging-in-Publication Data

---

Advances in mercury toxicology / edited by Tsuguyoshi Suzuki, Nobumasa
  Imura, and Thomas W. Clarkson.
       p.   cm. -- (Rochester series on environmental toxicity)
     "Proceedings of a Rochester International Conference in
  Environmental Toxicity, held August 1-3, 1990 in Tokyo, Japan"--T.p.
  verso.
     Includes bibliographical references and index.
     ISBN 0-306-44116-0
     1. Mercury--Toxicology--Congresses.   I. Suzuki, Tsuguyoshi, 1932-
  .  II. Imura, Nobumasa, 1935-   .  III. Clarkson, Thomas W.
  IV. Rochester International Conference on Environmental Toxicity
  (19th . 1990 : Tokyo, Japan)  V. Series.
     [DNLM: 1. Environmental Pollutants--adverse effects--congresses.
  2. Mercury Poisoning--congresses.   QV 293 A244]
  RA1231.M5A34  1991
  615.9'25--dc20
  DNLM/DLC
  for Library of Congress                                    91-45506
                                                                 CIP

Proceedings of a Rochester International Conference in Environmental Toxicity,
held August 1-3, 1990, in Tokyo, Japan

ISBN 0-306-44116-0

© 1991 Plenum Press, New York
A Division of Plenum Publishing Corporation
233 Spring Street, New York, N.Y. 10013

Printed in the United States of America

## Conference  Coordinators

### Seiichiro Himeno
Kitasato University
Tokyo, Japan

and

### Michael A. Terry
University of Rochester School of Medicine
Rochester, New York

## Sponsored  by

Japan Society for the Promotion of Science
Inoue Foundation for Science
University of Tokyo
National Science Foundation (INT - 8915129)
ICOH Subcommittee of Toxicology of Metals

We gratefully acknowledge additional support from the following contributors:

Banyu Pharmaceutical Company, Ltd.
Chugai Pharmaceutical Company, Ltd.
Eisai Company, Ltd.
Japan Immunoresearch Laboratories
Japan Industrial Safety and Health Association
Kanebo Ltd.
Kato Shinobu
Meiji Seika Kaisha, Ltd.
Mitsui Pharmaceuticals Inc.
Nihon Millipore Ltd.
Nippon Instruments Corp.
Nippon Kayaku Company, Ltd.
Nippon Roussel K.K.
Nitto Boseki Company, Ltd.
Sankyo Company, Ltd.
Sansou Seisakusho K.K.
Sapporo Breweries Ltd.
Schering-Plough K.K.
Sugiyama-Gen Iriki Company, Ltd.
Taiho Pharmaceutical Company, Ltd.
Terumo Corporation
Varian Instruments Ltd.
Yamamura Yukio
Yamanouchi Pharmaceutical Company, Ltd.

# DEDICATION

Tsuguyoshi Suzuki was born in Tokyo on 26 February, 1932. He graduated from the Faculty of Medicine at the University of Tokyo in 1955. He completed his doctor's course in the Division of Medical Sciences in the Graduate School of the University of Tokyo in 1960.

His professional career started with his appointment as Instructor in the Department of Public Health, Faculty of Medicine, University of Tokyo in 1960. Subsequently, in the same department, he rose to the position of Assistant Professor in 1966 and to Associate Professor in 1968. In 1971, he assumed the post of Professor of the Department of Public Health in the School of Medicine in the University of Tohoku, a position which he held until 1979 when he returned to the University of Tokyo as Professor of Human Ecology in the Faculty of Medicine. In addition to holding this professorship, Dr. Suzuki served as director of the Graduate Course in the School of Health Sciences, Division of the Medical Science from 1984 to 1989 and became Director of the Undergraduate Course in the same School from 1989 to 1991.

He has received numerous honors and acknowledgements of his scientific achievements and professional stature. These include election to the vice presidency of the Asia-Pacific Academic Consortium for Public Health from 1988 to 1989, his appointment to editorship of scientific journals including the Asia-Pacific Journal of Public Health, Nutrition Research and the Journal of Applied Toxicology. He has served on many prestigious national and international committees as well as acting as an expert advisor to the World Health Organization on public health issues.

His outstanding scientific reputation in public health and toxicology is built upon over 200 publications in peer reviewed scientific journals. He introduced the concept of human ecology into understanding how human populations respond to toxic agents in their environment. His article in this book describes his latest study comparing an industrialized to a non-industrialized population's response to the toxic stress from mercury. Throughout his career interest in human populations, he established a laboratory second to none in the application of analytical techniques to human biological monitoring. Biological monitoring focussed mainly on two elements, mercury and selenium as these elements played a central role in his studies in human ecology. In the case of mercury and its compounds, he pioneered human biological monitoring with applications to a wide variety of media including historical specimens, nails, hair, milk, placenta and autopsy tissues.

His human studies were complemented by highly productive experimental studies. A great deal of what we know today on the disposition of mercury vapor and methylmercury and their interaction with selenium is due to his experimental work. This unique combination of human and laboratory based research has made him a preeminent international authority on the toxicology of mercury.

Besides his great contribution to scientific progress, Dr. Suzuki has devoted himself to education of graduate and undergraduate students throughout his professional career. The unanimous opinion of the students whom he supervised is that Professor Suzuki is a severe but always encouraging teacher. His ingenuity and excellence as a teacher have been clearly proven by the fact that many of his students are now leading scientists not only in human ecology and toxicology, but also in other fields of environmental health sciences.

It was therefore appropriate that this conference on Advances in Mercury Toxicology was held at Professor Suzuki's home University of Tokyo. The conference and this book is an acknowledgement from the participants, authors and co-editors of his scientific accomplishments and his outstanding international scientific reputation.

# PREFACE

This book is based on an international meeting organized by the University of Tokyo and the University of Rochester, and is published as one belonging to the series of Rochester International Conferences in Environmental Toxicity. The meeting on "Advances in Mercury Toxicology" was held at the University of Tokyo on August 1 to 3, 1990. The invited papers are published in this book along with an "Overview" chapter that was written by the editors at a meeting held at the University of Rochester on August 1 to 2, 1991.

The purpose of the meeting was to assemble leading scientists to discuss their most recent findings on the toxicology of mercury. The time was opportune. Considerable progress has been made on the environmental fate and toxicology of mercury. Recent findings have given new insight into the global model for mercury. Transport in the atmosphere extends great distances resulting in pollution of lakes and rivers far distant from the source of mercury release. The process of methylation leads to accumulation of methylmercury in fish and thus in the human diet. New evidence indicates that acid rain and the impoundment of water for hydroelectric purposes affects the methylation and bioaccumulation processes resulting in higher levels of methylmercury in fish.

Studies on the disposition of mercury in the body are now moving from the descriptive phase to one of understanding the underlying mechanisms of transport and biotransformation. New hypotheses have been proposed on mechanisms of toxicity both with regard to the central nervous system and the kidney. Understanding may be at hand why the prenatal stage of the developing brain is more susceptible to methylmercury than the mature brain. Soon we may be in a position to identify the target molecules, at least for some forms of mercury poisoning.

Mechanisms of tolerance to mercury are also of considerable current interest. Included in this area is a consideration of the human ecology of mercury.

Thus the meeting offered an opportunity to evaluate these new developments and to open up new avenues of research. This book reflects not only the scientific material presented at the meeting but also, through the overview section, a general evaluation of the state of knowledge of mercury toxicology at time of increasing interest in this metal.

The editors wish to acknowledge the assistance to the planning and organization of this conference willingly and enthusiastically given by the staff and students of the Department of Human Ecology, Faculty of Medicine, University of Tokyo, and the Department of Public Health, School of Pharmaceutical Sciences, Kitasato University. We also thank Muriel

Klein, Joyce Morgan and Kathleen Amico for their skilled and tireless help in the preparation of this book.

T. Suzuki
N. Imura
T. Clarkson

# CONTENTS

## SESSION 3:   BIOTRANSFORMATION

Contents

## SESSION 4: MOLECULAR MECHANISMS OF TOXICITY

## SESSION 5: SELENIUM AS A MODIFYING FACTOR OF MERCURY TOXICITY

## SESSION 6: TOXIC EFFECTS OF MERCURY

**Contents**

## SESSION 7: CLINICAL AND EPIDEMIOLOGICAL ASPECTS

Contents

## SPECIAL LECTURE AND CLOSING REMARKS

## APPENDIX I.

## APPENDIX II.

# OVERVIEW

Tsuguyoshi Suzuki[1,] Nobumasa Imura[2], Thomas W. Clarkson[3]

[1]Department of Human Ecology, Faculty of Medicine,
University of Tokyo, Tokyo, Japan

[2]Department of Public Health, School of Pharmaceutical Sciences,
Kitasato University, Tokyo, Japan

[3]Environmental Health Science Center,
University of Rochester School of Medicine,
Rochester, NY, USA

This chapter is an overview of the papers presented at the Tokyo Conference in August, 1990. No references will be quoted as this text is based on the papers presented in this book. At times the names of individual investigators will be mentioned and appropriate references will be found in the symposium papers. This overview will follow the same sequence of topics as in the agenda of the Tokyo Conference. The fate of mercury in the environment will be followed by a discussion of the disposition of mercury in the body. This, in turn, will be followed by a closely related topic - the biotransformation reactions of mercury in mammalian tissues. The molecular mechanisms of action will then be considered as a prelude to the remainder of the book devoted to the toxic actions of mercury and how these may be modified by selenium. We conclude with a summary of the clinical and epidemiological aspects of mercury toxicity.

*Advances in Mercury Toxicology*, Edited by T. Suzuki *et al.*
Plenum Press, New York, 1991

Mercury exists in a number of physical and chemical forms in the environment and as a result of chemical syntheses for a variety of medical, agricultural, industrial and other purposes. Mercury has three stable oxidation states (Figure 1). In the ground or zero oxidation state, mercury exists as the metallic element. It is the only element that exists as a liquid at room temperatures and, in this form, was named 'quicksilver" by Aristotle over two thousand years ago. The loss of one electron gives the mercurous ion which actually consists of two atoms of mercury an ionic species that has two positive charges. This oxidation state is most commonly found as calomel or mercurous chloride, $Hg_2Cl_2$. It is still used in electrolytic reference cells. In previous centuries it was widely used in medicine. Even as recently as the 1950's calomel was a constituent of children's teething powders!

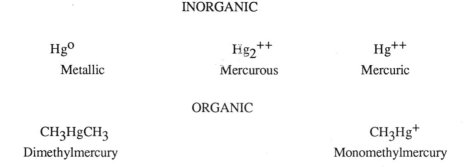

INORGANIC

| $Hg^o$ | $Hg_2^{++}$ | $Hg^{++}$ |
|---|---|---|
| Metallic | Mercurous | Mercuric |

ORGANIC

| $CH_3HgCH_3$ | $CH_3Hg^+$ |
|---|---|
| Dimethylmercury | Monomethylmercury |

Fig. 1  The major physical and chemical forms of mercury

The loss of two electrons results in the formation of the mercuric ion. In this oxidation state, mercuric mercury forms a large number of stable chemical compounds. Perhaps the best known of these is mercuric chloride, $HgCl_2$, sometimes called "corrosive sublimate". In the past it was commonly used as a disinfectant and is well known for its acute toxicity to humans.

The mercuric ion can form a number of organomercurial compounds in which the mercury atom is covalently linked to at least one carbon atom. Of special interest are the methyl organomercurials (Figure 1). Dimethylmercury is an uncharged, lipophilic compound that is highly volatile. Monomethyl mercury is a cation that forms a variety of compounds. The chloride salt, the form in which methylmercury is commonly available from chemical suppliers, is lipid soluble and has given rise to the erroneous impression that all forms of methylmercury are lipid soluble. This is not the case as many water soluble compounds are formed in nature.

Overall, the cations of mercury, for example inorganic mercuric mercury ($Hg^{++}$) and monomethylmercury ($CH_3Hg^+$), readily form stable complexes and chelate with organic ligands. However, the mercury cations have by far the highest chemical affinities for the sulfhydryl anions and for selenium in the selenide oxidation state, $Se^{2-}$. The bonds formed with the sulfhydryl ligand are rapidly reversible so that mercury cations will readily and rapidly transfer from one sulfhydryl group to another as depicted in the reaction involving the exchange of monomethylmercury between the sulfhydryl groups of albumin (Alb-SH) and cysteine (Cy-SH):

$$Al\text{-}S\text{-}Hg.CH_3 + Cy\text{-}SH = Alb\text{-}SH + Cy\text{-}S\text{-}Hg.CH_3$$

A number of physico-chemical properties of mercury play important roles in the environmental fate and toxicity of this metal: these include the high vapor pressure of $Hg^0$, the ease of conversion between $Hg^0$ and $Hg^{++}$, methylation of $Hg^{++}$ and demethylation of $CH_3Hg^+$ and the high chemical affinities of both $Hg^{++}$ and $CH_3Hg^+$ for sulfhydryl and selenide compounds. The various chemical and physical forms of mercury present a special challenge to the analytical chemist to speciate and measure each individual form as it exists at the extremely low concentrations in the environment (see the chapter by Akagi and Nishimura).

## Fate in the Environment

Mercury undergoes a global distribution (see Figure 1 in the chapter by Hecky et al.). It leaves the earth's surface, both terrestrial and aquatic, in the form of mercury vapor, $Hg^0$. Certain volcanoes are important sources of the natural release of mercury vapor to the atmosphere. Anthropogenic sources are also important and, indeed, may contribute more than half of the total emissions. The major "human" source is the burning of fossil fuels, especially coal. Other sources include metal smelter industries, cement manufacture and crematoriums. An increasingly important source is the incineration of municipal waste.

Mercury vapor has a residence time in the atmosphere of at least several months and probably more than one year (see chapter by Hecky et al.). It therefore travels over global distances. The northern hemisphere has average levels (about 4 $ng/m^3$) approximately double those of the southern hemisphere. Both the greater land area and the greater industrial activity in the North may account for this. Over 90% of the mercury in the atmosphere is in the form of gaseous mercury vapor. Less than 1% is in the particulate form but this form may be important in determining the amount of dry deposition to the earth's surface.

Mercury returns to the earth's surface in rain water in the form of $Hg^{++}$ and of $CH_3Hg^+$ compounds. The origin of the former is believed to be the oxidation of mercury vapor in the atmosphere whereas the origin of the methylmercury compounds probably comes from the atmospheric decomposition of dimethylmercury that has degassed from ocean and fresh water surfaces. Dry deposition of mercury compounds probably also occurs but its extent is not accurately known.

Once mercury has entered bodies of fresh or open water, it undergoes a variety of biotic and abiotic reactions. $Hg^{++}$ is reduced to $Hg^o$ which can volatilize back to the atmosphere. $Hg^{++}$ also undergoes methylation to both mono and dimethylmercury by certain microorganisms present in water and in sediments. Other classes of microorganisms are able to demethylate. The dimethyl form probably degasses to the atmosphere but the monomethyl form is avidly accumulated by aquatic organisms and is concentrated as it ascends the food chain. It attains its highest tissue concentrations in predatory fish at the top of the food chain. Most of the monomethyl mercury is located in the fish muscle, bound to protein and, in this form, enters the human diet. The bioconcentration factor from water to fish muscle may exceed values of one million fold.

In recent years, fish in lakes in areas remote from any known anthropogenic source have been found to have mercury levels in ranges similar to fish in polluted waters. It has been proposed (see Hecky et al.) that long distance transport from sources where large amounts of fossil fuels are burnt may account for these high values. Indeed, Hecky et al. describe an example where in one lake all the mercury in the fish can be accounted for by atmospheric deposition.

Another factor, at least for certain areas, is that remote lakes become acidified from acid compounds transported over long distances from power stations. Acidification results in higher levels of methylmercury in fish. Hecky et al. describe experiments that indicate that methylation is enhanced over demethylation reactions.

The impoundment of large bodies of freshwater for hydroelectric purposes also results in greater accumulation of monomethylmercury in fish tissues. Elevated mercury levels persist in predatory fish for at least 13 years after impoundment. Hecky et al. present evidence that this is due to the organic material that is flooded or otherwise entrained into the impounded water. This organic material serves as a substrate for the methylating bacteria.

The low levels of mercury in the atmosphere and in water require the development of ultra sensitive analytical methods that are also highly specific for the individual species of mercury. The chapter by Akagi and Nishimura describes such a method. They were able to measure methylmercury down to levels as low as 0.1 ng/l of seawater. They were able to show differences in levels of both total and methylmercury in samples from polluted and unpolluted ocean water in offshore areas near Japan. Methylmercury, on the average, accounted for between 7 to 16% of the total mercury.

The improvement in analytical techniques and in methods of collecting samples to avoid contamination has led to major revisions of the global cycle. For example, as recently as the 1970s, it was believed that the amount of mercury in the oceans was so great that human activity would make a negligible contribution. Now the new procedures for sampling and analysis reveal much lower amounts and therefore the anthropogenic contribution over the industrialized era is important.

## Disposition in the Body

The uptake, distribution and excretion of mercury, hitherto referred to as disposition, depends in the first place on the species of mercury.

### Monomethylmercury

The main features of the disposition of monomethyl mercury compounds (collectively referred to as MeHg ) in humans and animals have been well established over the past three decades. MeHg is efficiently absorbed from the gastrointestinal tract, about 95% of the ingested dose. It distributes to all parts of the body, readily crosses the blood-brain and placental barriers and is avidly accumulated in growing human hair or animal fur. In humans, distribution to tissues appears to be complete in about two days, except for the brain which requires about three days. Compared to the inorganic forms of mercury, the pattern of tissue deposition is relatively uniform.

Excretion from the body is a slower process than the rate of movement between tissues. Thus the mammalian body acts kinetically as a well mixed compartment so that MeHg excretion follows first order kinetics. A single biological half-life adequately describes the excretion rate for all practical purposes.

This may be somewhat puzzling as MeHg is slowly transformed in the body to inorganic mercury which has a more complex excretion pattern. However, except for kidney tissues, intact MeHg accounts for most of the mercury in the body and so half time components due to inorganic mercury may be masked when whole body half times are measured.

Four of the papers presented in the disposition session of the conference (Doi et al., Hirayama et al., Kostyniak, Naganuma et al.) focussed on factors affecting individual differences in disposition and on underlying mechanisms of transport and excretion of MeHg. The fifth paper (Kargacin and Kostial) in this session presented experiments on means to accelerate the elimination of mercury from the body.

An understanding of factors affecting disposition may give us a better understanding of differences in individual susceptibility to MeHg toxicity. If we know the basic mechanisms of elimination from the body, we may be able to develop better procedures for enhancing MeHg removal.

In all the discussions on transport and disposition of monomethylmercury, the key role of thiol compounds was apparent. In fact a general picture is emerging on the way in which MeHg enters and leaves mammalian cells. Studies in biliary secretion (see the chapter by Naganuma et al.) gave evidence that MeHg is secreted from liver cells into bile as a complex with reduced glutathione (GSH). Studies on MeHg disposition in the kidney (see chapter by Hirayama et al.) indicated that the same mechanisms apply to renal proximal cells. Given the millimolar levels of GSH in most mammalian cells, it may be that this is a common cellular exit pathway for MeHg.

Studies on the mechanism of MeHg passage across the blood brain barrier have indicated that MeHg enters the capillary endothelial cell as a complex with the amino acid cysteine. The structure of this complex is similar to that of the large neutral amino acid methionine. Thus, attached to the thiol group of cysteine, MeHg enters the cell on the large neutral amino acid carrier. More recent evidence indicates that this same mechanism may apply to entry into the proximal tubular cells of the kidney, the epithelial cells lining the bile duct, gall bladder and small intestines. Indeed, given the presence of the MeHg-cysteine complex in plasma, this may be a common entry pathway into all mammalian cells.

Several papers pointed to the importance of the enzyme, γ-glutamyl transpeptidase (γ-GT) in the intra-organ disposition and excretion of MeHg (Hirayama et al., Kostyniak, Naganuma et al.). MeHg is exported from liver cells into plasma as the GSH complex. After filtration at the glomerulus, the complex breaks down due to the hydrolysis of GSH to cysteinylglycine(Cys-gly) by γ-GT. MeHg may then be attached to the thiol group of Cys-gly. A second enzyme, a dipeptidase, completes the hydrolysis to cysteine and glycine leaving MeHg attached to the thiol group of cysteine. The latter is now free to enter the proximal tubular cells on the large neutral amino acid carrier.

Gamma-GT on the peritubular border of the proximal tubular cells is the first step in the hydrolysis of the MeHg GSH complex that is circulating in plasma as a result of export from the liver or other cells and tissues. MeHg cysteine is formed again leading to transport into the kidney cells. Intra-renal recirculation of MeHg thiol complexes probably occurs. The MeHg complex exported from the proximal tubular cells is hydrolyzed by γ-GT and dipeptidases to form MeHg-cysteine which once more enters the cell.

MeHg secreted into bile follows an analogous biochemical fate. MeHg-cysteine and MeHg-cysteinylglycine, produced in the biliary tree from the GSH complex, are subject to reabsorption in both the biliary tree and the small intestine thereby recycling MeHg back to the liver cells.

A number of genetic and sexual dimorphisms have been noted in the renal deposition and urinary excretion of MeHg in inbred strains of mice. Strain differences in the activity of γ-GT for genetic reasons or because of hormone action may account for these dimorphisms.

Binding to other thiol groups may account for other strain differences in the disposition of MeHg. For example, certain inbred strains of mice show different blood levels of mercury when similarly dosed with MeHg (see chapter by Doi et al.). It was found that these differences in blood levels were due to binding of MeHg to different molecular species of hemoglobin. Some hemoglobin molecules had extra cysteine moieties or cysteine-SH groups that were more accessible to MeHg. These compositional and structural differences in hemoglobin were genetically determined and correlated with the strain differences in mercury blood levels.

Thiol groups may underlie other genetic and sexual differences in the disposition of MeHg in mammalian tissues. Doi et al. have noted the wide range of biological half times in human populations. No doubt, genetic differences in thiol metabolism may underlie differences in excretion rates although many factors may contribute to the overall elimination rate from the body. Doi et al. have listed many of the factors affecting MeHg disposition. As we unravel these factors, we may be able to intervene to alter normal processes so as to increase the rate of elimination of MeHg from the body. For example, the discovery of the enterohepatic recirculation of MeHg led to the use of orally administered non-absorbable thiol agents that trapped MeHg in the small intestine thereby breaking the cycle and increasing fecal excretion.

Another important mechanism in regulating the disposition of mercury is the level of metallothionein (MT), which is related to the protection against the toxic effect of mercury. Tohyama et al. compared the immunohistochemical localization of MT in the kidney and brain after treatment with cadmium, inorganic mercury and MeHg in rats. Being different from the case of Cd, MT was detected in both the proximal and distal, and collecting tubular epithelium of the kidney after inorganic Hg administration. Of interest, in the brain was the existence of induced MT in ependymal cells, pia mater, arachnoid, vascular and endothelial cells and glial cells after MeHg treatment. This indicates the demethylation of MeHg and the capacity of MT induction by released inorganic mercury.

In the meantime, we can make use of the known interactions of mercury with synthetic complexing agents to accelerate removal from the body (see chapter by Kargacin and Kostial). Agents containing thiol groups are most efficacious, with two dithiol compounds showing the greatest promise. DMSA and DMPS have the advantage that they can be given orally and are generally more effective and less toxic than other synthetic thiols.

Kargacin and Kostial point out that the successful application of complexing agents must take into account such factors as dose of the agent, time of treatment and age of the subject. If these factors are neglected, a result (increased retention) opposite to that intended may occur. For example, a dose of a complexing agent that is too low may cause unwanted tissue redistribution including increase in brain levels.


Inorganic Mercury

Divalent inorganic mercury, on the average, is absorbed in the GI tract to the extent of about 7% of the ingested dose in adults. The range in individuals may rise has high as about 30%. Young children may experience higher absorption rates. Compounds of inorganic mercury may also be absorbed in the lung and across the skin.

Once absorbed, it distributes in the blood almost evenly between red cells and plasma. In plasma, it is bound to different proteins depending on dose, time and route of administration. It distributes to all tissues in the body but in a non-uniform manner. The kidney acquires by far the highest concentrations. Within a few days, about 30% of a single parenteral dose is found in the kidneys. Indeed, as much as 90% of the body burden may be found in this tissue after several weeks. It penetrates the blood-brain barrier only to a small extent, approximately 10 times less rapidly than does methylmercury. It is accumulated in placental tissue but crosses from mother to fetus only to a small extent as compared to methylmercury.

It is secreted from liver to bile and excreted in the feces. It is also excreted in urine. The urinary mercury derives from mercury already deposited in kidney tissue. Direct transport from plasma to urine via glomerular filtration is a minor pathway if it exists at all.

The whole body half time in human adults is, on the average, 37 days for females and 48 days for males. The half time in plasma is about 24 days and in red cells 28 days.

### Inhaled Mercury Vapor

Mercury vapor is retained in the lung to the extent of 80% of the inhaled dose. Most of this retained dose passes immediately into the bloodstream, the remainder being deposited in lung tissue. Mercury vapor can also cross the skin. Quantitatively, transport across the skin is much less important than pulmonary uptake in people exposed to mercury vapor.

On entering the blood compartment, it dissolves in plasma as the monatomic gas. It is lipid soluble and highly diffusible. It rapidly leaves the plasma to enter the red cells. It does, however, persist in plasma for a sufficient period to circulate to all parts of the body. It readily crosses the blood-brain and placental barriers. Once inside cells, it is oxidized to divalent inorganic mercury as discussed in the next section on biotransformation. The oxidation process results in virtually all of the inhaled vapor being converted to the divalent species.

Initially the pattern of tissue deposition differs from that of inorganic mercury. Much more deposition occurs in the brain. More is found in red cells than plasma. Placental transport is greater than that seen after inorganic mercury. However, within a few days after exposure, the pattern of deposition starts to transform to that of inorganic mercury as described in the previous section.

Excretion is predominantly in the form of divalent inorganic mercury in urine and feces but a small fraction of the inhaled dose (less than 10%) is exhaled. Dissolved vapor does appear in the urine during and shortly after exposure. The whole body half time is about 58 days, and in kidney virtually the same (64 days). The clearance from blood is described by two half times; one of 2-4 days accounting for about 90% of the initially deposited mercury, and another of 15-30 days accounting for most of the remainder.

## Biotransformation

The term "biotransformation" refers to biologically mediated changes in the oxidation state, to methylation or demethylation or to the formation of biologically inert complexes of mercury. The definition of "inert" forms of mercury implies the presence of active forms. Takahashi et al., in their chapter, refer to this active form as "ionizable mercury". The implication is that mercury in a chemical form that is capable of reversible reaction with tissue ligands is the active form. This active form will include the potentially many complexes formed with proteins, amino acids and, in particular, the reversible association with thiol groups.

### Monomethylmercury

The methylation of inorganic mercury, although an important reaction in the environmental fate of mercury, is to all practical purposes absent in mammalian tissues. Demethylation, however, does occur (see chapter by Takahashi et al.). Inorganic mercury has been found to account for a substantial amount of the total mercury in tissues in victims of Minamata Disease. These people were exposed to methylmercury in fish, thus, biotransformation must have taken place in their tissues.

This conclusion was confirmed by several investigators in animal experiments. The reaction takes place in macrophage cells and in intestinal microflora. In fact, conversion to inorganic mercury in the gut serves to break the enterohepatic cycle leading to excretion of inorganic mercury in the feces.

Takahashi et al. raise the question as to the biological activity on the inorganic mercury deposited in tissues as a result of demethylation of monomethylmercury (MeHg). They developed analytical methods to detect active (ionizable) inorganic mercury. By combining this method with their established method for measuring total mercury, they were able to measure "inert" or "stable" mercury by difference.

They were able to show that inert forms of mercury appear in liver and kidney tissue in a time dependent manner in rats injected with mercuric chloride. The process was affected by sulfur and selenium compounds added to the diet implying that these inert forms might involve sulfur or selenium compounds. Studies in Yugoslavia (cited by Suzuki) reported, a one-to-one atomic ratio of selenium to mercury in autopsy tissues of mercury miners and their families.

Takahashi et al. found that most of the inorganic mercury in brains of Minamata victims was in the inert form but a substantial amount of the inorganic mercury in liver and kidney was in the active form. These findings raise new questions about the role of inorganic mercury on the toxicity of MeHg. In agreement with previous experiments reported by Magos and colleagues, the results suggested that the neurological effects are due to the monomethyl radical. However, effects on other tissues such as kidneys, although rarely reported in humans, might involve the active form of inorganic mercury. Future epidemiological studies on MeHg exposed populations should take into account potential renal effects due to the released inorganic mercury.

## Inhaled Mercury Vapor

The oxidation of elemental mercury, $Hg^o$, to the divalent mercuric species, $Hg^{++}$, is one of the best understood biotransformation reaction of inorganic mercury (see chapter by Halbach). As far back as the 1930s, the famous German chemist, Stock, had noted that mercury vapor, in the presence of water and atmospheric oxygen, will oxidize to form mercuric oxide. The oxidation step was accelerated in the presence of blood. The oxidation step was suspected to be a purely abiotic chemical reaction with oxygen. However, Nielsen-Kudsk in 1965, made the chance observation that ethanol reduced the retention of the inhaled mercury vapor in humans. This finding led to Nielsen -Kudsk and other investigators to study the basis of the ethanol effect.

These studies have revealed that the mercury atom in the ground oxidation state, $Hg^o$, is a substrate for the enzyme catalase when the latter has formed a compound with one molecule of hydrogen peroxide as depicted in the reactions :

$$Cat.Fe\text{-}OH \quad + \quad H_2O_2 \quad = \quad Cat.Fe\text{-}OOH \quad + \quad H_2O$$

$$Cat.Fe\text{-}OOH \quad + \quad H_2O_2 \quad = \quad Cat.Fe\text{-}OH \quad + \quad O_2$$

$$Cat.Fe\text{-}OOH \quad + \quad Hg^0 \quad = \quad Cat.Fe\text{-}OH \quad + \quad HgO$$

Halbach has shown that the catalase/hydrogen peroxide reaction is the predominant and probably the only significant pathway of oxidation of mercury vapor in mammalian cells. His experiments, presented at this conference, also demonstrated that biochemical reactions that reduce the amount of endogenous hydrogen peroxide in cells also reduce the rate of oxidation of mercury.

This oxidation step is of key importance in the disposition of inhaled mercury vapor in the body. On entering the lung, most of the inhaled vapor passes directly into the bloodstream and dissolves in plasma. It rapidly diffuses into red blood cells where it is oxidized by the catalase pathway. However this removal by the red blood cells is not rapid enough to remove all the dissolved vapor before reaching the blood-brain barrier. Here the lipophilic vapor readily crosses the blood brain barrier to be oxidized to $Hg^{++}$ in the brain tissue. The oxidized form is assumed to be the proximate toxic species accounting for the well known effect on inhaled mercury vapor on the central nervous system.

## Molecular Mechanisms of Toxicity

It is well known that the toxic symptoms of mercury vary with its chemical forms. A large number of studies have been conducted to gain access to molecular mechanisms of toxic actions exerted by mercury compounds. The mechanism of neurotoxicity caused by methylmercury has been one of the major topics in heavy metal toxicology. Repeated outbreaks of intoxication by this metal alkyl that has occurred in various regions in the world clearly shows that the nervous system is the most sensitive tissue to methylmercury exposure. However, it still remains unclear which molecular species, the mercurials, or any radicals formed by these mercurials, is the proximate toxic species actually causing neurological injury, and which cellular molecule is the primary target of mercury toxicity. The mode of toxic action of mercury compounds on neuronal membranes, protein biosynthesis, microtubules and nucleic acid structure and the possible role of oxidative stresses in mercury-induced cellular damages were discussed in this session in order to approach the molecular mechanism of mercury cytotoxicity.

### Effects on Neuronal Membranes

Neuromuscular junctions are regarded as one of the major targets of mercury compounds and some neuromuscular preparations have served as a useful model for the study of mercury induced neurological damages. Narahashi and his coworkers using rat phrenic nerve-hemidiaphragm preparation demonstrated that methylmercury did not affect the frequency of spontaneous miniature end-plate potentials (MEPPs) at 4μM, but markedly increased it at 20 μM. Since 100 μM methylmercury shortened the latent period of MEPPs occurrence but did not increase its upper level, the accessibility of methylmercury to the site of action is rate limiting. Nerve-evoked end-plate potential (EPP) was influenced also by mercurials Some experiments with the rat muscle preparation revealed the suppression of EPP amplitude by methylmercury above 20 μM concentration which was ascribed to a decrease in the quantal content of EPP. Depolarization of nerve terminals is postulated as a possible underlying mechanism for the decrease in EPP amplitude and quantal content and for the increase in the frequency of MEPPs. Further, the recent experiments using a variety of mitochondrial function inhibitors and chemicals disrupting intraterminal $Ca^{++}$ buffering indicate that methylmercury stimulates release of $Ca^{++}$ from nerve terminal mitochondria, resulting in an increase in MEPP frequency.

Intracellular microelectrode and suction pipette method was applied to the mouse neuroblastoma NIE-115 cells treated with methylmercury, and demonstrated that resting potential, resting membrane resistance, and action potential were decreased by methylmercury above 20 μM. Especially, action potential was suppressed to 57% of the control even by 1 μM methylmercury. Voltage clamp experiments revealed that 60 μM methylmercury preferentially blocked the sodium current to potassium current.

Development of patch clamp techniques has made it possible to examine the effects of mercurials on any neuronal channel. Thus, the patch clamp techniques applied to the rat dorsal root ganglion neurons in primary culture demonstrated that mercuric chloride significantly increased GABA-induced chloride current at 1 μM concentration, while methylmercury, even at 100 μM, did not activate it (see Figures 7 and 8 of Narahashi's chapter). Since methylmercury and elemental mercury is known to be converted to inorganic mercury *in vivo*, the highly potent stimulating action of mercuric ion on the GABA-system is assumed to take an important role in manifestation of neurological symptoms by mercury compounds.

## Possible Roles of Free Radicals and Cellular Calcium Status

It has long been an interesting question what kind of molecular species actually contributes to the methylmercury induced cell damage. Verity and his collaborator attempted to examine the hypothesis that free radicals generated from methylmercury or those subsequently formed through hydrogen abstraction by the methylmercury radical may cause oxidative injury in the pathogenesis of methylmercury neurotoxicity. Using cerebellar granule cell suspensions as an *in vitro* model system, they observed an increase in membrane lipoperoxidation associated with neuronal cytotoxicity assayed by dye-exclusion and a substantial decrease in cellular level of reduced glutathione by 5-20 µM methylmercury.

However, early measurable cytotoxicity was detected prior to the rise in fluorescent signal from 2', 7'-dichlorofluorescein, an indicator for oxygen radical formation. Further, alpha-tocopherol inhibited lipoperoxidation but did not significantly protect neuronal cells from the methylmercury- induced damage of the membrane barrier and EDTA blocked lipoperoxidation caused by this metal alkyl without change in ability of cells to exclude trypan blue. These results demonstrate the generation of oxidative stress by methylmercury but reveal a dissociation of lipoperoxidation from cytotoxicity. The $Ca^{++}$ chelator EGTA, in contrast to EDTA, suppressed not only lipoperoxidation but also cytotoxicity exerted by methylmercury in cerebellar granule cell suspensions. This observation suggests that intracellular concentration of available $Ca^{++}$ may play a role in methylmercury neurotoxicity.

Considering the experimental results so far reported on the relation of cellular $Ca^{++}$ situation and free radical-induced cell damage, it is speculated that the combination and synergetic interaction of changes in intracellular $Ca^{++}$ homeostasis and disruption of - SH status initiated activation of phospholipases, proteases and endonucleases leading to neuronal injury. High sensitivity of the neuron to methylmercury toxicity can be elucidated by its relatively low glutathione content and/or the lack of its ability to efficiently supply this thiol compound when it is rapidly consumed by the stresses due to methylmercury exposure. To draw such an important conclusion as above, however, it seems to be necessary to assess the "cytotoxicity" by a more sensitive measure than the dye exclusion test.

## DNA as a Target of Free Radicals caused by Mercurials

DNA can also be a target for mercury toxicity. Summarizing the experimental results reported so far, Costa et al. demonstrated that mercuric mercury induced DNA single strand breaks in Chinese hamster ovary cells. Alkaline elution and CsCl density gradient analysis showed that mercuric chloride caused DNA strand breaks as in the case of X-ray irradiation. The DNA damage induced by mercuric chloride was probably due to superoxide radical formation which was confirmed by cytochrome C reduction and its depression in the absence of superoxide dismutase.

It is of interest that mercuric ion not only induces the DNA damage as above but also effectively inhibits DNA repair in contrast to X-ray irradiation. The incomplete repair of DNA strand breaks caused by mercuric mercury may account for the incapability of mercuric mercury to function either as a mutagen or as a carcinogen. Methylmercury was shown to be more potent in causing DNA strand breaks than inorganic mercury especially in nerve cells in which the strand breaks were produced by methylmercury even at concentrations as low as $10^{-6}$M.

This higher sensitivity of nerve cells in terms of DNA damage by methylmercury can also be explained by efficient uptake of methylmercury by nerve cells as in the case of its cytotoxicity towards mouse glioma cells (see the chapter by Miura and Imura). However, it is still ambiguous whether or not methylmercury is able to cause free radicals or, if this is the case, what species of radical is induced by methylmercury. Obviously, it is necessary also to know whether the DNA strand breaks precede the other cellular lesions on exposure to methylmercury.

## Protein Synthesis Machinery and Microtubules as Targets of Methylmercury

Protein synthesis has been recognized as one of the most sensitive targets of methylmercury neurotoxicity. Omata et al. employed an *in vitro* translation system of reticulocyte lysate to translate polyadenylated mRNA obtained from the brain of rats treated with methylmercury in the presence of $^{35}$S-methionine. They analyzed the $^{35}$S-labeled translation products by two-dimensional electrophoresis after mixing with those of the control labeled with $^3$H-leucine as an internal standard prior to application to the gel. Polyadenylated mRNAs were obtained by using oligo(dT) cellulose column from the rats administered 10 mg/kg of methylmercuric chloride daily for seven days and killed at an early, latent or symptomatic period of intoxication. The gels corresponding to protein spots on the electrophoretogram were cut out, extracted with SDS containing alkaline solution, and counted. The effect of methylmercury on the synthesis of individual proteins was carefully calculated by the radioactivity ratio of the spot from methylmercury treated rats to that from the control rats.

The results thus obtained revealed that the effect of methylmercury on protein synthesis in rat brain is not uniform for each protein species. The synthetic activity of some proteins were depressed by methylmercury, but biosynthesis of other proteins were stimulated by the exposure to methylmercury. Further, the pattern of these changes in protein synthesis was altered during the progress of the intoxication. Among 120 proteins separated only less than 10 species were identified (see Table 3 by Omata et al.). Of these identified proteins, actin and tubulin alpha and beta, i.e., cytoskeletal proteins, were decreased at symptomatic periods of intoxication. Further, it is of interest that synthesis of 14-3-3 proteins in the brain and dorsal root ganglia were reduced by methylmercury exposure, because the 14-3-3 protein family was recognized as regulators of neurotransmitter enzymes. Methylmercury also depressed six aminoacyl tRNA synthetase activities out of 12 enzyme activities examined, indicating the possibility that protein synthesis is modified by methylmercury at the aminoacyl tRNA formation process.

Considering these results together with the findings so far reported, methylmercury affects the translational step of protein synthesis to change protein levels almost uniformly on the one hand and, on the other hand, it modifies the pretranslational steps to cause specific changes in biosynthesis of individual proteins. Effects of methylmercury on tubulin gene expression described above may closely relate to the following finding by Miura and Imura that disruption of microtubules increases the tubulin pool, resulting in the inhibition of tubulin biosynthesis probably through mRNA unstabilization.

Miura and Imura have recently proposed that microtubules may serve as a primary target of methylmercury cytotoxicity. Electron microscopic observation, indirect immunofluorescence study and biochemical examination of nucleic acids and protein synthesis using mouse glioma cells revealed that microtubules were depolymerized by methylmercury of growth inhibitory concentration ($5 \times 10^{-6}$M) before the morphology of other cell organelles and cellular functions showed any detectable alteration. In contrast, inorganic mercury caused widespread lesions in various cell organelles including microtubules at the growth inhibitory concentration ($5 \times 10^{-5}$M). However, direct effects of inorganic mercury on *in vitro* tubulin polymerization were stronger than those of methylmercury.

These results demonstrate that the discrepancy in susceptibility of the cells to these two mercurials can be ascribed to the difference in their membrane permeability. In fact, the rate of methylmercury uptake by mouse glioma and neuroblastoma cells was approximately 10 times higher than that of inorganic mercury. The high susceptibility of microtubules to methylmercury indicated above was supported by the experimental results reported from several other laboratories using cultured cells and mice exposed to methylmercury *in utero*.

Besides the depolymerizing effect of methylmercury on microtubules, it has recently been confirmed that methylmercury depressed tubulin biosynthesis through increasing the cellular tubulin pool formed by depolymerizing microtubules. Mouse glioma cells were pulse-labeled with $^{35}$S-methionine for 15 min after treatment with methylmercury for 3 hr. Analysis of labeled proteins by two dimensional gel electrophoresis and subsequent autoradiography revealed that ß-tubulin synthesis was markedly reduced in methylmercury treated cells. At the growth inhibitory concentration ($5 \times 10^{-6}$M) methylmercury inhibited ß-tubulin synthesis by 50-70% (see Figure 8 by Miura and Imura).

On the other hand, radioactive bands of  proteins other than tubulin on gradient urea-PAGE gel remained unchanged. Northern blot analysis indicated that tubulin mRNA was decreased by methylmercury treatment, while the level of ß-actin mRNA used as a control was not changed. However, the relative transcriptional rate of ß-tubulin gene was shown to be unchanged by a nuclear run-on study using isolated nuclei from the glioma cells  with or without methylmercury treatment. Thus, the regulation of tubulin gene expression by methylmercury seems to be post transcriptional as in the case of colchicine treatment. The specific unstabilization of tubulin mRNA by the increase of the tubulin pool may be a cause for the decrease in its level observed by Northern blot analysis.

Considering a relatively large content of microtubules in the axoplasm of neuronal axons and its essential role in axonal transport, the disruption of microtubules and the successive inhibition of tubulin synthesis by methylmercury may suggest a possibility that the specific toxicity of this metal alkyl on microtubules is a key mechanism in its characteristic neurotoxicity.

Considering the papers and discussions in this session,  it may not  be an exaggeration to say that the identification of a primary site for methylmercury neurotoxicity appears to be close. However, it should be emphasized that to clearly understand the molecular mechanism of mercury compounds, the susceptibility of various cellular molecules to mercurials has to be strictly compared in the same strain of cells or animals and that the dose-effect relationship is to be clearly demonstrated. Otherwise, we cannot provide any reliable elucidation of the mechanism involved in the toxic actions exerted by mercurials.

## Selenium as a Modifying Factor of Mercury Toxicity

Several observations in the 1960s and 1970s indicated the potential for selenium compounds to antagonize the toxic effects of mercury. Parizek's laboratory first reported dramatic protection by selenite against the acute toxicity of inorganic mercury. Kosta and coworkers noted persistent high mercury levels with a one-to-one atomic ratio of selenium to mercury in autopsy tissues in mercury miners in Yugoslavia who had been retired for many years. Similar observations were made on retired Japanese mercury miners. Ganther and coworkers were the first to report that dietary selenium could delay the toxic effects of methylmercury in experimental animals.

These observations aroused great interest in mercury - selenium interactions. Two questions were of special interest: did selenium, naturally present in the diet, offer protection against the toxicity of mercury, and can we use selenium compounds therapeutically to treat cases of mercury poisoning? The papers by Imura and Naganuma and by Magos review the experimental work that arose out of these original observations. Since the interaction of selenium with divalent inorganic mercury ($Hg^{++}$) and methylmercury involve different mechanisms, the two will be treated separately.

Inorganic mercury

Compounds of selenite, when co-administered with mercuric chloride, protect against the renal toxicity of mercury. Selenium decreases the uptake of mercury by the kidneys but increases the overall residence time of mercury in the whole body.

Imura and Naganuma review their own work on the biochemical mechanism behind this protective effect. *In vivo* studies revealed the presence of high molecular weight complexes in plasma of experimental animals simultaneously dosed with mercuric chloride and sodium selenite. Mercury and selenium were present in a one-to-one atomic ratio. *In vitro* studies revealed that such complexes could be formed in both red blood cells and plasma and required the presence of reduced glutathione. A protein component in these complexes was confirmed by the use of enzymes that digest proteins. It was proposed that reduced glutathione reduces selenite to the selenide anion. The latter is able to form a selenium bridge between mercury and SH groups on proteins resulting in a stable mercury selenium complex.

The presence of these compounds in liver tissue could be explained by liver uptake from plasma. The large size of the complex prevented filtration in the kidneys thereby reducing both kidney levels and the urinary excretion of mercury.

Interaction with inhaled mercury vapor probably involves the same biochemical mechanism. Inhaled vapor is rapidly oxidized to divalent inorganic mercury which could then react with selenide anions.

The therapeutic or prophylactic use of selenium is unlikely. Selenium has to be given virtually simultaneously with mercury to exert its protective effect. When given

60 minutes ahead of mercury, toxicity is increased. When given only 20 minutes after mercury, its protective effect is already diminished.

On the other hand, the observations on human autopsy tissues suggest that selenium, naturally present in human tissues, can form a persistent complex with mercury. This complex may be the same protein complex as that seen in experimental animals. It may also be the breakdown product in the form of insoluble mercuric selenide. In either of these forms, it is assumed to be toxicologically inert. It is therefore of great interest for future studies to characterize the chemical nature of this mercury-selenium complex seen in human tissues.

However, one important caveat should be noted. Naturally occurring selenium does not protect as well against mercury as sodium selenite. Thus the protection against renal damage from mercuric chloride was greatest with selenite, less with selenomethionine and least with Se metabolically incorporated into liver. This sequence of efficacy was paralleled by the rate of *in vivo* formation of the mercury-selenium complex.


Methylmercury

That methylmercury can interact with selenium *in vivo*, there is no doubt. The uncertainties lie in the mechanisms of this interaction and its public health significance. Fish are the major source of human exposure to methylmercury and an important source of dietary selenium. Clearly, the possibility that selenium might attenuate the toxicity of methylmercury is an attractive idea.

Early experiments were encouraging. Selenite delayed the toxic effects of methylmercury. The time of administration of selenium relative to that of mercury was not as critical as in the case of inorganic mercury. However, selenium only delayed and did not prevent methylmercury toxicity. Thus its role in long term chronic exposure is not clear.

The "protective" effect of selenium, as in the case of inorganic mercury, appears to be greatest with selenite compounds and least with selenium naturally present in the diet. Thus selenium in tuna fish may be more than 50% the protective efficacy of selenite and possibly less.

Selenium does not produce consistent protective agents in the entire spectrum of prenatal toxicity of methylmercury, at least in experimental animals. Selenium deficiency is without effect on the teratogenicity of methylmercury. High doses of selenium protect against some prenatal effects but exacerbate others. Whether or not selenium can protect against prenatal effects is a critical public health question as the prenatal period is the most vulnerable stage of the life cycle.

The biochemical basis for the interaction of methylmercury with selenium has at least one step in common with the interaction involving inorganic mercury. Thus the first step involves the reduction of selenium to the selenide anion. The latter reacts with the methylmercury cation to form bis (methylmercuric) selenide. This compound is highly lipid soluble and may account for the fact that administration of selenium raises the brain levels of methylmercury. Whether or not bis (methylmercuric) selenide plays a role in the protective effect of selenium is unknown as it has been shown to be unstable.

Other pathways of interaction of methylmercury with the selenide anion may occur. Thus the methylation of selenium to the volatile dimethylselenide is increased by dosing animals with methylmercury. Interestingly, the rate of production of dimethylselenide increases with increasing doses of methylmercury. Thus the potential protective effect of selenium is lost as more and more selenium is lost by volatilization as methylmercury approaches toxic tissue levels (see the chapter by Magos).

The interaction of methylmercury with selenium is a two way phenomenon. Thus methylmercury can enhance the toxicity of selenite given to experimental animals. The mechanism is not known. Increased production of demethylselenide may be one factor. Another factor may be the production of bis (methylmercuric) selenide which could lead to greater tissue deposition of selenium.

The therapeutic application of selenium to methylmercury is as unlikely as it is for inorganic mercury. The protective effect is not well quantified. Selenium only delays the onset of symptoms in adults and appears to be ineffective for prenatal life (see the chapter by Magos).

That selenium, naturally present in human tissues, can protect against methylmercury in our diet still remains an intriguing but unproven hypothesis. Selenium naturally present in food appears to be less effective than inorganic selenium compounds artificially added to the diet. Nevertheless, this finding does not discount a protective role for selenium in the environment.

Not discussed in this symposium is the finding that marine mammals exposed to methylmercury in their diet have high levels of an inorganic mercury selenium complex in their livers. As will be discussed in the "Human Ecology" section of this review, it may be that selenium neutralizes the inorganic mercury split off from methylmercury in the tissues of these animals.

## Toxic Effects of Mercury

Toxic effects of mercury are expressed in different ways according to the chemical form of mercury, the dose and the route of exposure in various species of animals. However, there are the two major forms of toxic effects of mercury, i.e., nephrotoxic effects and neurotoxic effects, on which we have focussed in this conference. In both forms of toxicity, it was shown that an appropriate experimental model is useful for studying the mechanism of toxicity.

## Nephrotoxic Effects

Mercurials cause two types of renal injury: (1) tubular damage and (2) glomerular injury. In this conference the early, possibly the earliest sign of toxic action of inorganic mercury was on the proximal tubule segment of the nephron of the rat (Endou and Jung). When the isolated early portion of the proximal tubule was incubated in $HgCl_2$ at relatively high concentrations ($10^{-7}$ to $10^{-4}$M), the intracellular free calcium concentration, $[Ca^{++}]_i$, increased depending on the dose of mercury without treatment with angiotensin II (AII): the minimal dose was $10^{-7}$M. A quite similar pattern to this was observed with MeHg. At lower concentrations than $10^{-7}$M, response of $[Ca^{++}]_i$ in the early portion of proximal tubule to AII after incubation with $HgCl_2$ for 5 min was exaggerated. The maximum AII-induced increase of $[Ca^{++}]_i$ was observed at a concentration of $10^{-9}$M of $HgCl_2$. This stimulatory effect of $HgCl_2$ was completely inhibited by propranolol, an inhibitor of phospholipase. It is not clear why the early proximal tubule is so sensitive to mercury, even though this part of the nephron has an ability to respond to lower concentration of AII than the other parts. The concentration of $HgCl_2$, which was effective in modifying the response to AII, was very low. The physiological significance of this finding and comparative effects of MeHg at similar low dose levels will deserve further clarification.

Another type of renal injury, glomerular injury, was least understood until the effect of mercury on the immune system had been intensively studied with several experimental models which allowed an elucidation of the mechanism (see chapter by Druet). Various mercurials may induce autoimmune glomerular lesions, and other pathological conditions such as pink disease and possibly Kawasaki disease, an acute febrile mucocutaneous syndrome. The two diseases may involve disturbances in the immune system by mercury. Oral lichen planus may be of immune origin related to dental amalgam fillings in the patient.

In the study regarding the $HgCl_2$-induced autoimmunity in Brown Norway (BN) rats, the autoimmune abnormalities including a proliferation of CD4+ T cells and B cells in the spleen and lymph nodes, a hyperimmunoglobulinemia of IgE, and production of numerous antibodies to self (DNA, laminin, collagen and thyroglobulin) and non-self (2, 4, 6, -trinitrophenyl) were recognized with repeated subcutaneous dosings of 1 mg/kg. Antibodies to glomerular basement membrane (GBM) components, mainly laminin, are responsible for autoimmune glomerular nephritis, the first phase being the binding of circulating autoantibodies to the GBM. In the earlier studies, this was described as linear IgG deposits along the GBM and then renal and extra-renal granular deposits of IgG typical of glomerular nephritis. Heavy proteinuria and the nephrotic syndrome were seen, and 30-50% of the animals would die and the remaining would recover and become resistant.

The effects of much lower doses of $HgCl_2$ (12.5, 25, 50, 100 or 200 µg/kg, 3 times a week for 5 months) on the immune system were evaluated recently in rats by measuring the serum IgE concentration. An important finding was the existence of a dose-effect relationship between peak serum IgE concentrations and the amount of $HgCl_2$ injected. In rats that received 12.5 µg/kg of $HgCl_2$, the serum IgE concentration did not increase.

The autoimmune response to $HgCl_2$ has been confirmed to be genetically determined, e.g., highly sensitive BN rats have the RT-1n haplotype at the major histocompatibility complex while other resistant strains bore the RT-11 haplotype. In other species of animals, rabbit, guinea pig and mouse, immunological disregulation was observed in response to $HgCl_2$ administration. Depending upon the species or the strain tested, either autoimmunity or immunosuppression could be achieved. These effects, at least those of autoimmunity, are considered to be mediated by T cells reacting with class II molecules. The findings, reviewed in the chapter by Druet, indicate that it is only an opening scene for the more extensive study on epidemiology of immunological disorders, including glomerular nephritis, in relation to low doses of mercury in general populations.

Neurotoxic Effects

The central nervous system is one of the critical organs for exposure to mercury. The chemical forms of mercury most relevant to the neurotoxic effect is elemental mercury vapor or methylmercury. In this conference, all five papers presented in the session of characteristic toxicities have dealt with the neurotoxic effect of MeHg (Choi, Inoue, Sato and Nakamura, Satoh, Arito and Takahashi). Approaches adopted were behavioral, physiological, histopathological and biochemical. The species of animal used were mouse, rat, guinea pig, and monkey.

A common concern through these papers was to establish an appropriate animal model to simulate the health effects of MeHg on humans and to determine the minimum effective dose of MeHg. In cynomolgus monkeys (*Macaca fascicularis*) fed for one year at a daily dose of 0.02 to 0.04 mg/kg, ultrastructural changes in the brain were apparent without clinical symptoms except for abnormal nystagmus. One of the changes recognized was myelin-like membranous structures in the perikaryon of the neurons in the calcarine cortex. The peak concentration of mercury in the calcarine cortex was not measured, but 10 to 14 days after termination of MeHg feeding, it ranged from 1.6 to 8.0 µg/g (Sato and Nakamura).

Membrane structure of the cell may be the first target of MeHg in conjunction with lipid peroxidation (see chapter by Verity and Sarafian). In the occipital cortex of cats administered with a toxic or subclinical dose of MeHg, the histological changes were contrasted to the change in alpha-norepinephrine receptors (Sato and Nakamura). The capacity of the receptors in the occipital cortex was reduced even in the cats showing no histological abnormalities with brain Hg concentrations of around 10 µg/g wet weight.

Polygraphic (EEG, EMG, ECG and temperature recording with thermister) observations were conducted on rats administered orally with MeHg (Arito and Takahashi). The maximum brain mercury concentration of 10 µg/g was associated with the marked changes in sleep-wakefulness pattern and body temperature change, while this change in the vigilance state started to occur at concentrations of 1-3 µg/g. In accordance with this sleep-wakefulness change, the levels of 5-hydroxytryptamine, noradrenalin and 3-methoxy-4-hydroxy-phenylethyleneglycol in the frontal cortex were decreased during the dark period, indicating the lowered turnover of central monoamine metabolism due to MeHg.

## Toxic Effects in the Developing Brain

From the experiences in Minamata and Iraqi epidemics of MeHg poisoning, the developing brain has been considered to be the most sensitive to MeHg However, pathological observation on the patients affected *in utero* have been scarce (2 cases from Minamata and 2 cases from Iraq) and provides somewhat different features between the two epidemics (see chapter by Inoue). In Iraqi babies whose mothers consumed MeHg-contaminated bread during pregnancy, their brains showed developmental deviations characterized by disturbances in neuronal migration and laminar cortical organization within the cerebrum and cerebellum associated with diffuse astrocytosis of the white matter (see chapter by Choi). In Minamata, the brains of affected babies showed disorganization of the cerebral cortical architecture with degeneration of nerve cells and decrease in their number: the changes are basically similar to those found in adult Minamata disease sufferers (Inoue). As Choi pointed out in the conference, taking the complexity of brain development into account, precise information regarding the location and timing of insults in relation to ontogenic events is absolutely necessary when studying the effects of MeHg.

Another problem arises when studying the effects of MeHg on the developing brain using animal models, that is, the species difference problem. For instance, the cerebellum is one of the principal target sites for MeHg toxicity in humans, cats and rodents, but not in primates other than humans (Choi). This species-difference problem has a different aspect in relation to the balance between the maturation of the brain and the length of pregnancy. Newborn mice and rats have immature brains, thus, after birth their brains are developing and more susceptible to MeHg. The selection of experimental animals is very crucial in simulating the human experience. When guinea pigs, whose full term offspring have brains comparable to the human neonates' brains in histological development, were treated with MeHg in early pregnancy developmental disturbance of the fetal brain including abnormal neuronal migration was induced while treated in later pregnancy neurons of the cerebral cortex were degenerated. In the case of rats, exposure in the early postnatal period induced a widespread neuronal degeneration as observed in Minamata patients (Inoue).

In mice of C57BL/6J strain, the cellular and subcellular mechanism of the effect of MeHg on neuronal migration and on proliferative neuroepithelial germinal cells within the ventricular zone of telencephalic vesicles was investigated (Choi). With $^3$H-thymidine autoradiographic technique and ultrastructural studies, anomalous cytoarchitectonic patterning of the cerebral cortex was reconfirmed as a characteristic finding in prenatal MeHg poisoning. Additionally, Choi introduced the results of application of some new research techniques and tools in the toxicity study of MeHg on the developing brain. For instance, the failure of reorganization of mechanically dissociated embryonic rat cerebrum was observed in a rotation-mediated aggregation culture system in which the final MeHg concentration was 0.1, 0.2, or 0.5 μM. Taking the strong cytotoxic effects of MeHg into consideration, much lower concentrations of MeHg may deserve further study. Moreover, focus of interest has diverged into various aspects of brain development by Choi and his collaborators: 1) cell-to cell interaction, 2) neuronal cell adhesion, 3) astrocytes, 4) excitatory amino acid receptors, and others. Certainly, we now expect exciting new findings on the diverse pathological mechanism through which MeHg may affect the developing brain.

## Behavioral Teratology

Recent epidemiological studies in Iraq, Canada and New Zealand have suggested that *in utero* exposure to MeHg at lower doses than those toxic to adults may induce minor abnormalities which are only detectable by psychological and behavioral tests in the offspring. A need for further epidemiological studies to define the minimum toxic dose of MeHg for the human embryo and fetus has been commonly recognized. In connection with this problem, experimental behavioral teratology has come to fore again, since it has been known to have the most sensitivity to the abnormalities (see chapter by Satoh). Nevertheless, the dose effect and dose response relationship in behavioral teratology of MeHg has not been consistent. Based on the review in the literature, including his own unpublished studies, Satoh concluded that more consideration is necessary on the behavior to be observed, the dose to be given, the time of exposure, and the species to be examined.

## Clinical and Epidemiological Aspects

The four papers in this session covered the broad topics of the biological monitoring, clinical effects, diagnostic criteria and major epidemiological findings.

### Human Exposure and Biological Monitoring

The assessment of human exposure to mercury and its compounds is usually by biological monitoring. Measurement of mercury vapor concentrations in air has also been used to assess human occupational exposure. The papers by Yamamura and Yoshida and by Skerfving deal with biological monitoring and human exposure to mercury vapor and to methylmercury compounds. These are the two forms of mercury to which humans are most commonly exposed.

### Mercury Vapor

Human exposure to mercury vapor from the non-contaminated atmosphere is negligible compared to other sources. Occupational exposure in certain industries such as chloralkali manufacture and smelting of mercury containing materials can still lead to high risk levels despite the general improvement in occupational safety in most countries. Dental assistants are probably the largest group of people occupationally exposed to mercury vapor. A topic of considerably current interest is the role of dental amalgam fillings in human exposure to mercury vapor. Studies in the last decade have confirmed that amalgam fillings release mercury vapor into the air in the oral cavity. The rate of release is greatly exaggerated by chewing. The nature of the food also affects the release rate. Dental amalgams have been estimated to be the major source of background (non-occupational) exposure to mercury vapor.

As discussed in a previous section, the kinetics of disposition of inhaled mercury vapor are complex. The inhaled vapor dissolves in plasma from which it rapidly distributes to all tissues in the body. Once inside cells, it is oxidized to divalent inorganic mercury. Its disposition in tissues is highly nonuniform. Thus it is not surprising that the kinetics of disposition involve several compartments and half times. Yamamura and Yoshida found three half times for disappearance from plasma and red cells after chronic occupational exposure to mercury vapor. Urinary excretion also exhibited a triphasic pattern similar to that seen in blood. The longest half time was of the order of 100 days indicating that steady state will not be attained until after a year or more of exposure. Thereafter, one would expect to see a correlation between time weighted air concentrations and concentrations in blood and urine. Indeed such a correlation was found between time weighted air concentrations of mercury vapor and blood levels of inorganic mercury in workers employed for more than two years. Correlation between air and urine levels was not as good. However, most studies do find urine levels after chronic exposure to correlate with air levels on a group basis.

The situation during a rapidly changing exposure reflects the complex kinetics of disposition and is best illustrated by Skerfving's data on individuals who had their mercury amalgam fillings removed. The removal of the fillings causes a rise in mercury vapor levels in the oral cavity. Within one day, the plasma rises to a peak value and declines. The rise in urine levels is delayed for several days probably because the mercury must first be accumulated by the kidneys prior to excretion in the urine. Thereafter, the urine level also declines. Approximately half a year after the date of removal of the fillings, the urine and plasma levels approach a new steady state with levels about one fourth the original values.

In people with long established mercury amalgam fillings it was found that urine levels on the average increased with increasing number of fillings. The amount by which the urinary concentration increased was consistent with that calculated from the estimated amount of vapor absorbed based on the known kinetics of disposition of inhaled vapor.

An interesting addition to the use of urine as an indicator media is the specific measurement of dissolved vapor, first pioneered by Henderson. Yamamura and Yoshida showed that dissolved vapor in urine samples collected at the end of a work shift correlated closely with time weighted average air concentrations in exposed workers.

Skerfving's paper describes the application of a complexing agent, dimercaptopropane sulfonate (DMPS), to biological monitoring. Urine samples were collected 24 hours before and after giving a measured dose of DMPS in dental workers and in people with and without dental amalgam fillings. DMPS produced an increase in urinary excretion which was approximately proportional to the pretreatment urine level. By increasing urine mercury concentrations, it makes the mercury analysis easier and less prone to errors due to contamination.

### Methylmercury

The manufacture of methylmercury compounds for use as fungicides in agriculture has ceased. Human exposure to this form of mercury is now exclusively from consumption of fish and marine mammals. As described earlier, methylmercury derives from the methylation of inorganic mercury in the environment and its bioaccumulation in aquatic food chains, with the highest concentrations being found in large predatory fish and sea mammals. Skerfving points out that acidification of bodies of fresh water such as may be caused by "acid rain" results in higher levels in fish tissues.

Biological monitoring procedures for methylmercury are now well established and firmly based on the kinetics of disposition. Its high mobility in the body and relatively slow excretion ensures that concentration ratios between biological fluids and tissues are constant for any given animal species including human. Thus blood levels reflect levels in the target tissue, the brain. Maternal and cord blood levels parallel each other. As noted by Skerfving, most of the methylmercury in blood is located in the red blood cells making them a suitable indicator medium. Plasma is more appropriate for monitoring inorganic mercury as the latter distributes evenly between plasma and red blood cells. Skerfving illustrates the usefulness of red blood cells from data showing a linear relation between levels of methylmercury in red cells and the estimated intake from fish.

Methylmercury is avidly accumulated in hair. The concentration in newly formed human scalp hair is directly proportional to the simultaneous concentration in blood. Once incorporated into the hair fiber, its concentration is stable so that the hair may serve as a historical record of past blood levels.

## Clinical Signs and Symptoms and Diagnostic Criteria

Papers by Igata, Skerfving and Yamamura and Yoshida address this topic for both mercury vapor and methylmercury compounds. The severe effects of both forms of mercury are well known. Today the emphasis is on the detection of more subtle effects occurring at the lower levels of mercury in indicator media.

### Mercury Vapor

The classical signs of chronic mercurialism - erethism, tremor and gingivitis - have been described since antiquity. They are now rarely reported because of improved industrial hygiene measures. However, more subtle effects can still be detected arising from the action of inhaled vapor on both the nervous system and the kidneys as discussed in Skerfving's paper. The renal effects may, at least in part, be mediated by a direct effect of inorganic mercury on immunocompetent cells. Such cells are stimulated to produce antibodies that damage the glomerulus leading to proteinuria. Marked species differences have been seen in animals implying that individual susceptibility in humans may be high. Metallic mercury also produces skin allergies in susceptible individuals.

Skerfving reports studies to detect effects of inhaled vapor on endocrine function in occupationally exposed workers. Inorganic mercury is known to accumulate in both the pituitary and thyroid glands. However no effects could be detected in this study.

### Methylmercury

The diagnosis of severe cases of methylmercury poisoning is relatively straight forward. Effects are seen primarily in the nervous system such as cerebellar ataxia, and disturbances in sensory functions. The difficulty lies in detection of mild effects which tend to be nonspecific. This problem is discussed extensively in the paper by Igata. This paper is based on experience in developing objective criteria for Minamata Disease.

Minamata Disease can range in severity from full blown cases of poisoning, usually referred to as the Hunter-Russell syndrome, to mild cases, sometimes of late onset, that are difficult to distinguish from diseases of different etiologies.

Igata describes the development of a successful procedure for differential diagnosis of this disease in people living in the area of Minamata and claiming compensation. The first step was to establish objective medical criteria for signs and symptoms of the disease. Each patient received a series of medical examinations and the data subjected to a statistical procedure known as principal factor and discriminant analysis. This allows a graphical display of the clinical data and a dividing point can be identified between affected and non-affected individuals. Igata estimates that, using this method, misdiagnosis can be reduced to 2.5% of the claimants.

A considerable effort has been made to assure objective measurement of neurological impairment. Peripheral neuropathies are identified by biopsies, thermography and sensory nerve conduction velocities. Sensory disturbances of a central origin are assessed by somatosensory evoked potentials.

In addition to the usual neurological examination, cerebellar ataxia is studied by such techniques as the use of a gravidimeter, objective evaluation of the finger to nose test, analysis of voice, writing, and gait.

The CI scan has been introduced. Using morphological parameters for cerebellum, cerebrum and brain stem, a differential diagnosis of Minamata Disease can be made with an accuracy of over 80%.

Other organs such as liver and kidney are involved in severe cases of poisoning. The kidney effects such as increased excretion of microglobulins may result from the accumulation of inorganic mercury in kidney from the biotransformation of methylmercury. There is experimental evidence for effects of methylmercury on platelet cells. However, epidemiological studies have not found any association with arteriosclerosis or cerebral vascular disease.

The time course of Minamata Disease sometimes exhibits latent periods of several years. This has not been reported in other outbreaks of methylmercury poisoning and its mechanism is unknown. The chronic nature of the exposure in Minamata distinguishes it from other outbreaks. Thus the process of aging may uncover latent damage inflicted by mercury exposure in earlier years.

Congenital Minamata Disease due to prenatal exposure is characterized by diffuse damage to the central nervous system. A history of the mother's exposure during pregnancy is useful in diagnosis. The umbilical cord is frequently preserved in Japanese families and this can provide evidence of prenatal exposure.

Studies in other populations exposed to methylmercury were briefly discussed by Skerfving. Infants of mothers with exposures considerably lower than those seen in the severe cases in Minamata exhibited possible evidence of slight retardation of central nervous function. The lower limit for effects on the fetus has yet to be established. The most recent analysis of the Iraq population suggests that the threshold for fetal effects may be about five fold lower than the threshold for adults.

## Mercury in Human Ecology

The principal goal of human ecology is to understand human adaptive mechanisms to changes in the environment. The paper by Suzuki discusses human adaptation to mercury. Mercury is present in the environment not only as a naturally occurring element in the earth's crust but also as a result of human activities. Two series of studies are discussed; one on a population in Papua, New Guinea following a traditional way of life and the other on an industrialized society in Japan.

### Studies on a Traditional Population

Four villages in Papua were selected for study based on different dietary patterns. The average dietary intake was estimated for each village based on an exhaustive survey of total and organic (methyl) mercury in food items. Both total and organic mercury levels were highest in fish tissue, next in order were reptiles and the lowest levels were found in plants. A consistent finding in all foodstuffs was that the higher the total mercury, the higher the percentage of organic mercury. It was of special interest that one village had the highest average daily intake ever reported in any "non contaminated" population. This intake, approximately 80 µg per day of total and 70 µg per day of organic mercury, well exceeded the WHO tolerable limit of 30 µg per day.

The importance of fish consumption in determining human intake of mercury was further emphasized by measuring the ratio of the stable nitrogen isotopes. The ratio of N15 to N14 differs according to animal and fish species. Thus the ratio is highest in marine fish, intermediate in freshwater fish and reptiles, and low in terrestrial animals and birds. The mercury content of foods was significantly correlated with the nitrogen isotopic ratio.

The average human hair levels in the four villages did not correlate precisely with the estimated average dietary intake of mercury. The reason is believed to be that seasonal variations in average daily intake were not matched in time with the corresponding hair levels according to the time of hair sampling and the rate of hair growth.

The nitrogen isotopic ratio, in general, increases with increasing trophic levels of fish or animal species. Thus it was possible to confirm that mercury levels rose with the trophic level in both marine and freshwater fish. It was also of interest that, according to the nitrogen isotopic ratio, mercury levels in tissues of terrestrial animals also rose according to the trophic level.

A number of other elements were measured in the food of the Gidra villagers. The most important biological comparison is with selenium as this element protects against the toxicity of certain forms of mercury. In general, the selenium to mercury ratio in foodstuffs was always greater than unity but tended to approach unity when the mercury concentrations were highest. It was noted that the selenium to mercury ratio varied greatly between different populations.

<u>Studies on an Industrialized Population</u>

The study of an industrialized population focused on autopsy tissues collected in metropolitan Tokyo consisting of subjects with no known exposure to mercury.

Concentration of methylmercury fell within a narrow range in most tissues (to 50 ng/g wet weight). Liver was exceptionally high at 113 ng/g. The distribution of inorganic mercury was highly nonuniform. The levels in kidney and liver were much higher than those observed in cerebrum, cerebellum and spleen. The levels in autopsy hair samples were typical of the Japanese population. The average total mercury in hair was 4.7 microgram per gram with 87 to 99% in the form of organic mercury.

The relationship between mercury and selenium was also studied. Correlations of varying degrees of statistical significance were seen in a number of tissues between total, methyl and inorganic mercury on the one hand and with selenium on the other hand. The most striking correlation was found in autopsy kidney tissues between inorganic mercury and selenium. The atomic ratio of selenium to inorganic mercury was greater than unity at low mercury concentrations but approached unity at the highest mercury levels. It was noted that an atomic ratio of unity between inorganic mercury and selenium had been seen by other investigators in livers of sea mammals having a high dietary intake of methylmercury and in humans chronically exposed to high levels of mercury vapor.

CONCLUSIONS

Mankind has been exposed to both inorganic and methylmercury from time immemorial. From time to time and place to place, mercury levels may have risen to present an incipient toxic stress. Humans and other animals have adapted to this stress. In particular, populations dependent on fish, sea mammals and reptiles for a major part of their protein intake have experienced and continue to experience dietary intake of methylmercury far in excess of the general population. It is proposed that one adaptive mechanism is the biotransformation of methyl to inorganic mercury and subsequent excretion or inactivation of the inorganic mercury.  In particular, methylmercury is secreted mainly via the bile into the intestinal tract where it is demethylated by microflora.  Inorganic mercury is absorbed only to a small extent (about 7% of the oral intake) and is therefore eliminated in the feces.  Intact methylmercury is excreted to a negligible extent in either urine or feces.  Nevertheless, some inorganic mercury does reach the body tissues probably as a result of the small absorption from the intestine and also of other biotransformation processes that may be occurring in the body. Selenium may play a key protective role combining with inorganic mercury especially in the organ of highest accumulation, the kidney.  The induction of metallothionein by inorganic mercury may be an additional protective mechanism.

Adaptation to toxic stress from methylmercury may be least developed in the fetus and the neonate.  Methylmercury readily crosses the placenta and the fetal blood-brain barrier.  Microflora are absent in the fetus and, in the neonate, lack the capacity to demethylate.

In adults experiencing increasing exposures to mercury, the stores of available selenium will become depleted. Thus it is predicted that before mercury toxicity occurs, the individual will go into selenium deficiency.  Thus the selenium status could provide an early warning if incipient health risks from mercury occur.

# INCREASED METHYLMERCURY CONTAMINATION IN FISH IN

# NEWLY FORMED FRESHWATER RESERVOIRS

R.E. Hecky, D.J. Ramsey*, R.A. Bodaly, and N.E. Strange

Department of Fisheries and Oceans
Central and Arctic Region
Winnipeg, Canada R3T 2N6

*Agassiz North Associates
1214B Chevrier Blvd.
Winnipeg, Manitoba R3T 1Y3

## ABSTRACT

The formation of major new hydroelectric reservoirs in northern Canada in the last two decades was invariably followed by increased methylmercury concentrations in fish. Concentrations in piscivorous fish exceeded marketing limits and often approached mercury concentrations ($>5$ $\mu$g g-1 wet weight) in muscle formerly associated only with industrial mercury pollution. Experimental manipulations of large enclosures demonstrated that terrestrial vegetation and organic soils caused increased net methylation of mercury and bioaccumulation of methylmercury at low total mercury concentrations in water (1-2 ng L-1) and sediment (0.1-1.0 $\mu$g g-1 dry weight). Total mercury concentrations *per se* in water or sediments did not predict mercury concentrations in fish. Enhancement of microbial methylation relative to demethylation can be demonstrated in these new reservoirs and in reservoirs up to 60 years of age. Disruption of the natural microbially mediated mercury cycle accounts for the elevated Hg concentrations in fish, and indications are that it will be a persistent problem in boreal reservoirs. The reservoir experience emphasizes the critical role of microbial activity in mercury cycling. In natural lakes of the boreal forest, water temperature seems to be a critical variable controlling net mercury methylation by microbial activity.

*Advances in Mercury Toxicology*, Edited by T. Suzuki *et al.*
Plenum Press, New York, 1991

## INTRODUCTION

The Minamata tragedy alerted the world to methylmercury toxicity and the ability of the aquatic food chain to bioconcentrate methylmercury (D'Itri 1972). The sources of mercury and methylmercury at Minamata were industrial and local. Ever since then, industrial sources of mercury input have been sought whenever elevated levels of mercury concentration have been found in aquatic organisms. As a result of the Minimata experience, Canada, which has over 50% of the world's surface area of lakes, conducted a national mercury survey of its lake fisheries in the late 1960's. Canada also established acceptable marketing limits for mercury concentrations in fish products, currently 0.5 mg/kg, and instituted in 1970 a national inspection program on all commercial shipments of fish. This survey and inspection program did indeed identify a number of inland fisheries which were contaminated by industrial sources of mercury. But, it also identified a number of lakes far removed from industrial activity in sparsely populated northern Canada which had high concentrations of mercury in fish, i.e. in excess of marketable limits. The identification of the industrially contaminated fisheries led to the removal of mercury compounds from most industrial processes and to the introduction of waste water treatment for mercury removal. These industrially contaminated sites subsequently improved although not always to the extent desired (e.g. Rudd et al., 1983). The anomalous concentrations in remote lakes remained problematical, but the inspection program kept these fish from the market.

By the late 1970's the world's scientific community was alert to the hazards posed by long range transport of atmospheric pollutants. These pollutants arose primarily from the combustion of fossil carbon for energy production and from the smelting of base metals. These atmospheric pollutants can travel long distances from their sources and contaminate distant lakes. The best known and studied of these pollutants are the inorganic sulfur and nitrogen compounds which acidify rain and lakes. Mercury is also one of these pollutants (Joensuu 1971), and substantial concentrations of inorganic mercury can be found in rain, 2-20 ng/L, while methylmercury concentrations can be on the order of 0.15 ng/L (Bloom and Watras, 1989). An experimental acidification of Little Rock Lake (Wisconsin) was recently shown to increase methylmercury concentrations in fish (Wiener et al., 1990) while Fitzgerald and Watras (1989) have shown that atmospheric mercury deposition could account for the total mass of mercury in fish, water and accumulating sediment in Little Rock Lake. Long range transport of atmospheric mercury and acids could thus be contributing to the high mercury concentrations frequently found in fish in remote lakes. However, these processes should affect broad geographic regions equally, and there remains high lake to lake

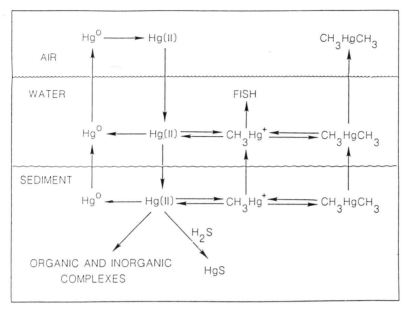

Fig. 1   The biogeochemical cycling of mercury in freshwater lakes (from Winfrey and Rudd, 1990).

variability in the mercury concentrations found in fish in any region. Explanation for this variability lies in the complexity of the aquatic mercury cycle (Figure 1).

Methylmercury is the compound of greatest concern in the aquatic mercury cycle. It is a powerful neurotoxin with 10 times the toxicity of the inorganic mercury compounds, and it bioconcentrates in fish muscle tissue where it strongly binds to sulfhydryl groups in muscle proteins. Methylmercury can be absorbed directly from water across the gills of fish, and it can be absorbed from food. These absorption processes have been shown to be sensitive to water quality and metabolic rates of fish (Rodgers and Beamish 1981, 1982, 1983).

Methylmercury can be produced by microbial activity (Jensen and Jernelov 1969), and it can be microbially degraded (Furukawa et al., 1969, Robinson and Tuovinen, 1984). These microbial activities can occur both within the water column, the sediments and even on and within fish (Rudd et al., 1980). Only the mercuric ion can be methylated probably through non-enzymatic transfer of methyl groups from methylcobalamin (Robinson and Tuovinen 1984). Abiotic methylation occurs but is thought to be of minor importance within the aquatic environment although it may be significant in terrestrial soil water (Berman and Bartha 1986). Demethylation is an enzymatically mediated reaction (Robinson and Tuovinen 1984). Because organisms that demethylate cannot methylate these processes are carried out by different microbes in the environment.

Methylmercury concentrations in fish consequently are dependent on a number of processes most important of which are the input of mercury, the net balance between methylation and demethylation of mercury, the absorption of methylmercury by fish and the growth rate of fish. Given the complexity of the aquatic mercury cycle which includes many independent abiotic and biotic processes, it is not surprising that there is a high variability among lakes of a region. To date environmental mercury research has largely focused on abiotic processes, but recent experience in new reservoirs in north temperate and subarctic regions has demonstrated the important role of biological processes particularly the balance of mercury methylation and demethylation processes in determining mercury concentrations in fish (Canada-Manitoba Mercury Agreement 1987). These studies will be reviewed below and their implications explored.

Temporal and spatial surveys on the Churchill River Diversion

In 1976 the level of Southern Indian Lake (northern Manitoba) was raised three meters to effect the diversion of 85% of the Churchill River flow from its natural course into the Nelson River watershed (Newbury et al. 1984). To control the southward diversion flow, a dam was built at Notigi Lake in the upper Nelson River watershed (Fig. 2). The impoundment of Southern Indian Lake resulted in marginal flooding of this large lake while the Notigi dam resulted in extensive flooding and the amalgamation of several smaller lakes into one large reservoir. There was also attendant flooding all along the diversion route because mean river discharge increased ten-fold.

These impoundments created extensive flooding of a permafrost affected, boreal forest terrain characterized by dense tree stands, peatlands and organic soils overlying glaciolacustrine clay deposits. The flooding and erosion which accompanied the impoundment and diversion caused numerous aquatic environmental impacts affecting the fisheries and aquatic resources utilized by aboriginal peoples at the town of South Indian Lake and the Nelson House reserve (Hecky et al., 1984). Southern Indian Lake was the site of a substantial commercial fishery and consequently there was a monitoring record for mercury concentrations in fish from the lake beginning in 1970.

By 1979 (Bodaly and Hecky 1979) it was evident in this monitoring record that mercury concentrations were increasing (Figure 3). Although Abernathy and Cumbie

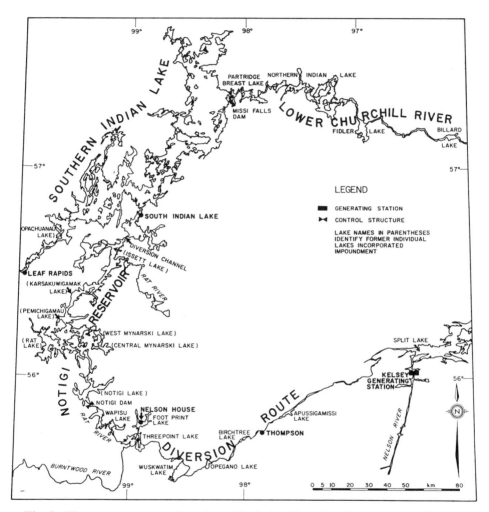

Fig. 2   The water systems of northern Manitoba (Canada) affected by the Churchill
River diversion (from Newbury et al., 1984).

(1977) had earlier inferred a possible link between flooding and mercury contamination in North Carolina reservoirs, Southern Indian Lake was the first record of mercury contamination of a fishery established through an observational time series extending from before to after impoundment. In addition, an attempt to open commercial fisheries in the Notigi Reservoir immediately ceased when quite high mercury concentrations were found in fish from this reservoir (Bodaly et al. 1984). In several watersheds draining into the Churchill River diversion, lakes flooded by higher water levels because of diversion were found to have higher mercury concentrations than unflooded lakes; and a general relationship between the degree of flooding and mercury concentrations in different fish species was evident (Canada-Manitoba Agreement 1987).

Surveys were conducted to determine if unusual soil or sediment mercury concentrations were the cause of the observed increases in fish mercury in the Churchill

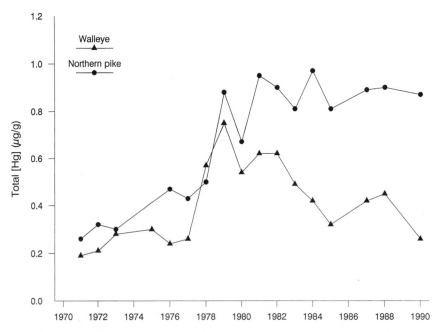

Fig. 3  Mean annual mercury concentrations in muscle tissue of the piscivorous northern pike and walleye caught in the commercial fishery of Southern Indian Lake. The federal mercury inspection program of commercial catches began in 1970. The lake was impounded in 1976 when its water level was raised 3 m above its long-term level.

River diversion reservoirs (Bodaly et al., 1987a). The flooded organic soils did have higher total mercury concentrations on a dry weight basis than the clay-rich lake sediments (Bodaly et al., 1984, 1987a), but the soils and sediments of the region had among the lowest mercury concentrations for these materials reported in the literature. A study of mercury deposition rates in a core of sediment from Southern Indian Lake confirmed that not only were sediment concentrations low but that deposition rates were also low, comparable to the Greenland ice cores (Bodaly et al., 1984). A survey of small fish species from numerous sites in the reservoirs and unflooded upstream lakes clearly showed that the reservoir fishes had higher mercury concentrations (Figure 4), but the considerable site-to-site variance could not be related to mercury concentrations in flooded or unflooded soils at the fishing sites (Bodaly et al., 1987).

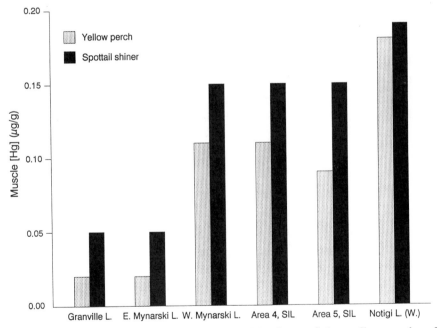

Fig. 4  Mercury concentrations in muscle tissue of the forage fishes yellow perch and spottail shiner in the unflooded Granville and East Mynarski Lakes compared to four flooded lakes. SIL = Southern Indian Lake where two different areas of this multibasin reservoir were sampled.

## Mesocosm studies

To unravel the complex of possible sources of mercury and the processes leading to its mobilization and bioaccumulation, a mesocosm approach was adopted. Ten meter diameter plastic enclosures were installed by SCUBA divers into 3-4 m of water in a protected bay of Southern Indian Lake. Hecky et al. (1987a) describe experimental designs and methods. The enclosures were open at the bottom and were installed over lake sediment except for one trial over flooded terrain. After removal of enclosed fish, one year old yellow perch (*Perca flavescens*) were introduced as assay organisms. In most experiments the radioisotope Hg-203 was added in a chloride solution in order to follow its uptake by fish under various experimental treatments. Because the Hg-203 was of low specific activity, a substantial amount of mercury was added with each radioisotope spike, 2-8 ng/L depending on the year and the volume of the mesocosm. Ambient mercury levels in the bay were generally less than 5 ng/L total mercury, the level of detection at that time (subsequent research indicated probable concentrations of 1-3 ng/L total Hg (Ramsey 1990b)). The experiments were normally run for sixty days with subsampling of fish at 1-3 week intervals. It was subsequently found that changes in non-radioactive mercury were easily detected in these fish, and the results reported below will emphasize the changes in total mercury in fish muscle. It was verified that 85-100% of the radioactive mercury taken into fish muscle was methylmercury. Experiments were initiated in four different years, and different treatments chosen and scaled to represent different reservoir conditions were studied. Control (untreated) mesocosms were established and monitored in each year. Details of treatments, observations and results are given in Hecky et al. (1987a).

The first year (1981) of experiments demonstrated that 63 kg (dry weight) of moss-peat material typical of the soils of the area had a dramatic effect on Hg-203 uptake in yellow perch (Figure 5).

In this mesocosm, mercury concentrations increased from 0.09 mg/kg to 0.48 mg/kg while concentrations in fish in the control enclosure actually fell from 0.09 mg/kg to 0.03 mg/kg. This treatment was repeated in three successive years and replicated in 1983 (Figure 6). In all years the moss-peat material stimulated mercury uptake. In 1982 it was found that there was a dose-response when a doubling of the amount of moss-peat added led to an approximate doubling of the uptake of mercury (Figure 7). In 1983 it was observed that the addition of bank clay materials, which were eroding into the lake, at concentrations of suspended sediments observed in the lake, did not modify the response of enhanced mercury uptake by perch in moss-peat treated limnocorrals (Figure 6). Although clay additions had been found to retard mercury bioaccumulation in similar experiments in the industrially mercury-contaminated English-Wabigoon River system (Rudd et al., 1983), several experiments with the bank clays eroding into Southern Indian Lake showed them to have little or a slightly stimulatory effect on mercury uptake by fish (Canada-Manitoba 1987). No mitigatory effect would be expected from clay additions short of burying all flooded terrain with a significant layer of bank clay.

All additions of terrestrial organic materials to the mesocosms resulted in enhancement of methylmercury uptake by fish. In 1984, moss-peat (53 kg d.w.), spruce boughs (31 kg d.w.) and prairie sod from southern Manitoba (610 kg d.w.) were compared in limnocorrals of similar volume (Hecky et al. 1987a). The sod

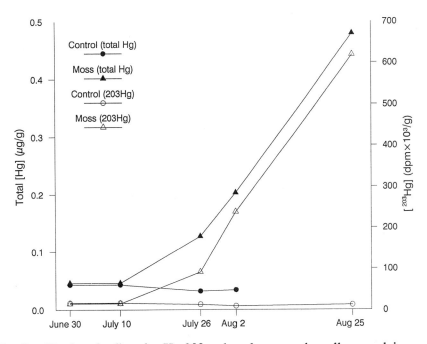

Fig. 5    Uptake of radioactive Hg-203 and total mercury by yellow perch in a mesocosm to which moss and peat have been added. The moss and peat were added with the radioisotope in late June.

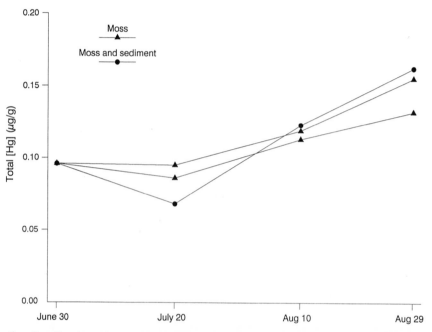

Fig. 6    Replication of mercury uptake into yellow perch muscle in two separate mesocosms receiving equal amounts of moss-peat material. Lake sediment was added in suspension to a third mesocosm receiving the same amount of moss-peat as the other two with little effect on mercury take.

treatment was meant to determine if similar mercury problems could be expected in different biophysical regimes such as the Canadian Prairies. In addition, the affect of the mercury added with the spike in previous experiments was investigated by adding an equivalent amount of moss-peat to a mesocosm without a radioisotope spike. Appropriate controls (with and without a Hg-203) spike were established. In an additional experiment to evaluate the role of mercuric ion concentration the amount of mercuric ion added by the spike was quadrupled in a mesocosm which received no other treatment. All the terrestrial organic materials stimulated mercury uptake (Figure 8).

The apparent lesser uptake of mercury by fish in the sod treated mesocosm resulted from much faster rates of growth in this mesocosm. Yellow perch in the sod-treated mesocosm grew from 1.7 g (initial mean weight of individual fish added) to 7.1 g over

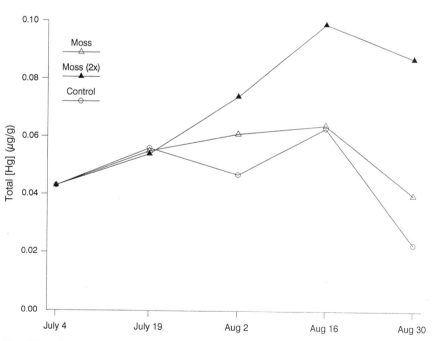

Fig. 7    The effect on uptake of mercury into yellow perch muscle tissues of doubling the amount of moss-peat material added during 1982 experiments.

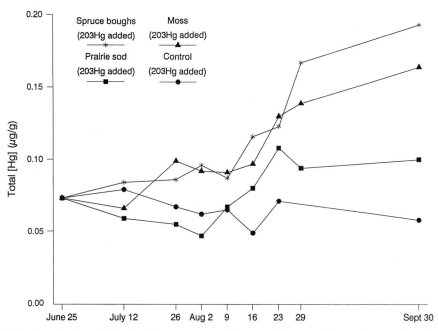

Fig. 8    Effect of adding different terrestrial organic materials on the uptake of mercury into yellow perch muscle.

the course of the experiment, compared to growth to a mean size of 2.9 to 4.4 g at the end of the experiments in the other mesocosms. This was the only treatment in the four years of experiments to elicit a significantly different fish growth rate. In 1984, the body burdens of methylmercury in fish at the end of the experiments were very similar for all three terrestrial organic treatments indicating similar rates of methylmercury production in all three mesocosms (Hecky et al. 1987a). The moss-peat treatment which did not receive a radioisotope spike, had mercury uptake rates comparable to the moss-peat mesocosm which received a spike, demonstrating that increased mercury uptake by fish was not dependent on receiving additional mercuric ion beyond what was available from natural sources in the mesocosm (Figure 9). Increasing the mercury spike concentration four-fold in an untreated mesocosm did not substantially affect it relative to standard controls (Figure 9).

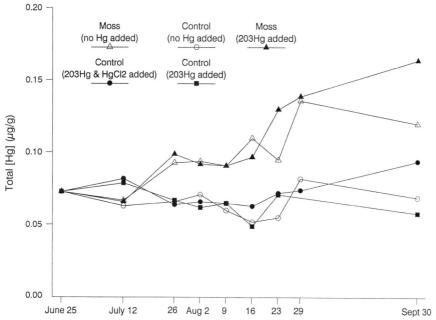

Fig. 9     Effect of different initial concentrations of added $HgCl_2$ on uptake of mercury into yellow perch muscle. Total mercury concentrations due to the initial spike were 0 ng L$^{-1}$ (as added $HgCl_2$) in one moss and one control mesocosm, 2.5 ng L$^{-1}$ with radioisotope spike and 8.0 ng L$^{-1}$ when $HgCl_2$ added.

The 1984 experiments were run for three months rather than two months during which the specific activity of muscle mercury apparently reached an asymptote indicating equilibrium with the natural sources of mercuric ion in the mesocosms. This equilibrium condition allows the calculation of the amount of mercury being mobilized within the mesocosms because the specific activity of the original radioisotope addition is known. The dilution of the specific activity of the spike indicates the size of the methylatable mercuric ion pool in the mesocosm (Table 1).

TABLE 1

| Limnocorral | Treatment (dry weight) | Total Hg in treatment | Total Hg added in spike | Spike S.A. | Fish S.A. | Available treatment Hg |
|---|---|---|---|---|---|---|
| | | mg (1) | mg (2) | $10^8$ Bq(mg Hg)$^{-1}$ (3) | (4) | mg (5) |
| SP84-3 | 31 kg spruce | 0.9 | 0.66 | 2.11 | 0.359 | 3.9 |
| SD84-4 | 610 kg sod | 36.6 | 0.66 | 2.11 | 0.293 | 4.8 |
| LM84-5 | 53 kg moss-peat | 2.1 | 0.66 | 2.11 | 0.156 | 8.9 |

There was a four-fold range in the size of this pool and in two cases the size of the pool was larger than the added materials could have provided. The latter cases demonstrate that the *in situ* lake sediments enclosed by the mesocosms can readily supply mercuric ion for methylation. The total mercury content of the upper two cm of the 75 square meter mesocosms would contain about 15 mg of mercury. Because the fish had similar body burdens of mercury at the end of the experiments, it is assumed that fish in the mesocosms were taking up methylmercury at similar rates. However, this similarity in methylmercury uptake, and its production at similar rates in these organically enriched mesocosms would not be predicted either by the incremental mercury added with the treatments or by the size of the methylatable mercury pool in the mesocosms. Table 1 shows that the moss-peat supplied more mercuric ion than the spruce boughs, the sod or the natural lake sediments, but this availability of mercuric ion did not significantly affect methylmercury concentrations in fish. This result is consistent with the general lack of effect of the Hg-203 spike or quadrupling mercury concentrations on results in treatments or controls, and this suggests that mercuric concentrations *per se* do not explain the enhanced mercury concentrations in fish in new reservoirs.

Studies of the methylation-demethylation balance

The lack of a relationship between mercuric ion concentrations and methyl mercury uptake in fish focused attention on the balance of the methylation and demethylation processes. Radioisotope methods were developed (Furutani and Rudd 1980, Ramlal et al., 1986) to allow estimation of the specific rates of these individual processes without resorting to prolonged incubations or high mercury concentrations which may be toxic to some bacteria. Because ambient concentrations of the relevant substrates, mercuric ion and methylmercuric ion, are not determinable in sediments, results are reported relative to the mercury compound added and the weight of sediment assayed. These are called specific rates of methylation (M) and demethylation (D). These rates are expressed as % (of added compound) processed per gram dry weight of sediment per hour. They are used in a comparative fashion as they are not true tracer experiments. The methods do enable evaluation of the factors which separately affect mercury methylation and demethylation rates (e.g. Ramlal et al., 1986, Xun et al., 1987).

The ratio of specific methylation to specific demethylation rates (M/D) has been used as a surrogate for net methylmercury production (Winfrey and Rudd 1990) because until recently, the ambient concentrations of aqueous methylmercury in these newly flooded reservoirs were below detection levels. Investigation of numerous flooded sites within the Churchill River diversion and other reservoirs has consistently shown that the flooded forest floor gives higher M/D ratios than offshore clay-rich sediments or the littoral (shallow water) environment in upstream unflooded lakes (Ramsey 1990a). This balance is independent of total mercury concentrations in sediment. One site on Southern Indian Lake has been monitored since 1983 and the difference between flooded soils and offshore sediments has been consistent over that time. Recently (Ramsey 1990b) the site with higher M/D has been shown to have higher methylmercury concentrations in water than the offshore site (0.034 ng/L vs 0.017 ng/L) despite the flooded site having lower concentrations of total mercury (1.2 ng/L vs 1.60 ng/L).This 3-fold difference in the ratio of methyl to total mercury in water between offshore (1%) and flooded sites (3%) results from the higher M/D ratio at flooded sites. These recent results support the use of the M/D ratio as being indicative of net methylation processes. The M/D ratio in sediments is highly seasonal with maximum rates usually observed in late June and August. The August period coincides with the time of maximum mercury uptake in our mesocosm experiments which were begun too late to observe any June peak. While winter methylation rates are consistently low, demethylation rates seem to be less sensitive to temperature (Ramlal et al. 1987). This strong seasonality also is indicative of the microbial control of the methylmercury cycle as the pattern is not readily explained by inorganic processes.

The M/D balance has been used to explore the question of the duration of elevated mercury concentrations in reservoirs. In Southern Indian Lake concentrations of mercury in piscivorous northern pike (*Esox lucius*) are as high 13 years after impoundment as they have been at any time over that period (Figure 3). Lake whitefish (*Coregonus clupeaformis*), a deep water dwelling benthivore exhibited an early maximum, within five years of impoundment, and its mercury concentrations have returned to near their pre-impoundment levels (Bodaly et al. 1984). The dietary and habitat preferences of the pike likely account for this difference in time course. The pike is piscivorous, and prefers shallow water environment with debris cover. These traits will maximize its exposure to the methylmercury being produced in flooded areas, and its concentrations are likely to stay high as long as the M/D balance of the flooded areas is elevated.

A study has been conducted on reservoirs, in northern Manitoba, ranging in age from that of Southern Indian Lake up to nearly 60 years in age. There is a general logarithmic decay of the mean seasonal value of M/D with age in these reservoirs (Figure 10; Ramsey 1990a), but very long periods of time, in excess of several decades, are required for the flooded areas of these boreal reservoirs to return to M/D rates comparable to those found in shallow waters of unflooded lakes. This is indeed an ominous conclusion for the people dependent on these fishery resources. Canada has a greater number and area of reservoirs than any country in the world, with the exception of the Soviet Union, and these numbers are increasing rapidly. Between 1975 and 1985, the aggregate area of new reservoirs built in Canada, is comparable to that of Lake Ontario, one of the world's great lakes. The next 10 years of reservoir construction could double that total. That aggregate lake would be one of the most mercury polluted water bodies in the world, and it will remain so for a long time.

CONCLUSION

The studies on the mercury contamination of new reservoirs calls into question the traditional view that new sources of mercury, usually industrial, are necessary to

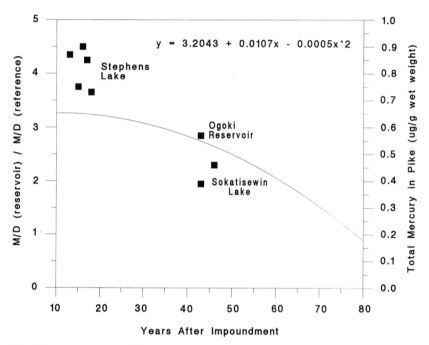

Fig. 10    The time course of the elevation of M/D in flooded areas above reference
(natural) values (solid line, left ordinate) and trend in fish mercury
concentrations (squares, right ordinate) in older reservoirs in northern
Manitoba.  The equation is the best fit to the ratio values which are calculated
from individual linear regressions for microbial methylation and microbial
demethylation on reservoir age.

account for increased mercury concentrations in fish (Hecky 1987). Rather, disruption of the natural cycling of mercury appears to be at fault based on our studies. Increased organic loading of terrestrial materials was the *sine qua non* for enhancement of mercury bioaccumulation. Increased availability of mercury or mercuric ion *per se* did not cause elevated mercury concentrations in mesocosm experiments which were designed to be as natural as possible. The results emphasize the role of microbial activity as controlled by decomposable, organic substrate and temperature as well as the role of mercury inputs in defining net methylation rates as Wright and Hamilton (1983) have previously shown in laboratory studies. These studies implicated the alteration of the microbially controlled balance of methylation and demethylation leading to higher methylmercury concentrations in water which then is bioaccumulated to high concentrations in fish muscle. Demonstration that an upset of the microbial balance can cause elevated mercury concentrations in fish should lead to a reconsideration of our paradigm of mercury contamination which emphasizes increased deposition of mercury as the ultimate cause of mercury contamination. A better understanding of the microbial ecology of methylating and demethylating bacteria is urgently required as other environmental perturbations aside from terrestrial organic loading may disrupt the balance of methylation and demethylation. It may be time to revisit those anomalous situations which have not recovered as planned after the elimination of industrial mercury or which are far removed from industrial activity and investigate the role of the aquatic microbial mercury cycle in the persistence of those problems. Modification of microbial populations might also provide an effective remedial action where other methods have been ineffective.

## REFERENCES

Abernathy, A.R. and Cumbie, P.M., 1977. Mercury accumulation by largemouth bass (*Micropterus salmoides*) in recently impounded reservoirs. Bull. Environ. Contam. Toxicol.. 17:595-602.

Berman, M. and Bartha, R., 1986. Levels of chemical versus biological methylation of mercury in sediments. Bull. Environ. Contam. Toxicol. 36:402-404.

Bloom, N.S. and Watras, C.J., 1989. Observations of methylmerjcury in precipitation. Sci. Total Environ. 87/88:199-207.

Bodaly, R.A., Hecky, R.E. and Fudge, R.J.P., 1984. Increases in fish mercury levels in lakes flooded by the Churchill River diversion, northern Manitoba. Can. J. Fish. Aquat. Sci. 41:682-691.

Bodaly, R.A. and Hecky, R.E., 1979. Post-impoundment increases in fish mercury levels in the Southern Indian Lake reservoir. Manitoba Can. Fish Mar. Serv. Manuscr. Rep. 1531:iv+15 p.

Bodaly, R.A., Strange, N.E., Hecky, R.E., Fudge, R.J.P. and Anema, C., 1987. Mercury content of soil, lake sediment, net plankton, vegetation, and forage fish in the area of the Churchill River diversion, Manitoba, 1981-82. Can. Data Rep. of Fish. Aquat. Sci. 610:iv+33 p.

Canada-Manitoba Mercury Agreement. 1987. Summary Report. Canada Manitoba Agreement on the Study and Monitoring of Mercury in the Churchill River Diversion. Winnipeg, Canada.

D'Itri, F.M., 1972. The Environmental Mercury Problem, The Chemical Rubber Co., Cleveland, Ohio.

Fitzgerald, W.F. and Watras, C.J., 1989. Mercury in surficial waters of rural Wisconsin lakes. Sci. Total Environ. 87/88: 223-232.

Furukawa, K., Suzuki, T. and Tonomura, K., 1969. Decomposition of organic mercurial compounds by mercury-resistant bacteria. Agric. Biol. Chem. 33:128-130.

Furutani, A. and Rudd, J.M.W., 1980. Measurement of mercury methylation in lake water and sediment samples. Appl. Environ. Microbiol. 40:770-776.

Hecky, R.E., 1987. Methylmercury contamination in northern Canada. Northern Perspectives 15(3):8-9.

Hecky, R.E., Bodaly, R.A., Ramsey, D.J., Ramlal, P.S. and Strange, N.E., 1987b. Evolution of limnological conditions, microbial methylation of mercury and mercury concentrations in fish in reservoirs of northern Manitoba, Canada-Manitoba Agreement on the Study and Monitoring of Mercury in the Churchill River diversion. Summary Report. Appendix 3:53 p. + 5 appendices.

Hecky, R.E., Bodaly, R.A., Strange, N.E., Ramsey, D.J., Anema, C. and Fudge, R.J.P., 1987a. Mercury bioaccumulation in yellow perch in limnocorrals simulating the effects of reservoir formation. Can. Data Rep. Fish Aquat. Sci. 628:v+158 p.

Hecky, R.E., Newbury, R.W., Bodaly, R.A., Patalas, K. and Rosenberg, D.M., 1984. Environmental impact prediction and assessment: the Southern Indian Lake experience. Can. J. Fish Aquat. Sci. 41:720-732.

Jensen, S. and Jernelov, A., 1969. Biological methylation of mercury in aquatic organisms. Nature (London) 223:753-754.

Joensuu, O.I., 1971. Fossil fuels as a source of mercury pollution. Science 172:1027-1028.

Newbury, R.W., McCullough, G.K. and Hecky, R.E., 1984. The Southern Indian Lake impoundment and Churchill River diverison. Can. J. Fish Aquat. Sci. 41:548-557.

Ramlal, P.S., Rudd, J.W.M. and Hecky, R.E., 1986. Methods of measuring specific rates of mercury methylation and degradatio and their uses in determining factors controling net rates of mercury methylation. Appl. Environ. Microbiol. 51:110-114.

Ramlal, P.S., Anema, C., Furutani, A., Hecky, R.E. and Rudd, J.W.M., 1987. Mercury methylation and demethylation studies at Southern Indian Lake, Manitoba: 1981-1983. Can. Tech. Rep. Fish Aquat. Sci., 1490:v+35 p.

Ramsey, D.M., 1990a. Experimental studies of mercury dynamics in the Churchill River diversion, Manitoba. Collection Environment et Geologie, 9:147-173.

Ramsey, D.M., 1990b. Measurements of methylation balance in Southern Indian Lake, Granville Lake, and Stephens Lake, Manitoba, 1989. Northern Flood Agreement Federal Environmental Monitoring Program Rep. (in press).

Robinson, J.B., and Tuovinen, O.H., 1984. Mechanisms of microbial resistance and detoxification of mercury and organomercury compuonds: physiological, biochemical, and genetic analyses. Microbiol. Reviews 48:95-124.

Rodgers, D.W. and Beamish, F.W., 1981. Uptake of waterborne methylmercury by rainbow trout (*Salmo gairdneri*) in relation to oxygen consumption and methylmercury concentration. Can. J. Fish Aquat. Sci. 38:1309-1315.

Rodgers, D.W. and Beamish, F.W., 1982. Dynamics of dietary methylmercury in rainbow trout (*Salmo gairdneri*), Aquat. Toxicol. 2:271-290.

Rodgers, D.W. and Beamish, F.W., 1983. Water quality modifies uptake of waterborne methylmercury by rainbow trout (*Salmo gairdneri*). Can. J. Fish Aquat. Sci. 40:824-828.

Rudd, J.W.M., Furtani, A. and Turner, M., 1980. Mercury methylation by fish intestinal contents. Appl. Environ. Microbiol. 40(4):777-782.

Rudd, J.W.M., Turner, M.A., Furutani, A., Swick, A. and Townsend, B.E., 1983. A synthesis of recent research with a view towards mercury amelioration. Can. J. Fish Aquat. Sci. 40:2206-2217.

Wiener, J.G., Fitzgerald, W.F., Watras, C.J. and Rada, R.G., 1990. Partitioning and bioavailability of mercury in an experimentally acidified lake. Environ. Toxicol. Chem. 9:909-918.

Winfrey, M.R. and Rudd, J.W.M., 1990. Environmental factors affecting the formation of methylmercury in low pH lakes: a review. Environ. Contam. Toxicol. 9:853-869.

Wright, D.R. and Hamilton, R.D., 1982. Release of methylmercury from sediments: effects of mercury concentration, low temperature, and nutrient addition. Can. J. Fish Aquat. Sci. 39:1459-1466.

Xun, L., Cambell, N.E.R., and Rudd, J.W.M., 1987. Measurements of specific rates of net methylmercury production in the water column and surface sediments of acidified and circumneutral lakes. Can. J. Fish Aquat. Sci. 44:750-757.

# SPECIATION OF MERCURY IN THE ENVIRONMENT

Hirokatsu Akagi[1] and Hajime Nishimura[2]

[1] National Institute for Minamata Disease
Minamata, Kumamoto 867, Japan

[2] Faculty of Engineering, University of Tokyo
Bunkyo-ku, Tokyo 113, Japan

## ABSTRACT

Despite a large number of investigations concerning the distribution of mercury in the aquatic environment, our understanding of the overall dynamics of environmental mercury is still unsatisfactory. This may be mainly because little information is available on the distribution of mercury in the methylated form that is more biologically available for aquatic organisms.

Hence, the work reported here was undertaken to establish a highly sensitive and reliable method for the determination of methylmercury as well as total mercury in various biological and non-biological environmental samples by the combination of dithizone extraction and ECD-gas chromatography. These samples include natural water containing mercury down to parts per trillion levels. The method for methylmercury is based on the fact that methylmercuric dithizonate in the final sample solution is converted into methylmercuric chloride as soon as it is subject to ECD-gas chromatography.

The differences in the composition of the samples determine the pretreatment of the sample for the highest extraction efficiency. Thus, the biological sample, suspended solids and sediment were treated with KOH in ethanol, whereas the water sample was treated with $KMnO_4$ and sulfuric acid. After pretreatment, methylmercury in the sample

*Advances in Mercury Toxicology*, Edited by T. Suzuki *et al.*
Plenum Press, New York, 1991

was extracted with dithizone in benzene and then back-extracted into alkaline Na2S solution. The excess sulfide ions were removed by acidification with HC1 and purging with $N_2$ and methylmercury was re-extracted with a small portion of dithizone in benzene. The extract was washed with NaOH solution and subjected to the conventional ECD-gas chromatography.

## INTRODUCTION

It is generally known that almost all fish contain at least trace amounts of mercury mostly in the form of methylmercury. So far, a large amount of research has been reported on where the methylmercury comes from and how it is accumulated in fish, but our understanding of the overall dynamics of environmental mercury is still unsatisfactory.

The conversion of inorganic mercury into methylmercury in the sediment has long been recognized as a critical step in the environmental behavior of this metal. Thus, it is reasonable to assume that methylmercury, produced in the bottom sediment is released into the overlying water and ultimately accumulated in fish tissue. Although much information on total mercury levels in various environmental samples has been accumulated in many countries, there is very little data on methylmercury that is more biologically available for aquatic organisms. Therefore, the speciation of mercury in the environment is of great importance in surveying the pathway of methylmercury accumulated in fish. In particular, the levels of total mercury in natural water is extremely low and its accurate determination is still a major problem. The purpose of this study is to establish simplified, sensitive, and reliable methods for the determination of both total and methylmercury found in various biological and non-biological samples, including natural water.

Analytical Method for Total Mercury

The most commonly used method for determining total mercury in biological and environmental samples is cold vapor atomic absorption spectrometry. The method involves sample digestion with strong acids followed by reduction to elemental mercury, aeration and measurement of mercury absorption.

This procedure is certainly sensitive but both the apparatus and digestion method are insufficient for accurate determination of the background level of mercury in the environment.

Apparatus

Figure 1 shows the most widely used apparatus for the determination of total mercury. In this system, elemental mercury vapor generated by the reaction with tin (II) compound is continuously circulated and absorbence is measured when equilibrium of mercury vapor between gas phase and liquid phase is reached. It is apparent that mercury vapor is diluted invariably by air during the circulation and subsequently the sensitivity is reduced.

Another problem is the loss of mercury by vaporization. In this technique, reducing agent is added to the sample solution before the closed system is formed. During this operation, loss of mercury was found to occur, unless addition of the reducing agent was made with much care. Therefore, the reducing agent should be added after the reaction vessel is attached to the closed circulating system to prevent such a mercury loss.

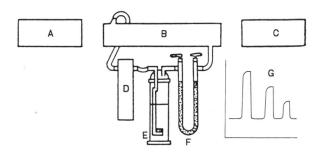

Fig. 1.        Closed system for atomic absorption spectrometric analysis of mercury.
               A, Mercury lamp; B, Absorption cell;, C, Detector; D, Pump;
               E, Reaction vessel; F, Drying agent; G, Recorder response.

The mercury vapor introduction system as improved in our laboratory, is shown in Figure 2. This apparatus consists of an air circulation pump, a reaction vessel, an acidic gas trap and a four-way stop-cock, connected by tygon tubing. A known volume of sample solution is placed in the reaction vessel and made up to 20 ml with mercury free distilled water. The vessel is stoppered tightly and then attached to the closed circulation system. After 1 ml of tin (II) chloride solution is injected from the syringe, the air is circulated through the four-way stop-cock to allow the concentration of mercury vapor to come to an equilibrium. During this circulation, acidic gases leaving from the sample solution, are removed by sodium hydroxide solution.

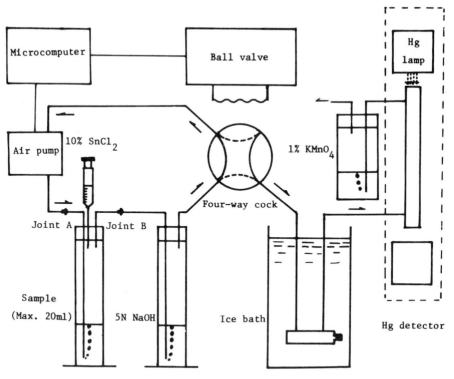

Fig. 2   Cold vapor introduction apparatus

After a definite time of circulation, usually 30 seconds, the mercury vapor is introduced to the absorption cell by turning the four-way stop-cock 90°, and the mercury absorption recorded.  When the maximum peak height has been attained, the reaction vessel is removed and then joints A and B are connected directly to purge the residual mercury vapor.  With this procedure, the whole determination is completed in one minute. This manual system can be easily improved to a semi-automated system which is more convenient to operate.  This could be achieved by attaching an electrically driven ball valve which could turn 90° at one time to the four-way stop- cock, and the operations of both ball valve and air circulation pump being controlled by a microcomputer.

The results of repeated measurements for standard solution were consistent and sufficiently sensitive, as shown in Figure 3. With this mercury analyzer, around 0.5ng mercury can be accurately determined.

Sample Digestion Procedure

For accurate analysis of total mercury, complete digestion of organic materials in the sample is essential. A number of wet digestion procedures have been proposed but most of them involve time consuming operation steps and require considerable numbers of reagents and careful handling during digestion. There is a need for a simple, rapid and effective method which is applicable to various environmental samples.

As a major problem in the sample preparation procedure is the loss of mercury during the digestion process, we attempted to prevent the loss of mercury with various combinations of acids and oxidizing agents. The results for standard methylmercury dithizonate (50 ng as Hg) digested on the hot plate at 250°C for 0.5-1hr are listed in Table 1. An apparent loss of mercury was observed in nitric acid and in a mixture of nitric acid and sulfuric acid, but excellent results were obtained in the mixtures with perchloric acid or potassium permanganate. These results show that the presence of oxidizing agents prevent complete mercury loss even under severe heating conditions. In this experiment, the vessel used for the preparation of the sample solution is a 50ml volumetric flask with a thick wall. A condenser, commonly used for preventing the loss of mercury, was not used. The complete prevention of mercury loss is presumably due to the long neck of the volumetric flask used for sample digestion.

Table 1   Losses of mercury during wet-digestion under various conditions

| System | Heating time (hr) | Mercury lost (%) |
|---|---|---|
| $HNO_3$ 10ml | 0. 5 | 6 1. 7 |
| $HNO_3$ 10ml $H_2SO_4$ 10ml | 0. 5 1. 0 | 8. 8 4 4. 2 |
| $HNO_3$ 10ml $H_2SO_4$ 10ml $HClO_4$ 2ml | 0. 5 1. 0 | 0 0 |
| $HNO_3$ 10ml $H_2SO_4$ 10ml 5% $KMnO_4$, 1ml | 0. 5 1. 0 | 0 0 |

Dithizone-benzene extract (10ml) containing 50ng Hg was evaporated to dryness in a 50ml volumetric flask with thick wall and then digested with each system on hot plate at 250 ° . After cooling, the digested sample was made up to 50ml with water and analysed by atomic absorption spectrometry.

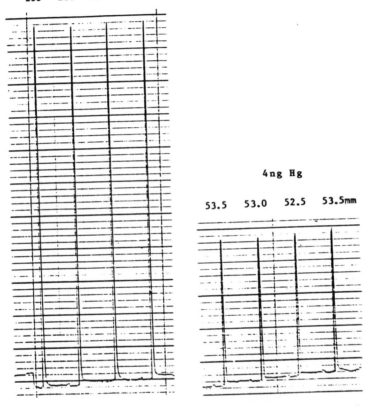

Fig. 3  Reproducibility of mercury measurements by the semi-automated system.
Sample: 100-250ng Hg/50 ml
Sample size: 2 ml

Of these two oxidizing agents, perchloric acid was chosen because the use of potassium permanganate required one more reagent to reduce it. In order to obtain an optimum condition for the rapid digestion of various environmental samples, the effects of temperature on sample digestion were examined using fish tissues. As shown in Figure 4, total mercury detected increased with increase in the temperature of the hot plate. At 250°C the value of total mercury detected did not change with heating time, indicating that the digestion process was completed in 10 minutes.

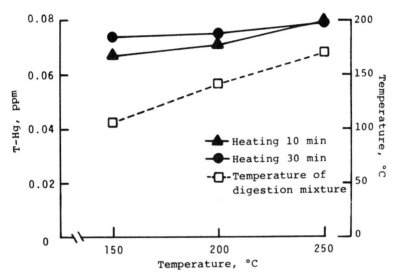

Fig. 4. Effect of digestion temperature on total mercury determination in fish tissues.
Digestion mixture: $HNO_3$ - $H_2SO_4$ - $HClO_4$ (2+10+2)
Sample: Mackerel

The temperature of the digestion mixture increased with increasing temperature of the hot plate and reached to 170°C at the hot plate temperature of 250°C. Another experiment indicated that the recovery of total mercury decreased slightly with the increase in the amounts of nitric acid added. The highest total mercury value obtained in a short time is presumably due to the increased temperature of the digestion mixture, leading to enhancement of the digestion process.

The final procedure for biological samples is shown in Chart 1. A known amount of sample is placed in a 50ml volumetric flask, to which 2 ml of nitric acid, 10ml of sulfuric acid and 2ml of perchloric acid are added and heated on a hot plate at 250°C for 30 min. After cooling, the digested sample is made up to 50 ml with water and analyzed by cold vapor atomic absorption spectrometry. The digestion procedure was successful with fish samples; the recoveries were 98.98-103.60% for two species of fish fortified with 20-100ng of methylmercury, as shown in Table 2. This procedure is simple and rapid but it is not suited for the digestion of a large amount of tissue sample because of violent frothing on heating. For organic samples like fish tissue, 1g or less of the sample should be used for analyzing total mercury. This procedure is also applicable to environmental samples such as sediment, suspended solids, plankton, etc., except for water samples. In sediment samples, up to 5 g of the sample were used for the digestion and successful recoveries were obtained similar to fish tissues. This procedure is simple and can analyze accurately as many as 100 samples a day.

The concentration of mercury in water is extremely low and its measurement requires some preconcentration. Of the methods reported in the literature for preconcentration of mercury in water samples, dithizone extraction technique described by Chau and Saitoh (1970), was supposed to be suitable for this purpose. However, this technique was found to be insufficient to extract total mercury at extremely low levels in water samples. Therefore, different pretreatments of water samples prior to dithizone extraction were examined to obtain satisfactory recoveries. As a result, pretreatment with sulfuric acid and potassium permanganate was found to be essential for extraction of total mercury present in water samples as shown in Figure 5.

Chart 1   Analytical procedure for total mercury
in biological and sediment samples

```
Sample (in 50ml Measuring flask)
    |   HNO₃, 2ml
    |   H₂SO₄, 10ml
    |   allowed to stand for 5min.
    |   HClO₄, 2ml
    |   Heated at 250°C for 30min.
    |   cooled down to room temperature
Digested Sample
    |   H₂O
    |   made up to 50ml
Sample Solution, 0~20ml
    |   10% SnCl₂, 1ml
    A.A.S.
```

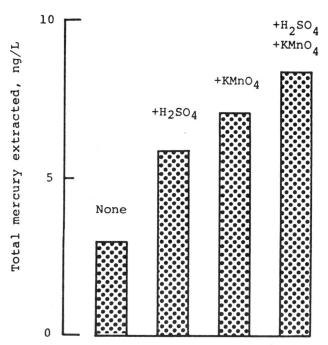

Fig. 5   Effects of pretreatment with $H_2SO_4$ and $KMnO_4$ on dithizone extraction of
mercury in seawater.
   $H_2SO_4$: 10ml of 20N $H_2SO_4$ / 2L of seawater
   $KMnO_4$: 5ml of 1% $KMnO_4$ /2L of seawater

Table 2  Recoveries of mercury added to fish samples

| Sample | Mercury added (ng) | Mercury found (ng) | Recovery av. % |
|--------|------|------|------|
| Mackerel (1g) | 0 | 72.27 | —— |
|  | 0 | 71.44 |  |
|  | 20 | 92.78 | 103.01 |
|  | 20 | 92.17 |  |
|  | 50 | 119.07 | 98.98 |
|  | 50 | 123.61 |  |
| Horse mackerel (1g) | 0 | 142.05 | —— |
|  | 0 | 142.05 |  |
|  | 50 | 190.26 | 103.60 |
|  | 50 | 197.44 |  |
|  | 100 | 243.08 | 103.08 |
|  | 100 | 247.18 |  |

Methylmercury–cysteine solution was added to each sample.

The reason for this action is presumably due to the liberation of ionized mercury compounds from binding sites on organic material and particulate matter in water samples by oxidation. In this pretreatment, the selection of the concentrations of these two reagents and the pretreatment time is of great importance, because acidic permanganate reacts with chloride ions to produce chlorine which results in the oxidation of dithizone when it is applied to seawater. The preconcentration method was finally obtained and subsequent procedures are shown in Chart 2. Two liters of a water sample in a separatory funnel was mixed with 10ml of 20N sulfuric acid and 5ml of 0.5% potassium permanganate and allowed to stand for 5 minutes. The treated sample was then neutralized with 20ml of 10N sodium hydroxide and 5ml of 10% hydroxylamine hydrochloride and allowed to stand for 20 minutes. After addition of 5ml of 10% ethylenediaminetetraacetic acid tetrasodium salt, the sample was extracted with 20ml of purified 0.01% dithizone in benzene. The sample was left until it separated and the water layer was discarded. A known volume of benzene layer (usually 10ml) was transferred into a 50ml volumetric flask with a thick wall and evaporated to dryness using a rotary evaporator. The residue in the volumetric flask was digested with nitric acid - sulfuric acid - perchloric acid (2+10+2) system, diluted and measured by cold vapor atomic absorption spectrometry in the same manner as described above.

In this procedure, carrying mercury standard solutions through the procedure is needed for the determination of mercury, because the volume of the benzene layer decreases considerably by partitioning.

To make sure of the validity of this method recovery tests were carried out. The results are shown in Table 3. Quantitative recoveries were obtained for seawater with added mercury at levels of 1-2ng/l. The detection limit of this procedure was estimated to be around 0.2ng/l for 2L of the water sample.

## Analytical Methods for Methylmercury

### Pretreatment and Extraction

The most widely used procedure originated by Westöö (1966), for the determination of methylmercury in environmental samples such as fish and sediment has the disadvantages of poor extraction efficiency and of requiring a long time. The major problem encountered is that most biological samples form thick emulsion at the initial benzene extraction stage. This requires high speed centrifugation and repeated extraction with benzene as well.

Previously, we have developed a highly sensitive radiochemical technique for studying the behavior of mercury compounds in a given environment and in experimental animals using radioactive tracer (Akagi et al., 1979; Czuba et al., 1981). This technique involves extraction of total mercury with dithizone benzene, separation of the resulting inorganic and organic mercury dithizonate by thin layer chromatography followed by gamma-counting of each fraction for 203-mercury. In the course of this study, we have found that the pretreatment of biological and sediment samples by potassium hydroxide in ethanol is very effective to extract mercury with dithizone benzene solution.

On the other hand, Westöö (1966) has studied gas chromatographic methods for estimating methylmercury compounds in fish and showed, incidentally, that methylmercuric dithizonate had the same retention time as its chloride or bromide. This finding has strongly suggested that the determination of methylmercury in various samples by the combination of dithizone extraction and gas chromatographic separation is possible.

The chemical form of methylmercury compound detected by gas chromatography of its dithizonate was identified as methylmercury chloride by gas chromatography-mass spectrometry as shown in Figure 6 indicating that methylmercuric dithizonate was converted into its chloride as soon as it was subjected to gas chromatography.

Thus, this study was designed to establish a systematic, sensitive and reliable method for the determination of methylmercury in various biological and non-biological

Table 3 Recoveries of total mercury added to seawater. Methylmercury-cysteine solution was added to each sample

|  | Added (ng/L) | Found (ng/L) | Recovered (ng/L) | Recovery (%) |
|---|---|---|---|---|
| Sample A | none | 1.74 | —— | —— |
|  | 1.0 | 2.68 | 0.94 | 94.0 |
|  | 2.0 | 3.77 | 2.03 | 101.5 |
| Sample B | none | 2.80 | —— | —— |
|  | 1.0 | 3.76 | 0.96 | 96.0 |
|  | 2.0 | 4.84 | 2.04 | 102.0 |

Methylmercury-cysteine solution was added to each sample.

Chart 2 Analytical procedure for total-Hg in seawater

```
Seawater, 2L
    | 20N H₂SO₄, 10ml
    | 0.5% KMnO₄, 5ml
    | allowed to stand for 5min.
    | 10N NaOH, 20ml
    | 10% NH₂OH·HCl, 5ml
    | allowed to stand for 20min.
    | 10% EDTA(-4Na), 5ml
    | extracted with 0.01% Dithizone in benzene, 20ml
 _____|_____
|                       |
Aq. layer      Bz layer, 10ml (in 50ml measuring flask)
               | evaporated to dryness
               Residue
               | HNO₃ - HClO₄ (1+1), 4ml
               | H₂SO₄, 10ml
               | heated at 250°C for 30min.
               | cooled down to room temperature
               | H₂O, made up to 50ml
               Sample solution (Max. 20ml)
               | 10% SnCl₂, 1ml
               Hg
               |
               A.A.S.
```

Chart 3  Analytical procedure for methylmercury in biological samples

```
        Sample (0.5-1g)
             |  1N KOH-EtOH, 20ml
             |  heated at 100°C for 1hr.
    Digested  sample
             |  1N HCl, 20ml
             |  20% EDTA(-4Na), 2ml
             |  washed with n-hexane, 5ml
    Aqueous  layer
             |  extracted with 0.01% dithizone in benzene, 10ml
    Benzene  layer
             |  washed with H₂O, 20ml
             |  washed twice with 1N NaOH, 10ml
    Benzene  layer, 4ml
             |  back-extracted with 0.02% Na₂S
             |                     in 0.2N NaOH-EtOH(1 +1), 4ml
    Aqueous  layer
             |  washed with benzene, 2ml
    Aqueous  layer
             |  slightly acidified with 1N HCl
             |  N₂-bubbling, 50ml/min, 5min.
             |  Walpole's Buffer (pH3.00), 5ml
             |  extracted with 0.05% dithizone in benzene, 2ml
    Benzene  layer
             |  washed twice with 1N NaOH, 2ml
             |  washed with H₂O, 4ml
             |  acidified with 1N HCl, 5drops
         GLC
```

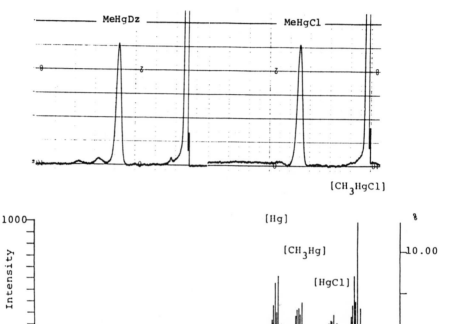

Fig. 6  Gas chromatography - mass spectrometry spectrum of methylmercuric dithizonate.

environmental samples with the combined techniques of dithizone extraction and gas chromatography. The outline of the analytical methods for methylmercury in various samples is shown in Figure 7. Alkaline digestion is advantageous for analysis of methylmercury, particularly in biological samples. The digests are clear and do not form any emulsion on initial solvent extraction, due to the breakdown of proteinacious materials in the sample matrics during digestion. Subsequently, an efficient recovery of methylmercury can be obtained.

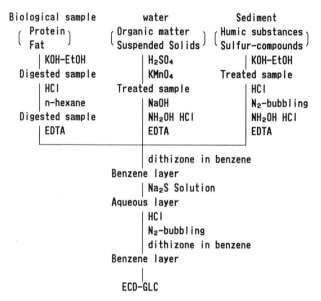

Fig. 7. Outline of the analytical methods for methylmercury in various biological and non-biological samples.

For alkaline digestion, 1N potassium hydroxide ethanol solution, widely used for analyzing PCB in tissues, was used.

This alkaline treatment was also applied to the pretreatment of sediment samples. Sediment samples often contain humic substances and sulfide ions having a great affinity for mercury compounds. The humic substances are soluble in alkaline solution and thus methylmercury bound to humic substances can be liberated from sediment by alkaline treatment, while the elimination of sulfide ions can be achieved by bubbling air or nitrogen through the treated sample solution after slightly acidified with hydrochloric acid.

Chart 4   Analytical procedure for methylmercury in sediments

```
Sample  1-5g  (wet weight)
       | 1N KOH-EtOH,  20ml
       | shaken for 15min.
       | 2N HCl,  10ml
       | N₂-bubbling with stirring,  100ml/min,  5min
Digested  sample
       | 20% NH₂OH · HCl,  2ml
       | 20% EDTA(-4Na),  2ml
       | shaken for 10min.
       | extracted with 0.05% dithizone in benzene,  10ml
Benzene  layer,  8ml
       | washed twice with 1N NaOH,  10ml
Benzene  layer,  4ml
       | back-extracted with 0.02% Na₂S
       |                 in 0.2N NaOH-EtOH (1+1),  4ml
Aqueous  layer
       | washed with benzene,  2ml
Aqueous  layer
       | slightly acidified with 2N HCl
       | N₂-bubbling,  50ml/min,  5min.
       | Walpole's Buffer (pH3.00),  5ml
       | extracted with 0.05% dithizone in benzene,  2ml
Benzene  layer
       | washed twice with 1N NaOH,  2ml
       | washed with H₂O,  4ml
       | acidified with 2N HCl,  5drops
     GLC
```

A known amount of a finely-chopped biological sample (0.5-1g) was placed in a 50ml centrifuge tube and digested with 20ml of 1N potassium hydroxide solution at 100°C for 1 hr. Sediment samples did not require high temperatures. They were simply shaken with 20 ml of 1N potassium hydroxide for 15 minutes using a mechanical shaker.

The pretreatment procedure for water samples is the same as that in the total mercury analysis in water described above.

The digested biological sample was slightly acidified with 20ml of 1N hydrochloric acid and shaken with 5ml of n-hexane to extract fat and fatty acids. After the hexane layer was removed, the sample was mixed with 2ml of 20% ethylenadiamine tetraacetic acid tetrasodium salt and then extracted with 10ml of purified 0.05% dithizone benzene solution.

Chart 5  Analytical procedure for methylmercury in water

```
Water  Sample, 2L (X3)
        |   20N H₂SO₄, 10ml
        |   0.5% KMnO₄, 5ml
        |   allowed to stand for 5min.
        |   10N NaOH, 20ml
        |   10% NH₂OH·HCl, 5ml
        |   allowed to stand for 20min.
        |   10% EDTA(-4Na), 5ml
        |   0.01% dithizone in benzene, 20ml
Benzene  layer combined
        |   washed twice with 0.5N NaOH, 60ml
Benzene  layer, 40ml
        |   back-extracted with 0.02% Na₂S
        |                   in 0.2N NaOH-EtOH (1+1), 10ml
Aqueous  layer
        |   washed with benzene, 5ml
Aqueous  layer
        |   slightly acidified with 2N HCl
        |   N₂-bubbling, 100ml/min, 5min.
        |   Walpole's Buffer (pH3.00), 5ml
        |   extracted with 0.01% dithizone in benzene, 0.4ml
Benzene  layer
        |   washed twice with 0.5N NaOH, 2ml
        |   washed with H₂O, 2ml
        |   acidified with 2N HCl, 2drops
        GLC
```

The treated sediment sample with 1N potassium hydroxide was bubbled with nitrogen gas through the solution for 5 min at a flow rate of 100ml/min., after acidified with hydrochloric acid. The sample was then mixed with 2ml of 20% hydroxylamine hydrochloride and 2ml of 20% ethylenadiamine tetraacetic acid tetrasodium salt and extracted with 10ml of purified 0.05% dithizone benzene solution. The purpose of the addition of ethylenediamine tetraacetic acid terasodium salt and hydroxylamine hydrochloride is to protect dithizone from unnecessary consumption by the other metal ions and oxidation, respectively. For the analysis of methylmercury in water, 6L of the water sample was used for dithizone benzene extraction, considering the extremely low concentration of methylmercury present in the water sample.

## Clean-up Procedure

In  most cases, cleanup of the initial extract is effected by conversion of methylmercury into a water soluble form such as cysteine or glutathione complex. Sporak (1956) has described a method for back-extraction of organomercury compound from benzene with aqueous sodium sulfide. Although sulfur compounds are known to be poisonous to electron capture detectors, sodium sulfide has a great affinity for mercury compounds and the excess of this reagent can easily be removed as hydrogen sulfide by bubbling with air or nitrogen through the aqueous layer after acidification with hydrochloric acid. In order to seek the most efficient back-extraction procedure, various kinds of sodium sulfide solutions were tested for methylmercury dithizonate and methylmercuric chloride in benzene. The results are illustrated in Figure 8. The extraction efficiency of methylmercuric dithizonate from benzene solution by neutral solutions of sodium sulfide were insufficient, but satisfactory recoveries were obtained in alkaline ethanol solution of sodium sulfide. A similar tendency was observed for methylmercuric chloride in benzene, although the values were relatively low. Loss of methylmercury from each extract by bubbling nitrogen was examined. Each back-extract was acidified slightly with 1N hydrochloric acid and nitrogen gas was bubbled through the solution for 5 minutes at a flow rate of 50ml/min. No significant loss of methylmercury was observed, as shown in Figure 8.

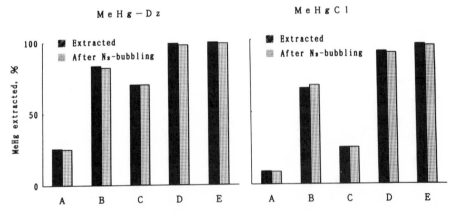

Fig. 8 Extractability of methylmercury from benzene solution of MeHgDz and MeHgCl by various 0.02% $Na_2S$ solutions and the loss of methylmercury from each solution by $N_2$-bubbling (50ml/min, 5 min) after acidification with HC1.

A: $H_2O$, B: 0.1N NaOH, C: $H_2O$-EtOH (1+1), D: 0.1N NaOH-EtOH, E: 0.2N NaOH-Etoh (1+1)

Based on these experimental results, 0.02% sodium sulfide in 0.2N sodium hydroxide-ethanol (1+1) was adopted as a reagent for back-extraction. The dithizone benzene extract from biological and non-biological samples was washed twice with 1N sodium hydroxide to remove the excess of dithizone and a portion of benzene layer was back-extracted into 0.02% sodium sulfide in 0.2N sodium hydroxide-ethanol (1+1). In the fish and sediment samples, 4 ml out of 10 ml of extract was transferred into a 10 ml centrifuge tube and back-extracted with 4 ml of 0.02% sodium sulfide solution. In water samples, however, dithizone benzene extracts from three 2L of water were combined and washed twice with 0.5N sodium hydroxide. Then 40 ml aliquot of the extract was transferred into a 50ml centrifuge tube and back-extracted with 10 ml of 0.02% sodium sulfide solution.

Re-extraction Procedure

After the benzene layer was completely drawn out by suction, 1N hydrochloric acid was added dropwise to the aqueous layer until a blue color appeared with bubbling with nitrogen through the aqueous layer and the bubbling was continued for another 5 min. Five ml of Walpole's buffer (pH 3.0) was added to the aqueous layer and the mixture was re-extracted with dithizone benzene solution. The benzene layer was washed twice with diluted sodium hydroxide and the benzene layer was transferred to another 10 ml centrifuge tube and washed with distilled water. After the water layer was removed, the benzene layer was acidified with a small volume of diluted hydrochloric acid. Ten µl of this benzene layer were injected into a conventional gas chromatograph with an electron capture detector. The final sample solution should be stored in the dark, as the complex is labile to light.

The final analytical procedures for methylmercury in biological samples, water, and sediment are shown in charts 3-5.

In order to test the reproducibility of the present method, six replicate samples of horse mackerel were analyzed. The mean values and standard deviations were 155.35ng/g and 3.41ng/g, respectively. The validity of each analytical procedure was examined by checking the recoveries of methylmercury. Fish, seawater and sediment samples were spiked with methylmercury at low concentrations close enough to individual background levels. The results are shown in Table 4.

Excellent recoveries were observed for all kinds of samples even in the water sample with added methylmercury at a level of 0.1ng/l. The present method for biological samples was also evaluated by comparing with the conventional benzene extraction method. Three different fish containing mercury at levels of 318.5, 179.4 and 32.6ng/g

Table 4   Recovery of methylmercury from fish, sediment and seawater

| Sample | Methylmercury added(ng) | Methylmercury found(ng) | Recovery(%) |
|---|---|---|---|
| Sardine, 1g | 0 | 73.5 | — |
| | 50 | 121.0 | 95.0 |
| | 100 | 176.5 | 103.0 |
| | 200 | 266.0 | 96.3 |
| Rockfish, 1g | 0 | 63.0 | — |
| | 50 | 111.0 | 96.0 |
| | 100 | 157.5 | 94.5 |
| | 200 | 268.5 | 102.8 |
| Sediment A, 2g | 0 | 13.57 | — |
| (Organic: 4.3%) | 25 | 39.07 | 102.0 |
| | 50 | 60.93 | 94.7 |
| Sediment B, 2g | 0 | 56.20 | — |
| (Organic: 10.7%) | 25 | 80.17 | 95.9 |
| | 50 | 105.79 | 99.2 |
| Seawater A, 6L | 0.00 | 0.59 | — |
| | 0.60 | 1.22 | 105.0 |
| | 1.50 | 2.05 | 97.3 |
| Seawater B, 6L | 0.00 | 0.95 | — |
| | 3.00 | 4.06 | 103.6 |
| | 6.00 | 6.75 | 96.7 |

Methylmercury-cysteine solution was added to each sample.

as total mercury were analyzed for methylmercury by these two methods. The comparison results are given in Table 5. The present method yielded methylmercury values close to total mercury levels regardless of mercury concentrations in fish, but in the benzene extraction method, methylmercury values were relatively low and the methylmercury ratios to total mercury decreased with decrease of total mercury contents in fish, indicating that it is difficult to extract total methylmercury from fish containing low levels of mercury.

Table 5  Comparison of results obtained by conventional benzene extraction method and the present method for methylmercury in fish

|  |  | methylmercury | | | |
| --- | --- | --- | --- | --- | --- |
|  | Total-Hg | Method A | | Method B | |
| Sample | ng/g | ng/g | % MeHg | ng/g | % MeHg |
| Argyrosomus argentatus | 333. 3 | 284. 3 | 89. 5 | 329. 4 | 102. 5 |
|  | 303. 7 | 285. 7 |  | 323. 5 |  |
| Trachurus trachurus | 178. 8 | 128. 4 | 70. 9 | 170. 4 | 95. 7 |
|  | 179. 9 | 128. 4 |  | 172. 1 |  |
| Stolephorus japonicus | 33. 0 | 18. 7 | 59. 8 | 32. 7 | 99. 2 |
|  | 32. 1 | 20. 3 |  | 32. 0 |  |

Method A : Benzene extraction method          Method B : The present method

Figure 9 shows gas chromatograms from sediment samples with or without added methylmercury. As seen in these chromatograms, although several peaks other than methylmercury appeared, these did not interfere with the methylmercury determination, and no significant difference in the methylmercury peak heights could be seen between the duplicate determinations.

The gas chromatograms of methylmercury extracted from two different seawater samples are shown in Figure 10, together with methylmercury standard solution and reagent blank. No interference from all reagents used in this procedure was observed. With this procedure, methylmercury was clearly detected and the peak was well separated from any interferences. The methylmercury concentrations were evaluated as 0.13 and 0.29ng/l, respectively. The present methods for total and methylmercury were also applied to seawater samples collected from various parts of Japan: the offing of Hachijo Island, the central part of Osaka Bay, the coast of Yatsusiro Sea and Minamata Bay.

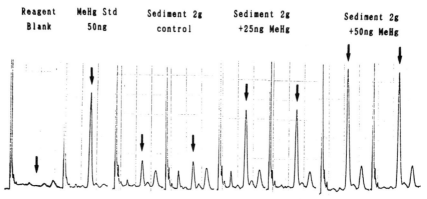

Fig. 9 Typical gas chromatograms of methylmercury extracted from sediment
Organic content: 4.3%
Total mercury content: 0.053 ppm (dry wt basis)

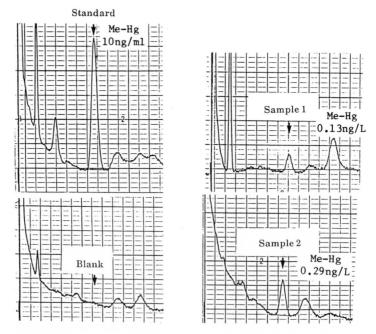

Fig. 10 Gas chromatograms of methylmercury extracted from seawater samples by the
present method.

Table 6  Analyses of total mercury and methylmercury in seawaters

| Station | T-Hg (ng/l) | Me-Hg (ng/l) | Me-Hg/T-Hg (%) |
|---|---|---|---|
| The offing of Hachijou Islands (Surface) | 0.877 | 0.080 | 9.14 |
| The central part of Osaka Bay (Mid-depths, 10m) | 4.345 | 0.325 | 7.48 |
| Yunoko Bay, Yatsushiro Sea (Surface) | 1.199 | 0.197 | 16.43 |
| | 1.820 | 0.210 | 11.54 |
| Minamata Bay, Yatsushiro Sea (Surface) | 8.640 | 0.582 | 6.74 |

Hachijo Island is a small island lying 300km to the south from Tokyo and its offing is in the middle of Kuroshio current. Yunoko Bay and Minamata Bay face towards Yatsushiro Sea.

The results of analyses are shown in Table 6. Total mercury concentrations ranged from 0.877 to 8.64ng/l and the lowest value was observed in Kuroshio current. The highest level was observed in Minamata Bay and the value in Osaka Bay was about the half of that in Minamata Bay.

Methylmercury level in Kuroshio current was also the lowest and the level was around 0.1ng/l which corresponded to approximately 10% of the total mercury. The highest level of 0.58ng/g was observed in Minamata Bay but this value was much lower than expected from its highly polluted sediment.

CONCLUSION

Simplified and highly sensitive methods for the determination of both total mercury and methylmercury in various environmental samples have been developed. For rapid and accurate analysis of total mercury, the mercury vapor introduction system as well as sample digestion procedures were improved and the sample preconcentration procedure by dithizone extraction was also developed for water samples.

Analytical procedures for methylmercury were achieved by the combined techniques of dithizone extraction and electron capture gas chromatography. The differences in the composition of samples determine the pretreatment of samples for the highest extraction efficiency. Thus, biological samples and sediment were treated with potassium hydroxide in ethanol, whereas water samples were treated with potassium permanganate and sulfuric acid. The cleanup of methylmercury in the dithizone benzene extracts from various environmental samples was achieved by means of alkaline sodium sulfide solution. These methods were shown to be applicable to the various types of environmental samples containing mercury at background levels.

## ACKNOWLEDGEMENTS

The authors wish to express their sincere appreciation to Professor T. Suzuki, University of Tokyo, and Professor N. Imura, Kitasato University, for their valuable advice and strong support throughout the course of the present study.

## REFERENCES

Akagi, H., Mortimer, D. C., and Miller, D.R., 1979, Mercury methylation and partition in aquatic systems, Bull. Environ. Contam. Toxicol., 23: 372.

Chau, Y. K., and Saitoh, H., 1970, Determination of submicrogram quantities of mercury in lake waters, Environ. Sci. Technol., 4: 839.

Czuba, M., Akagi, H., and Mortimer, D. C., 1981, Quantitative analysis of methyl- and inorganic-mercury from mammalian, fish and plant tissue, Environ. Pollut. Ser. B, 2: 345.

Sporak, K. F., 1956, Determination of mercury in fungicidal preparations containing organo-mercury compounds, Part I. The determination of organo-mercury compounds by direct titration procedures, Analyst, 81: 474.

Westöö, G, 1966, Determination of methylmercury compounds in foodstuffs, Acta Chem. Scand., 20: 2131.

# INDIVIDUAL DIFFERENCE OF METHYLMERCURY METABOLISM IN ANIMALS AND ITS SIGNIFICANCE IN METHYLMERCURY TOXICITY

Rikuo Doi

Yokohama City University
Department of Hygiene, School of Medicine
Yokohama 236, Japan

## ABSTRACT

Various factors affect the absorption, distribution, biotransformation, excretion of methylmercury (MeHg), and consequently its toxicity. The most important is the influence of genetically determined factors such as species, sex, and the molecular structure of binding substances. Environments, e.g. nutrition, coexisting metals, etc., are also important for MeHg toxicity. It is well known that the greater part of MeHg exists in erythrocytes in the form bound to hemoglobin. The molecular structures of hemoglobin shows the most remarkable effect on the binding of MeHg. The number and position of cysteinyl residues in the hemoglobin determine the affinity of hemoglobin to MeHg and the distribution of MeHg in an erythrocyte. The role of molecular structures of the binding substances of MeHg in the brain and other organs remains to be elucidated in the future.

Sex has a role on the tissue distribution and half-time of MeHg. Usually female mice showed higher tissue MeHg levels and longer half-times than male mice when they were given the same dose of MeHg per body weight. The role of the sex hormone to the organ distribution of MeHg has been reported by Hirayama et al. (1987). The age of the animals also has a role on the distribution of MeHg in mice. Usually young mice have a lower blood MeHg level than the older mice at the same dose.

Physiological conditions, such as local blood flow and erythrocyte count have some effect on the distribution and elimination of MeHg in the blood, brain and liver.

*Advances in Mercury Toxicology*, Edited by T. Suzuki *et al.*
Plenum Press, New York, 1991

Polyerythrocytemic mice due to long term exposure to CO, showed a higher organ MeHg distribution and a shorter half-time of MeHg. Thus local ischemia, anemia or polycytemia may affect the tissue distribution of MeHg.

The individual differences of MeHg metabolism in man remain to be studied (Al-Shahristani and Shihab, 1974).

INTRODUCTION

Individual differences in the half time of metabolism of methylmercury (MeHg) have been described by various researchers (Birke et al., 1972; Miettinen, 1973; Al-Shahristani and Shihab, 1974; Kershaw et al., 1980). Miettinen (1973) reported that the biological half time (BHT) of MeHg was in the range from 52 to 98 days when 6 women and 9 men were administered p.o. with a tracer dose of $^{203}$Hg-MeHg. The average half time was longer in males (79$\pm$3 days) than females (71$\pm$ 6 days). In the patients of MeHg poisoning in Iraq, BHT of MeHg was estimated to be in the range of 35 to 189 days by segmental mercury analysis of hair (Al-Shahristani and Shihab, 1974). There are few studies, however, which intend to elucidate the cause of individual difference itself.

We considered that we will be able to develop more effective methods and better social systems for the prevention of MeHg poisoning if we know the precise basis of individual differences. We have carried out several experiments to elucidate the basis of individual differences using inbred strains of mice (Doi and Kobayashi, 1982; Doi and Mochizuki, 1982; Doi and Tagawa, 1983; Doi et al., 1983; Doi and Tanaka, 1984; Doi, 1986). Inbred mice were mainly used in these studies because they have genetically defined characteristics.

Heredity of strain difference in blood mercury level

We tried to elucidate the difference in the organ distribution and BHT of MeHg with some mouse strains. Figure 1 shows the changes of blood mercury levels in 3 strains of inbred mice, BALB/c, C3H/HeN and C57BL/6N, and a random-bred strain (ICR) after i.p. MeHg administration at a dose of 1 mg MeHg per kg of body weight. BALB/c and C3H showed 2 times higher blood mercury levels as compared with those of C57BL/6N and ICR on the first day after MeHg injection.

The half-time in the brain was longer than that in the other organs in all strains, and ICR strain showed the shortest half time in all organs examined.

Table 1   Biological half-time of methylmercury in the organs of the different strains of mice.

| Strain | Blood | Brain | Liver | Kidney |
|---|---|---|---|---|
| | | Biological half-time (days) | | |
| BALB/C | 5.03 | 14.9 | 6.03 | 8.73 |
| C3H | 5.52 | 14.5 | 7.92 | 7.73 |
| C57BL | 7.79 | 16.3 | 9.49 | 7.47 |
| CD-1 (ICR) | 3.81 | 9.3 | 4.36 | 4.54 |

Ref.: Doi and Kobayashi (1982)

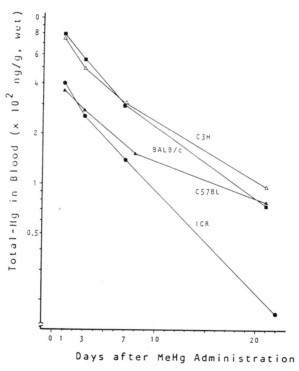

Fig. 1   Strain difference in chemobiokinetics of methylmercury: whole blood.
Ref.: Doi and Kobayashi (1982)

Figure 2 shows the results of cross-breeding experiments between 2 strains of
-mice, C3H and C57BL/6N. C3H mice had 2 times higher blood mercury levels than
that of C57BL/6N on the 1st day after i.p. MeHg administration. The blood mercury
level in $F_1$ mice was at the middle of their parents. In $F_2$ mice, blood mercury levels
had 3 distinct peaks corresponding to those of C3H, C57BL/6N and $F_1$, and in back-
cross mice 2 major peaks of blood mercury levels which corresponded to $F_1$ and C3H
were observed. These results suggest strongly that the difference in blood mercury
levels is controlled principally by hereditary factors.

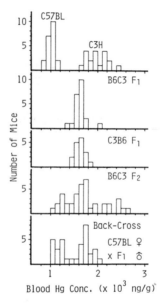

Fig. 2    Frequency distribution of blood mercury concentration of mice 24 hr after
MeHg injection i.p. MeHg dosage was 3 mg/kg for all strains
Ref.: Doi et al. (1983)

Figure 3 shows mercury concentrations in the blood and brain of various strains
of mice on the first day after i.p. MeHg administration. Blood mercury was
approximately 2 times higher in 5 inbred strains, i.e. C3H, AKR, CBA, NZB/JIc,
NZB/San, and a Japanese wild mouse, *Mus musculus molossinus*, than those of other
7 inbred strains, i.e. C57BL/6N, C57BL/10Sn, DDD, HTH, KK, NC and SS.

Mercury levels in the brain were approximately the same in all strains, though an apparent sex difference was seen in a strain of C57BL/6N.

The major difference between these 2 groups of mouse strains is the molecular structure of their hemoglobins. The group of mice with higher mercury levels has 2 cysteinyl residues on a beta chain of hemoglobin, while the other group has only one cysteinyl residue on a beta chain (Dayhoff et al., 1972).

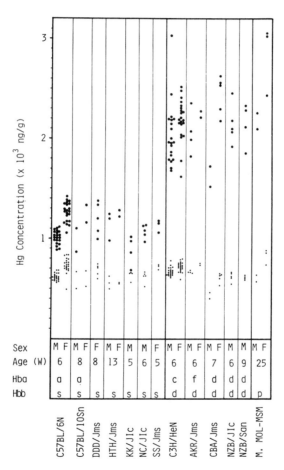

Fig. 3     Mercury concentrations in blood and brain of four strains of mice 24 hr after MeHg injection i.p. at the dose of 3 mg/kg. Large points show the blood mercury concentration and small points show the brain mercury concentration. Ref.: Doi et al. (1983)

81

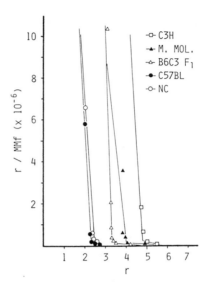

Fig. 4    Scatchard plots for methylmercury chloride to mouse hemoglobins. The x intercept "*r*" shows moles of MeHg bound per mole of hemoglobin. MMf means molar concentration of free MeHg in Hb-MeHg mixture. Final concentration of hemoglobin was $5 \times 10^{-6}$M in the Hb-MeHg mixtures of all strains and of all species in Figure 5.
Ref.: Doi and Tagawa (1983)

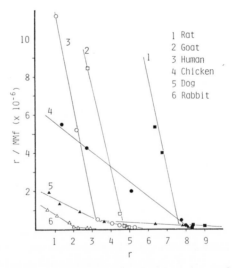

Fig. 5    Scatchard plots for methylmercury chloride to hemoglobins of six species of animals. Ref.: Doi and Tagawa (1983)

Biochemical and biological behavior of methylmercury

Figure 4 is a Scatchard plot for MeHg chloride to mouse hemoglobins. The x intercept, "r", shows mole of MeHg bound per mole of hemoglobin. Two types of binding sites, primary and secondary binding sites ($N_1$ and $N_2$), were supposed to be on the hemoglobins of all strains examined since the plots were hyperbolic. Figure 5 is a Scatchard plot for MeHg to the hemoglobins of various species.

Table 2 is the summary of Scatchard plots analysis in Fig. 4 and 5. The number of primary binding sites ($N_1 = 4.8$) on the hemoglobin of C3H strain was twice the number of $N_1$ on the hemolobin of C57BL ($=2.4$). These results corresponded well with the blood mercury concentrations in these mice after MeHg administration at the equal dosage and to the number of cysteinyl residues on the hemoglobin beta-chain.

Table 2  Summary of binding experiments of methylmercury chloride to hemoglobins

| Species | N1 | $K_1$ $(\times 10^6)$ | N2 | $K_2$ $(\times 10^4)$ |
|---|---|---|---|---|
| C3H | 4.8 | 16 | 5.8 | 24 |
| B6C3 F1 | 3.4 | 30 | 4.9 | 7.5 |
| C57BL | 2.4 | 17 | 4.3 | 4.3 |
| NC | 2.5 | 15 | 4.0 | 7.3 |
| M.MOL. | 4.0 | 9.5 | 5.7 | 8.0 |
| Rat | 7.7 | 4.9 | 11 | 7.0 |
| Human | 3.3 | 4.9 | 5.8 | 13 |
| Goat | 4.7 | 4.3 | --- | --- |
| Dog | 4.2 | 0.50 | 10 | 6.7 |
| Rabbit | 2.3 | 0.61 | 3.2 | 8.0 |
| Chicken | 8.2 | 0.77 | --- | --- |

Note: $N_1$, Number of primary binding sites; $N_2$, number of secondary binding sites; $K_1$, association constant for the primary site; $K_2$, association constant for the secondary site. ( Ref.: Doi and Tagawa, 1983).

Table 3 is a summary of the molecular structure of hemoglobins. Small "c" means a cysteinyl residue in the contact junction of hemoglobin molecule, and it has lower reactivity with MeHg. Large "C" means a cysteinyl residue out of the junction, and it can easily bind to MeHg because it is located on the outer surface of the hemoglobin molecule.

TABLE 3

Cysteinyl Residues on Hemoglobin alpha- and beta-chain

| | Positions on polypeptide chains[b] | | | | | | | | | | Number of residues[c] | | |
|---|---|---|---|---|---|---|---|---|---|---|---|---|---|
| | Alpha-chain | | | | Beta-chain | | | | | | | | |
| Species[a] | 13 | 104 | 111 | 130 | 13 | 23 | 93 | 112 | 125 | 126 | c | C | Total |
| Mouse-s | A | c | S | A | G | V | C | I | A | A | 2 | 2 | 4 |
| Mouse-d | A | c | S | A | C | V | C | I | A | A | 2 | 4 | 6 |
| Rat | C | c | c | A | G | V | C | I | c | A | 6 | 4 | 10 |
| Human-A | A | c | A | A | A | V | C | c | P | V | 4 | 2 | 6 |
| Goat | A | S | c | A | G | V | C | V | L | L | 2 | 2 | 4 |
| Dog | T | c | c | A | G | V | C | c | Q | V | 6 | 2 | 8 |
| Rabbit | A | c | N | A | A | V | C | I | Q | V | 2 | 2 | 4 |
| Chicken | I | c | A | C | G | C | C | I | E | C | 2 | 8 | 10 |

Note: Abbreviations:

c = cysteinyl residue in the $\alpha 1 \beta 1$ contact junction;

C = cysteinyl residue outside of the $\alpha 1 \beta 1$ contact junction;

A = alanine; E = glutamic acid; G = glycine; I = isoleucine; L = leucine;
N = asparagine; P = proline; Q = glutamine; S = serine; T = threonine; and V = valine.

[a]Mouse-s and Mouse-d mean the strains of mice with Hbb[s] or Hbb[d] allele, respectively. Human-A means human adult Hb (Hb A).

[b]Positions from the N terminal.

[c]Number of cysteinyl residues in a Hb tetramer.
  Ref.: Doi and Tagawa (1983).

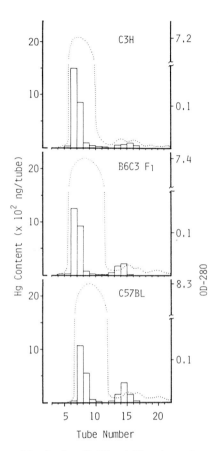

Fig. 6    Chromatogram of Sephadex G-25 gel filtration of stroma-free hemolysates of mice.   Ref.: Doi and Tagawa (1983)

Figure 6 shows the amount of MeHg bound to hemoglobin and a low molecular weight fraction like glutathione after Sephadex G-25 gel filtration of the hemolysates. In the erythrocytes of C57BL/6N, more than 3 times of MeHg was bound to the low molecular weight fraction as compared with that of C3H.   In the erythrocytes of $F_1$ mice the amount of MeHg bound to the low molecular weight fraction was at the middle of C3H and C57BL/6N.

Figure 7 shows the release rate of MeHg from the erythrocytes of mice, man and rat.   Methylmercury was released most rapidly from the erythrocytes of man, and most slowly from those of rat.   With the erythrocytes of mice, those of C57BL/6N showed the highest release rate, those of C3H the lowest and those of $F_1$ mice was at the middle of their parents.   These results correspond well to the number of binding sites on the hemoglobins of the animal species and strains.

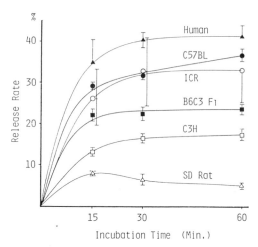

Fig. 7   Release of methylmercury from erythrocytes into BSA solution. Each point represents $x \pm$ SD for at least three animals. Ref.: Doi and Tagawa (1983)

It is well known that there is no significant difference in the amount of glutathione between the erythrocytes of different species of animals (Beutler et al., 1963; Jocelyn, 1967; McNeil and Beck, 1968; Srivastava and Beutler, 1968; Tietze, 1969; Hsu et al., 1982).   Therefore, the results which were shown in Figures 6 and 7 suggest that hemoglobin is the determinant factor for the binding of MeHg to erythrocytes.

## Other Factors for the Strain Difference of MeHg Metabolism

Table 4 shows an example of sex difference of MeHg distribution in mice. Female mice had higher mercury concentrations in blood and brain than those of male mice especially in C57BL/6N strain.  Female mice also had a longer half time of MeHg than male mice (Figure 8).  The sex difference in MeHg distribution has also been reported by Magos et al. (1981) and Hirayama and Yasutake (1986).  Mercury levels in the brain, liver and blood were higher in females than males in these reports, and Magos et al. (1981) emphasized the importance of difference in growth after MeHg administration.  Hirayama et al. (1987) suggested that MeHg metabolism may be regulated by sex hormones through regulating GSH metabolism by sex hormones. Inouye et al. (1986) reported that the brain mercury levels were higher in female fetuses of mice than males after transplacental exposure to MeHg. Sager et al. (1984) reported that the antimitotic effects of MeHg were greater in the cerebellum of male mice than females after MeHg administration to 2-day-old pups of BALB/c strain.

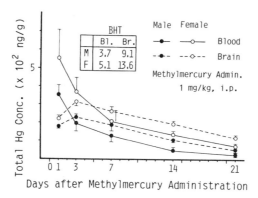

Fig. 8    Sex difference in the distribution and biological half-time of methylmercury in the blood and brain of ICR mice.  Ref.: Doi and Mochizuki (1982)

Table 4    Sex Difference in Mercury Concentrations in the Blood and Brain of Mice 24 hrs after Methylmercury Administration   (ng/g, Mean ± SD)

| Species | Sex | n | Blood | Brain |
|---------|-----|---|-------|-------|
| C57BL/6N | M | 21 | $1008 \pm 68$ | $612 \pm 37$ |
|  | F | 21 | $1259 \pm 95 \overset{*}{*}$ | $730 \pm 76 \overset{*}{*}$ |
| C3H/HeN | M | 20 | $2041 \pm 308$ | $669 \pm 40$ |
|  | F | 21 | $2188 + 215$ | $715 \pm 47$ [*] |

significant: * p $\leq$ 0.005,  $\overset{*}{*}$p $\leq$ 0.001
Methylmercury Adm.: 3 mg/kg, i.p.

Ref.: Doi and Mochizuki (1982)

Figure 9 shows the difference of MeHg distribution according to age of mice. At 24 hours after MeHg administration, the older the age, the higher the blood mercury levels were in C3H and C57BL/6N strains. An extremely slow clearance of mercury was reported in rats (Thomas et al., 1982), in infant monkeys (Lok, 1983) and in mice (Hirayama and Yasutake, 1986) which were exposed to MeHg at the earliest stages of age. Lok (1983) showed a sudden drop of the blood mercury levels accompanied by the marked increase of fecal excretion of mercury after weaning of the monkeys. On the other hand, Hirayama and Yasutake (1986) reported an increased urinary excretion of mercury in mice at the stages after weaning, while the increase of fecal mercury excretion was not so significant as that of urinary excretion. These findings seem to suggest that there is a wide variation of the mechanisms in MeHg excretion between species of animals.

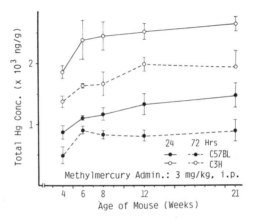

Fig. 9    Mercury levels in the blood of mice of various ages: 24 and 72 hrs after methylmercury administration. Ref.: Doi and Mochizuki (1982)

Figures 10 and 11 show MeHg distribution in mice which were pre-exposed to carbon monoxide at the concentration of 350 ppm for a week prior to MeHg injection. Hemoconcentration was observed in these mice within several days of CO exposure and mercury levels were relatively higher in the blood and organs of CO exposed mice than control mice at the earliest stage after MeHg injection. If the mice were exposed to CO immediately after MeHg administration, hemoconcentration occurred after several days. The mercury concentrations in the blood, liver and kidneys were lower in CO exposed mice than those of control mice at the first and third day after MeHg administration (Figure 12 and 13). Blood mercury levels exceeded, however, the levels of control mice after several days. On the other hand, the mercury level was higher in CO exposed mice than in control mice at the first day. These changes, such as

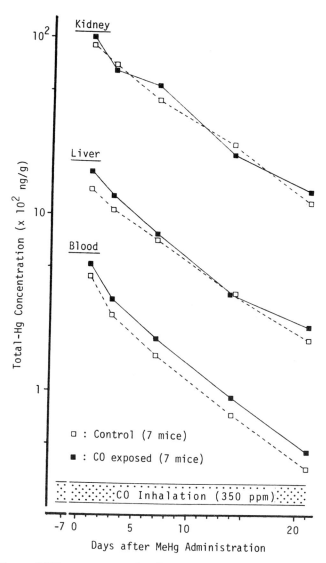

Fig. 10  Effects of CO exposure on the distribution of methylmercury in male mice:
Blood, Liver, Kidney    Ref.: Doi and Tanaka (1984)

in the mercury levels will be explained by compensatory reactions against CO. These are hemoconcentration and the increase in cardiac output, heart rate and cerebral blood flow (Paulson et al., 1973; Doblar et al., 1977; Pitt et al., 1979). An extremely decreased physical movement at the earliest stage of CO exposure might be one of the causes of decreased blood mercury level in CO exposed mice. Accompanied by gradual hemoconcentration, physical activities were recovered day by day and mercury concentration increased in the blood of these mice despite continuous CO exposure.

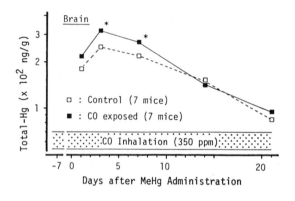

* : statistically significant at $p < 0.05$.

Fig. 11   Effects of CO exposure on the distribution of methylmercury in male mice: Brain   Ref.: Doi and Tanaka (1984)

Figure 14 shows the excretion of $^{203}$Hg through feces and urine after the admnistration of $^{203}$Hg-MeHg to 3 strains of mice. C3H excreted more than 40% of the dose into feces. ICR mice excreted nearly twice more $^{203}$Hg into urine as compared to the other 2 strains, while ICR mice excreted the least $^{203}$Hg through feces among these strains. As a result, retention of $^{203}$Hg decreased most rapidly in ICR, next C3H and slowest in C57BL/6N. Yasutake and Hirayama (1986) reported similar strain difference in mercury excretion in MeHg administered mice. It is supposed that the strain difference of glutathione secretion into the bile might be one of the principal causes of the strain difference in $^{203}$Hg excretion.

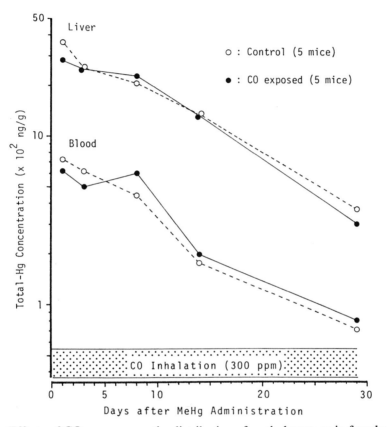

Fig. 12. Effects of CO exposure on the distribution of methylmercury in female mice: Blood, Liver. Ref.: Doi and Tanaka (1984)

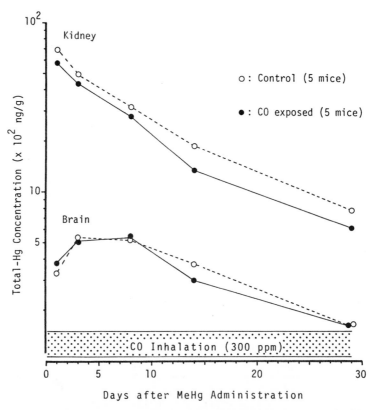

Fig. 13 Effects of CO exposure on the distribution of methylmercury in female mice: Brain, Kidney Ref.: Doi and Tanaka (1984)

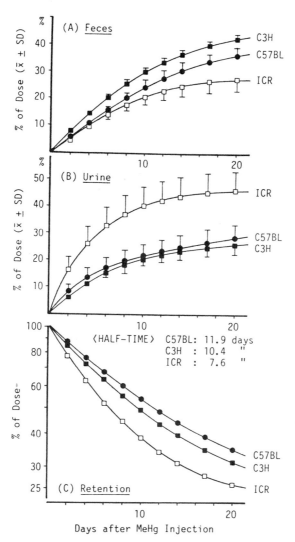

Fig. 14 Excretion of $^{203}$Hg via feces (A) and via urine (B), and whole body retention of $^{203}$Hg in mice (C) after a single i.p. injection of $^{203}$Hg-CH$_3$HgCl at a dose of 1 mg/kg    Ref.: Doi (1986)

CONCLUSION

Three major groups of factors might be supposed as the factors determining individual difference of MeHg metabolism, i.e. hereditary and environmental factors and time or age (Table 5). Hereditary factors might be the most important in these three major groups. These include molecular structure of the binding substances of MeHg (Doi and Tagawa, 1983; Doi et al., 1983), sex, anatomical structures, physiological activities, etc. The role of biliary excretion of glutathione is discussed by Naganuma et al.

Table 5   Factors influencing individual differences in methylmercury metabolism

1. **Hereditary (Genetic) Factors**
   a) **Molecular Structure of the Binding Substances:**
      Hemoglobin and Other Cellular/Plasma Proteins with -SH
   b) **Sex of the Animals/Man:**
      Sex Hormones
   c) **Anatomical Structure:**
      Distribution of Blood Vein
      Blood-Brain Barrier
      Placental Barrier
   d) **Physiological Activities:**
      Blood Stream
      Production of -SH Substances: Glutathione, Hemoglobin, Albumin, etc.
      Red Blood Cells: Anemia, Polycytemia
      Bile Secretion: Glutathione

2. **Environmental Factors**
   a) **Nutrition:**
      Protein: Fish, Meat, Egg, etc.
      Dietary fibers
   b) **Intestinal Bacterial Flora:**
      Methylation of Hg/Demethylation of MeHg by Intestinal Bacterial Flora
   c) **Drugs and Chemicals:**
      Drugs/Chemical with -SH
      Alcohol, Selenium, etc.
   d) **Temperature**
   e) **Others**

3. **Time/Age**
   a) **Time-dependent Changes of Physiologic Activities:**
      Sex Hormones
      Blood Stream, Production of -SH Substances, Bile Secretion
      Intestinal Bacterial Flora
      Nutrition
      Others

Environmental factors include nutrition (Landry et al., 1979; Rowland et al., 1984), dietary fibers (Rowland et al., 1986), ethanol (Tamashiro et al., 1985), intestinal bacterial flora (Rowland et al., 1977, 1980 and 1984; Seko et al., 1981), drugs and chemicals especially those with thiol groups in the molecule (Clarkson et al., 1973; Takahashi et al., 1975; Klaassen, 1975; Gabard, 1976; Suzuki et al., 1976; Hirayama, 1980 and 1985; Yonaga and Morita, 1981; Thomas and Smith, 1982), environmental temperature

(Nomiyama et al., 1980) etc. These will affect MeHg metabolism through changing physiological activities of animals. The role of selenium as a modifying factor of mercury toxicity is discussed by Imura and also by Magos.

Time/age may change MeHg metabolism through changing physiological activities of man and animals.

Individual difference in the metabolism of MeHg might be controlled by the combinations of these 3 major factors, and species and strain differences are also controlled by these factors. Therefore, species difference, strain difference and individual difference are basically common in their mechanisms, though there might be wide variations in the combinations of factors.

We have not elucidated, to our regret, the whole aspect of individual difference in MeHg metabolism even now, but it will be done more precisely and extensively in the near future and we will have more effective methods for preventing MeHg poisoning.

## REFERENCES

Al-Shahristani, H. and Shihab, K.M., 1974. Variation of biological half-life of methyl-mercury in man. Arch. Environ. Health 28: 342-344.

Beutler, E., Duron, O., and Kelly, B.M., 1963. Improved method for the determination of blood glutathione. J. Lab. Clin. Med. 61: 882-888.

Birke, G., Johnels, A.G., Plantin, L-O., Sjostrand, B., Skerfving, S. and Westermark, T., 1972. Studies on humans exposed to methylmercury through fish consumption. Arch. Environ. Health 25: 77-91.

Clarkson, T.W., Small, H. and Norseth, T., 1973. Excretion and absorption of methylmercury after polythiol resin treatment. Arch. Environ. Health 26: 173-176.

Dayhoff, M.O., Hunt, L.T., McLaughlin, P.J. and Barker, W.C., 1972. Globins. In: Atlas of Protein Sequence and Structure. M.O. Dayhoff, ed. Vol. 5, D-51-85. National Biomedical Research Foundation, Washington, D.C.

Doblar, D.D., Santiago, T.V. and Edelman, N.H., 1977. Correlation between ventilatory and cerebrovascular responses to inhalation of Co. J. Appl. Physiol. 43:456-462.

Doi, R., 1986. Strain differences in excretion of methylmercury in mice. Bull. Environ. Contam. Toxicol. 36: 500-505.

Doi, R. and Kobayashi, T., 1982. Organ distribution and biological half-time of methyl-mercury in four strains of mice. Jpn. J. Exp. Med. 52: 307-314.

Doi, R. and Mochizuki, Y., 1982. Chemobiokinetics of methylmercury: Effects of age and sex of mice to the distribution and excretion of methylmercury. Jpn J. Hyg. 37: 143 (in Japanese).

Doi, R. and Tagawa, M., 1983. A study on the biochemical and biological behavior of methylmercury. Toxicol. Appl. Pharmacol. 69: 407-416.

Doi, R., Tagawa, M., Tanaka, H. and Nakaya, K., 1983. Hereditary analysis of the strain difference of methylmercury distribution in mice. Toxicol. Appl. Pharmacol. 69: 400-406.

Doi, R. and Tanaka, H., 1984. Effects of carbon monoxide exposure on the distribution of methylmercury in mice. J. Toxicol. Sci. 9: 11-22.

Gabard, B., 1976. Treatment of methylmercury poisoning in the rat with sodium 2,3-dimercaptopropane-1-sulfonate: Influence of dose and mode of administration. Toxicol. Appl. Pharmacol. 38: 415-424.

Hirayama, K., 1980. Effect of amino acids on brain uptake of methyl mercury. Toxicol. Appl. Pharmacol. 55: 318-323.

Hirayama, K., 1985. Effects of combined administration of thiol compounds and methylmercury chloride on mercury distribution in rats. Biochem. Pharmacol. 34: 2030-2032.

Hirayama, K. and Yasutake, A., 1986. Sex and age differences in mercury distribution and excretion in methylmercury-administered mice. J. Toxicol. Environ. Health 18: 49-60.

Hirayama, K., Yasutake, A. and Inouye, M., 1987. Effect of sex hormones on the fate of methylmercury and on glutathione metabolism in mice. Biochem. Pharmacol. 36: 1919-1924.

Hsu, J.M., Rubenstein, B. and Paleker, A.G., 1982. Role of magnesium in gluthione metabolism of rat erythrocytes. J. Nutr. 112: 488-496.

Inouye, M., Kajiwara, Y. and Hirayama, K., 1986. Dose- and sex-dependent alterations in mercury distribution in fetal mice following methylmercury exposure. J. Toxicol. Environ. Health 19: 425-435.

Jocelyn, P.C., 1967. An assay for glutathione in acid solution. Anal. Biochem. 18: 493-498.

Kershaw, T.G., Clarkson, T.W. and Dhahir, P.H., 1980. The relationship between blood levels and dose of methylmercury in man. Arch. Environ. Health. 35: 28-36.

Klaassen, C.D., 1975. Biliary excretion of mercury compounds. Toxicol. Appl. Pharmacol. 33: 356-365.

Landry, T.D., Doherty, R.A. and Gates, A.H., 1979. Effects of three diets on mercury excretion after methylmercury administration. Bull. Environ. Contam. Toxicol. 22: 151-158.

Lok, E., 1983. The effect of weaning on blood, hair, fecal and urinary mercury after chronic ingestion of methylmercuric chloride by infant monkeys. Toxicol. Letters 15: 147-152.

Magos, L., Peristianis, G.C., Clarkson, T.W., Brown, A., Preston, S. and Snowden, R.T., 1981. Comparative study of the sensitivity of male and female rats to methylmercury. Arch. Toxicol.. 48: 11-20.

McNeil, T.L., and Beck, L.V., 1968. Fluorometric estimation of GSH-OPT. Anal. Biochem. 22: 431-441.

Miettinen, J.K., 1973. Absorption and elimination of dietary mercury ($Hg^{2+}$) and methylmercury in man. In: Mercury, Mercurials and Mercaptans. M.W. Miller and T.W. Clarkson, eds. C.C. Thomas, Springfield, IL, pp. 233-243.

Nomiyama, K., Matsui, K. and Nomiyama, H., 1980. Effects of temperature and other factors on the toxicity of methylmercury in mice. Toxicol. Appl. Pharmacol. 56: 392-398.

Paulson, O.B., Parving, H-H., Olsen, J. and Skinhoj, E., 1973. Influence of carbon monoxide and of hemodilution on cerebral blood flow and blood gases in man. J. Appl. Physiol. 35: 111-116.

Pitt, B.R., Radford, E.P., Gurtner, G.H. and Traystman, R.J., 1979. Interaction of carbon monoxide and cyanide on cerebral circulation and metabolism. Arch. Environ. Health 34: 354-359.

Rowland, I., Davies, M. and Grasso, P., 1977. Biosynthesis of methylmercury compounds by the intestinal flora of the rat. Arch. Environ. Health 32:24-28.

Rowland, I.R., Davies, M.J. and Evans, J.G., 1980. Tissue content of mercury in rats given methylmercuric chloride orally: Influence of intestinal flora. Arch. Environ. Health 35: 155-160.

Rowland, I.R., Robinson, R.D. and Doherty, R.A., 1984. Effects of diet on mercury metabolism and excretion in mice given methylmercury: Role of gut flora. Arch. Environ. Health 39: 401-408.

Rowland, I.R., Mallett, A.K., Flynn, J. and Hargreaves, R.J., 1986. The effect of various dietary fibres on tissue concentration and chemical form of mercury after methylmercury exposure in mice. Arch. Toxicol. 59: 97-98.

Sager, P.R., Aschner, M. and Rodier, P.M., 1984. Persistent, differential alterations in developing cerebellar cortex of male and female mice after methylmercury exposure. Devel. Brain Res. 12:1-11.

Seko, Y., Miura, T., Takahashi, M. and Koyama, T., 1981. Methyl mercury decomposition in mice treated with antibiotics. Acta Pharmacol. Toxicol. 49: 259-265.

Srivastava, S.T. and Beutler, E., 1968. Accurate measurement of oxidized glutathione content of human, rabbit and rat red blood cells and tissues. Anal. Biochem. 25: 70-76.

Suzuki, T., Shishido, S. and Ishihara, N., 1976. Different behaviour of inorganic and organic mercury in renal excretion with reference to effects of D-penicillamine. Brit. J. Indust. Med. 33: 88-91.

Takahashi, H., Hirayama, K. and Ikegami, Y., 1975. Resin for methyl mercury elimination in rats. Kumamoto Med. J. 28:: 60-62.

Tamashiro, H., Arakaki, M., Akagi, H., Murao, K. and Hirayama, K., 1985. Factors influencing methyl mercury toxicity in rats - Effects of ethanol. Jpn. J. Pub. Health 32: 397-402 (in Japanese with English abstract).

Thomas, D.J. and Smith, J.C., 1982. Effects of coadministered low-molecular-weight thiol compounds on short-term distribution of methyl mercury in the rat. Toxicol. Appl. Pharmacol. 62: 104-110.

Thomas, D.J., Fisher, H.L., Hall, L.L. and Mushak, P., 1982. Effects of age and sex on retention of mercury by methyl mercury-treated rats. Toxicol. Appl. Pharmacol. 62: 445-454.

Tietze, F., 1969. Enzymic method for quantitative determination of nanogram amounts of total and oxidized glutathione. Anal. Biochem. 27: 502-522.

Yasutake, A. and Hirayama, K., 1986. Strain difference in mercury excretion in methylmercury-treated mice. Arch. Toxicol. 59: 99-102.

Yonaga, T. and Morita, K., 1981. Comparison of the effect of N-(2,3-dimercaptopropyl) phthalamidic acid, DL-penicillamine, and dimercaprol on the excretion of tissue retention of mercury in mice. Toxicol. Appl. Pharmacol. 57: 197-207.

# MECHANISMS OF URINARY EXCRETION OF METHYLMERCURY (MM)

Paul J. Kostyniak

Department of Pharmacology and Therapeutics
    and the Toxicology Research Center
University of Buffalo
School of Medicine and Biomedical Research
Buffalo, New York  14214

## ABSTRACT

A two-fold difference in elimination rates of methylmercury (MM) between two genetic variant strains of mice was due to a five-fold greater rate of urinary excretion in the fast-excreting strain. Further investigation indicated that the difference could not be accounted for by biotransformation of MM to the inorganic form. However, several factors that could conceivably influence MM renal excretion were altered, including an increased plasma total glutathione (GS) concentration, and an increased rate of output of GS in the urine. To assess the role that GS may play in the elimination of methylmercury in the urine, the $\gamma$-GTP inhibitor, L-[$\alpha$s,5s]-$\alpha$-Amino-3-chloro-4,5-Dihydro-5-Isoxazoleacetic Acid (AT-125), was administered to CBA/J mice previously treated with $^{203}$Hg-MM to determine whether increasing urinary GS could result in a simultaneous increase in urinary excretion of MM. Increasing doses of AT-125 resulted in a dose-dependent increase in urinary GS. MM excretion, however, did not increase in a direct proportion to increasing concentrations of GS.

Earlier studies in whole blood indicated that millimolar plasma concentrations of low molecular weight thiols were required to redistribute MM from cellular binding sites into plasma. Similarly, increased urinary MM excretion coincided only with the period following AT-125 administration when urinary thiol was elevated to 300 to 500 times normal levels to the millimolar concentration range. These studies suggest that changes in GS elimination into urine may only be of consequence with regard to MM excretion once GS levels have been elevated to the millimolar concentration range.

*Advances in Mercury Toxicology*, Edited by T. Suzuki *et al.*
Plenum Press, New York, 1991

## INTRODUCTION

Studies on the excretion of MM have focused primarily on the fecal route which in several species serves as the primary route of excretion (Norseth, 1971; Norseth and Clarkson, 1971). Little attention was paid to the renal elimination of MM where the mechanisms responsible for elimination of the metal have remained elusive. Our interest in the urinary pathway of excretion was sparked by an observation that the CFW mouse strain eliminated mercury at a rate which was approximately twice the rate of excretion in the CBA/J strain.

### Strain Differences in Excretion Rate

As indicated in Table 1, the whole body half-time for MM in the CBA/J strain was twice that of the CFW strain. This is equivalent to a two fold greater excretion rate in CFW mice when compared to the CBA/J strain. The possible mechanisms for the difference in elimination of $^{203}$Hg between the two strains could not be accounted for by an altered biotransformation of MM to the inorganic form (Hg$^{++}$), since both strains eliminated the same relative percentage of $^{203}$Hg in the inorganic form in urine (Mulder and Kostyniak, 1985c). Similarly, when dosed with $^{203}$Hg labeled mercuric chloride, the elimination rates of $^{203}$Hg in the strains show a considerably different kinetic picture than for methylmercury. The rates of excretion for inorganic mercury were virtually identical for the two strains. The half-time for excretion in the CBA/J strain was $3.68 \pm .09$ versus $3.83 \pm .18$ in the CFW strain. There were only minor differences in inorganic mercury disposition between the two strains as indicated in Table 2, and kidney levels in particular were not significantly different.

### The Role of Endogenous Thiols

Since it is well known that MM has an avid affinity for sulfhydryl groups, a series of studies was designed to address the role that endogenous thiols may play in the excretion of MM in urine. Initial studies focused on identifying MM complexes which are formed in urine. It was evident that problems which had plagued early investigators trying to identify biliary complexes of MM were also in effect identifying the urinary MM complexes and peptides. The pattern of mercury binding to urinary proteins

Table 1    Mercury excretion in two strains of mice after dosing with $^{203}$Hg labeled methylmercury*

| Parameters | $^{203}$Hg Excretion CBA/J | CFW |
|---|---|---|
| t 1/2 ** | 7.4 ± 0.2 days | 3.0 ± 0.2 days |
| Feces *** | 6.8 ± 0.4% | 7.3 ± 0.4% |
| Urine *** | 2.4 ± 0.2% | 13.1 ± 0.7% |

\* Data from Kostyniak, 1980

\** Biological halftime (mean ± SE) obtained from whole body counting data analyzed according to a first order kinetic model.

\*** Percent body burden of $^{203}$Hg excreted per day (mean ±SE).

Table 2    Tissue mercury concentrations at necropsy 7 days after dosing with $^{203}$Hg labeled mercuric chloride

| | Mercury Concentration (ngHg/g)* CBA/J | CFW |
|---|---|---|
| Blood | 1.8 ± .1 | 1.6 ± .5 |
| Brain | 5.1 ± .5 | 2.8 ± .4** |
| Lung | 8.4 ± 1.1 | 6.2 ± 1.9 |
| Liver | 30 ± 2 | 20 ± 2** |
| Kidney | 323 ± 11 | 364 ± 21 |
| Spleen | 9.4 ± .5 | 4.1 ± .7** |
| Muscle | 1.4 ± .1 | 1.7 ± .4 |
| Bone | 3.9 ± .4 | 3.8 ± .4 |
| Pelt | 5.0 ± .1 | 5.4 ± 2.4 |

\* Mean ±S.E.
\** Significantly different from CBA/J at p = .0005

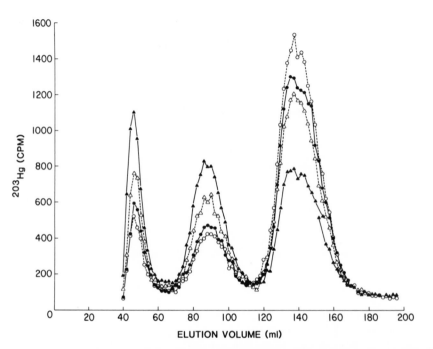

Fig. 1 Sephadex G-75 gel exclusion chromatography of CFW mouse urine spiked with
$^{203}$Hg-MM (5 µM final) and incubated at room temperature for the following
time periods: 1 hour, ○-○; 6 hours, ●-●; 12 hours, △-△; 24 hours, ▲-▲
(from Mulder, K.M., 1985).

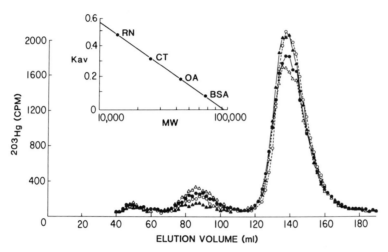

Fig. 2 Sephadex G-75 Chromatography of mouse urine collected into 0.2M EDTA and incubated with $^{203}$Hg-MM for time periods indicated: 1 hour, ▲-▲; 6 hours, ○-○; 12 hours, ●-●; 24 hours, △-△. The insert displays the column calibration with the following standards: Ribonuclease A (RN; 13,700), chymotrypsinogen A (CT: 25,000), ovalbumin (OA; 43,000), and bovine serum albumin (BSA; 67,000), from Mulder, K.M., 1985).

utilizing Sephadex G75 column chromatography is given in Figure 1. This profile shows large molecular weight proteins in the column void volume, a peak at approximately 18,000 daltons, and a lower molecular weight fraction less than 3,000 daltons. The multiple curves in this figure show a time dependent shift in $^{203}$Hg binding from lower molecular weight peptides to high molecular weight proteins over a 24 hour time course. This illustrates the dynamic nature of MM binding among the changing thiol pools within urine. This shift in binding occurred over a larger time course than that observed for the loss of glutathione from urine which exhibited a half time of about 11.5 minutes in the same system. The glutathione loss from urine can occur through several pathways including formation of mixed disulfides with proteins and metabolism to its constituent amino acids.

The shift in binding of MM from low to high molecular weight components could be largely prevented, as demonstrated in Figure 2, by the addition of EDTA to urine. Similarly, the acidification of urine would also prevent this shift in mercury binding and also prevent the formation of mixed disulfides, thus stabilizing binding to low molecular weight thiols (Mulder and Kostyniak, 1985). When subjected to high performance liquid chromatography in order to separate the low molecular weight MM complexes, the primary low molecular weight complex identified in urine co-eluded with a MM-cysteine standard as depicted in figure 3. This peak was indeed distinguishable from MM-glutathione. The presence of EDTA or acidification of urine had no effect on the MM-complex peak.

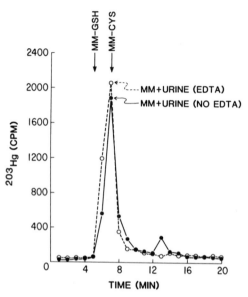

Fig. 3 Reverse-Phase HPLC profiles of $^{203}$Hg-MM (70 $u$M) added to mouse urine collected either with or without EDTA and to GSH (1.2 mM) and cysteine (1.9M) standards. Time O represents the time of injection of each sample, (from Mulder, K.M., 1985).

Glutathione could indeed play a role in the renal excretion of MM even though it was not found as a complex with MM in mouse urine. In the intact animal, glutathione undergoes a continuous cycling within the renal proximal tubule cell as depicted in Figure 4. The tubule cell secretes glutathione into the lumen of the proximal tubule. Glutathione is not reabsorbed as a tripeptide, but rather it must be metabolized by a two step process into its constituent amino acids before they are reabsorbed. The first step in this degradation involves attachment by $\gamma$-glutamyl transpeptidase, a brush border enzyme which results in the formation of the cysteinyl-glycine dipeptide. This dipeptide can be metabolized further by a dipeptidase releasing the amino acids cysteine and glycine. The primary amino acids are then reabsorbed. Within the cell the synthetic machinery exists to re-synthesize glutathione from its constituent amino acids. This is necessary to maintain cellular glutathione at millimolar concentrations within the cell. Glutathione can undergo additional chemical and enzymatic reactions within the cell with reactive chemicals and metabolites to form glutathione conjugates. This is a key pathway in the detoxification of a variety of xenobiotics. Glutathione can also form dimers through a disulfide linkage resulting in an oxidized form of glutathione. This oxidized form can be reduced enzymatically by glutathione reductase to allow for homeostatic control of intracellular glutathione, such that the majority of glutathione within the cell (>90%) is in the reduced form. Reduced glutathione can also interact with protein thiols forming mixed disulfides with proteins. The reduced glutathione, with a free sulfhydryl group, exists intracellularly in millimolar concentrations, and is a likely candidate for transport of MM across the lumenal membrane and out of the cell. Similarly, the sulfhydryl of GSH may also be responsible for the reabsorption of MM across the lumenal membrane

We hypothesized that if methylmercury was excreted in the kidney as the MM-GSH complex and reabsorbed as a cysteine complex, the inhibition of γ-GTP, the first step in glutathione catabolism would result in an inability of the secreted MM-GSH complex to be reabsorbed and thus an increased excretion of MM should occur in consert with the glutathionuria established through γ-GTP inhibition. AT-125 (acivicin) is a specific inhibitor of γ-GTP. This drug causes a dramatic dose dependent increase in urinary GSH concentration as indicated in Table 3. The low dose of AT-125 (7.5mg/Kg) causes a 150 fold increase in urinary total glutathione concentration. The intermediate dose (15 mg/Kg) caused a 300 fold increase in urinary glutathione and the highest dose resulted in a 500 fold increase in urinary glutathione into the millimolar range. The substantial magnitude of these increases in urinary glutathione was not, however, paralleled by a proportional increase in the total excretion of MM from the animal. As indicated in the table, there was no significant change in total $^{203}$Hg excretion after any dose of AT-125 in spite of a more than 500 fold increase in urinary glutathione excretion. The only effect on mercury excretion was to produce up to a two fold increase in urinary excretion of mercury, thus simply shifting a portion of the mercury normally excreted in the feces into the urine. Thus the manipulated increase in GSH excretion did not produce a proportional response in mercury output into urine.

Earlier studies indicated that the methylmercury tended to equilibrate among thiol pools in the body. In a whole blood system it is possible to mobilize methylmercury

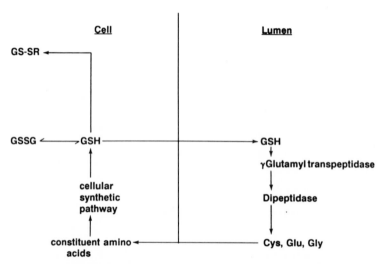

Fig. 4 Depiction of glutathione (GSH) transport across the renal proximal tubule cell
   membrane as influenced by enzymatic breakdown into constituent amino acids
   cysteine (Cys), glutamate (Glu) and glycine (Gly) in the lumen and the
   resynthesis into GSH within the cell. Mixed disulfides (GSSR) from the
   interaction with proteins and the oxidized dimer (GSSG) are in equilibrium with
   the reduced form of glutathione in the cell.

from protein binding sites in both red cells and plasma by establishing concentrations of monothiol chelating agents above the millimolar rage *in vitro* (Kostyniak, et al., 1975) and *in vivo* (Kostyniak et al., 1977). The mobilization can be rapid enough to affect a dramatic removal of methylmercury from the poisoned patient, utilizing an extracorporeal complexing hemodialysis procedure in which 50% of the methylmercury in blood can be cleared on a single pass through the dialyzer (Kostyniak, et al., 1977, Al-Abbasi, et. al., 1978). Methylmercury in tissues will re-equilibrate rapidly enough with the blood compartment to affect a total clearance that can exceed the original blood

Table 3  Effects of AT-125 on glutathione and methylmercury excretion in the mouse[+]

| AT-125 mg/Kg | UrinaryTGS uM | Total Excretion* of [203]Hg (% Body Burden/24h) | % Total [203]Hg Excreted in Urine |
|---|---|---|---|
| 0 | 3 | 7.1 ± .4 | 23.3 ± 1.1 |
| 7.5 | 450 | 6.4 ± .6 | 22.5 ± 2.3 |
| 15 | 900 | 6.8 ± .4 | 29.9 ± 2.8** |
| 30 | 1500 | 7.6 ± .5 | 44.8 ± 2.9*** |

[+] Data from Mulder and Kostyniak, 1985b.
* Total excretion of [203]Hg in urine and feces (mean ± SE for a minimum N = 6)
** Significantly different from control or 7.5 mg/Kg (p<.05)
*** Significantly different from the 15 mg/Kg group (p<.05)

mercury content by upwards of a factor of five (Al-Abbasi, et. al., 1978). Similarly, methylmercury can be rapidly mobilized from tissue stores by direct infusion of selective thiol chelators such as dimercaptosuccinic acid. With high infusion rates methylmercury output into urine can rival that obtained by the direct removal from blood by the extracorporeal complexing hemodialysis procedure (Kostyniak, 1983). Thus, the ability of thiols to affect methylmercury redistribution is a concentration dependent phenomenon, and requires thiol concentrations in the millimolar range and above. Even though AT-125 can produce a five hundred fold increase in urinary glutathione concentration, this concentration is just within the millimolar range and produces only a doubling of methylmercury output in urine. Thus the mobilization of mercury into urine appears to be highly dependent upon thiol concentrations established in the urine and not just enhancement of glutathione movement across the lumenal membrane. Even the ability to increase urinary glutathione by a factor of 300 causes no significant alteration in urinary mercury output. The rapid equilibration of methylmercury across intracellular and extracellular membrane barriers to equilibrate among thiol pools is a likely explanation for the lack of its urinary excretion increase in spite of the establishment of a profound glutathionuria. Further studies into this problem may be addressed using an *in vitro* proximal tubule cell culture system which has been characterized in our laboratory (Aleo et. al., 1989, Aleo et. al., 1990).

## CONCLUSION

A two-fold difference in elimination rates of MM between two genetic variant strains of mice was due to a five-fold greater rate of urinary excretion in the fast-excreting strain. Further investigation indicated that the difference could not be accounted for by biotransformation of MM to the inorganic form. However, several factors that could conceivably influence MM renal excretion were altered, including an increased plasma total glutathione (GS) concentration, and an increased rate of output of GS in the urine. To assess the role that GS may play in the elimination of methylmercury in the urine, studies were performed to identify the specific urinary peptides and amino acids capable of forming complexes with MM in urine. MM was bound as a low molecular weight complex with co-eluted with a MM-cysteine standard. Additional studies manipulated urinary GSH excretion *in vivo* using the γ-GTP inhibitor, L-($\alpha s, 5s$)-$\alpha$-Amino-3-chloro-4,5-Dihydro-5-Isoxazoleacetic Acid (AT-125). AT-125 was administered to CBA/J mice previously treated with [203]Hg-MM to determine whether increasing urinary GS could result in a simultaneous increase in urinary excretion of MM. Increasing doses of AT-125 resulted in a dose-dependent increase in urinary GS. MM excretion, however, did not increase in a direct proportion to increasing concentrations of GS.

Earlier studies in whole blood indicated that millimolar plasma concentrations of low molecular weight thiols were required to redistribute MM from cellular binding sites into plasma. Similarly, increased urinary MM excretion coincided only with the period following AT-125 administration when urinary thiol was elevated to 300 to 400 times normal levels to suggest that changes in GS elimination into urine may only be of consequence with regard to MM excretion once GS levels have been elevated to the millimolar concentration range.

ACKNOWLEDGEMENT

This work was supported in part by grants GM 25329 and GM07145.

REFERENCES

Al-Abbasi, A.H. Kostyniak, P.J., and Clarkson, T.W., 1978, An extracorporeal complexing hemodialysis system for the treatment of methylmercury poisoning. III. Clinical applications. J. Pharmacol. Exp. Ther. 207: 249-254.

Aleo, M., Taub, M.L., Nickerson, P.A., and Kostyniak, P.J., 1989, Primary cultures of rabbit renal proximal tubule cells: I. Growth and Biochemical Characteristics. In Vitro 25: 776-783.

Aleo, M.D., Taub, M.L. and Kostyniak, P.J., 1990, Primary cultures of rabbit renal proximal tubule cells: Selected Phase I and Phase II Metabolic Capacities. In Vitro 4:6 727-773.

Kostyniak, P.J., Clarkson, T.W., Cestero, R.V., Freeman, R.B. and Al-Abbasi, A.H., 1975, An extracorporeal complexing hemodialysis system for the treatment of methylmercury poisoning. I. *In vitro* studies of the effects of four complexing agents in the distribution and dialyzability of methylmercury in human blood. J. Pharmacol. Exp. Ther. 192:260-269.

Kostyniak, P.J., Clarkson, T.W., and Al-Abbasi, A.H., 1977, An extracorporeal complexing hemodialysis system for treatment of methylmercury poisoning. II. *In vivo* applications in the dog. J. Pharmacol. Exp. Ther. 203: 253-263.

Kostyniak, P.J., 1980. Differences in elimination rates of methylmercury between two genetic variant strains of mice. Toxicol. 6: 405-410.

Kostyniak, P.J., 1983, Methylmercury removal in the dog during infusion of 2,3-dimercapto-succinic acid. J. Toxicol. Environ. Health 11: 947-957.

Mulder, K.M., 1985, Investigation of the mechanism of a strain variation in renal elimination of methylmercury in mice. PhD Thesis, University of Buffalo.

Mulder, K.M., and Kostyniak, P.J., 1985a, Stabilization of glutathione in urine and plasma: Relevance to urinary metal excretion studies. J. Anal. Toxicol. 9, 31-35.

Mulder, K.M., and Kostyniak, P.J., 1985b, Effect of L-($\alpha$S,5S)-$\alpha$-amino-3-chloro-4,5-dihydro-5-isoxazoleacetic acid on urinary excretion of methylmercury in the mouse. J.Pharmacol. Exp. Ther. 234: 156-160.

Mulder, K.M. and Kostyniak, P.J., 1985c, Involvement of glutathione in the enhanced renal excretion of methylmercury in CFW Swiss mice. Toxicol. Appl. Pharmacol. 78:451-457.

Norseth, T., 1971, Biotransformation of methyl mercuric salts in the mouse studies by specific determination of inorganic mercury. Acta. Pharmacol. et Toxicol. 29: 375-384.

Norseth, T. and Clarkson, T.W., 1971, Intestinal Transport of 203Hg-Labeled Methyl Mercury Chloride: Role of biotransformation in rats. Arch. Envir. Health 22:568-577.

# ROLE OF GLUTATHIONE IN MERCURY DISPOSITION

Akira Naganuma, Toshiko Tanaka, Tsutomu Urano and
Nobumasa Imura

Department of Public Health, School of Pharmaceutical Sciences
Kitasato University, Tokyo 108, Japan

## ABSTRACT

A complex of methylmercury (MeHg) with glutathione (GSH) has been found in several tissues such as brain, liver, erythrocytes and bile. Although the binding of MeHg to GSH is reversible, the complex formation may play an important role in the transport of MeHg.

Species difference in the biliary secretion rate of MeHg reflected total concentration of GSH and its degradation products such as cysteine (Cys) and cysteinylglycine (CysGly) in the bile. It has been reported that hepatic GSH is secreted into bile, and then decomposed enzymatically to CysGly and Cys. These facts suggest that the GSH transport system from liver to bile is one of the determinants of the biliary secretion of MeHg. The main form of MeHg in the bile was found to be MeHg-GSH in mice and hamsters and MeHg-CysGly in rats and guinea pigs. These complexes of MeHg were reabsorbed from the intestine more efficiently than MeHgCl, and transported into the kidneys and liver.

On the other hand, substantial amounts of GSH is released directly from the liver into plasma as a major source of plasma GSH. Specific depletion of hepatic GSH by pretreatment of mice with 1,2-dichloro-4-nitrobenzene (DCNB) reduced renal accumulation of MeHg. The renal uptake of MeHg in mice receiving MeHg-GSH intravenously was significantly higher than that in mice receiving MeHgCl. Although

*Advances in Mercury Toxicology*, Edited by T. Suzuki *et al.*
Plenum Press, New York, 1991

MeHg-GSH has hardly been detected in plasma after administration of MeHgCl, our results described above suggest the possibility that the GSH released from the liver into plasma plays an important role in the renal accumulation of MeHg. Inhibition of γ-glutamyltranspeptidase (γ-GTP) by acivicin pretreatment also reduced renal MeHg uptake and increased urinary excretion of MeHg and GSH. These facts indicate that MeHg is transported to the kidney as GSH complex and the MeHg is incorporated into the kidney by a γ-GTP dependent system. Sex and strain differences and developmental changes in the renal uptake of MeHg can at least partially be explained by the difference in renal γ-GTP activity.

INTRODUCTION

Glutathione (GSH) is considered to be the most important sulfhydryl compound in the detoxification and metabolism of xenobiotics. A methylmercury-GSH complex has been detected in several tissues such as brain (Thomas and Smith, 1979), liver (Omata et al., 1978), erythrocytes (Naganuma and Imura, 1979; Naganuma et al., 1980) and bile (Ballatori and Clarkson, 1983; Naganuma and Imura, 1984). Although the binding of methylmercury to GSH is reversible (Naganuma et al., 1980), the complex formation may play an important role in the transport of methylmercury.

Role of GSH in biliary secretion of methylmercury

It has been reported that over 10% of the injected methylmercury was excreted in rat bile during the 24 hr period following its administration (Norseth, 1973), and about 90% of this methylmercury was then reabsorbed from the gut (Clarkson et al., 1973). These experimental results suggest that biliary excretion and enterohepatic circulation of methylmercury are important factors in determining the disposition and toxicity of methylmercury in mammals. We compared biliary secretion of methylmercury in several species to obtain information on the role of GSH in the biliary secretion of methylmercury in animals (Naganuma and Imura, 1984; Urano et al., 1988b). Table 1 shows biliary concentrations of mercury in rats, mice, hamsters, guinea pigs and rabbits 2 to 3 hr after administration of methylmercury (Urano et al., 1988b). The concentration of mercury in the bile of guinea pigs and rabbits were markedly lower than those of rats, mice and hamsters. Ballatori and Clarkson (1983) suggested that the biliary transport system for GSH might be a determinant of biliary secretion of methylmercury in rats.

Table 1   Species difference in biliary excretion of methylmercury. Excretion of
mercury into bile of animals 2-3 hr after administration of methylmercury (5
μmol/kg, iv) was determined.

| | Bile flow (ml/hr/kg) | Hg in bile (% of dose) |
|---|---|---|
| Rat | $4.3 \pm 0.7$ | $0.74 \pm 0.23$ |
| Mouse | $2.3 \pm 1.0$ | $0.44 \pm 0.16$ |
| Hamster | $2.3 \pm 0.5$ | $0.60 \pm 0.19$ |
| Guinea pig | $10.3 \pm 2.7$ | $0.05 \pm 0.01$ |
| Rabbit | $3.7 \pm 0.6$ | $0.02 \pm 0.01$ |

Since GSH secreted from the liver into bile canaliculi is metabolized to CysGly and
Cys by the actions γ-glutamyl transpeptidase and dipeptidase located on biliary ductular
epithelium (Abbott and Meister, 1986), concentrations of GSH, CysGly and Cys in the
bile of these animals were determined by HPLC. Total concentrations of GSH,
CysGly and Cys in the bile of guinea pigs and rabbits were significantly lower than
those of rats, mice and hamsters (Table 2). These results suggest that species
difference in biliary secretion of methylmercury reflect total concentration of GSH and
its degradation products in the bile. Thus the GSH transport system from the liver to
bile may be one of the determinants of biliary secretion of methylmercury as speculated
by Ballatori and Clarkson (1983). Furthermore, efficacy differences in hepatobiliary
transport of GSH among the species may play an important role in the species
difference in the biliary secretion of methylmercury.

Figure 1 shows Sephadex G-15 gel filtration patterns of bile obtained from rat,
mice, hamster or guinea pigs receiving methylmercury (Urano et al.,1988a; Urano et
al., 1988b). Both bile collection and gel filtration were performed in the presence of
EDTA to prevent oxidation of sulfhydryls. The major form of Hg in the bile was found
to be methylmercury-GSH in mice and hamsters and methylmercury-CysGly in rats
and guinea pigs. In the case of rabbit bile, an inadequate mercury separation pattern
was obtained by Sephadex G-15 filtration and the recovery of mercury from the
column was extremely low, probably because of the low concentration of mercury in
the rabbit bile.

113

Table 2  Species difference in concentrations of glutathione and its degradation products in bile

| | Concentration (nmol/ml bile) | | | |
|---|---|---|---|---|
| | Total NPSH[a] | GSH | CysGly | Cys |
| Rat | 3790 ± 770 | 3040 ± 720 | 640 ± 100 | 100 ± 20 |
| Mouse | 3297 ± 242 | 3253 ± 245 | 11 ± 2 | 34 ± 22 |
| Hamster | 2885 ± 608 | 2800 ± 240 | 84 ± 17 | N.D. [b] |
| Guinea Pig | 106 ± 24 | N.D. | 106 ± 24 | N.D. |
| Rabbit | 35 ± 7 | N.D. | 31 ± 6 | 4 ± 1 |

[a] NPSH, non-protein sulfhydryls.  [b] N.D., not detected.

These S-conjugates of methylmercury undergo biliary secretion into the intestine and are available for reabsorption. Figure 2 shows *in situ* absorption of methylmercury from ligated intestinal segment of rats (Urano et al., 1990). The methylmercury complexes with GSH and CysGly in bile were absorbed from the intestine more efficiently than methylmercuric chloride, and can be transported into organs such as the kidneys and liver. The methylmercury thus reabsorbed and transported to the liver completes the enterohepatic circulation which has been thought to be important for the disposition and toxicity of methylmercury.

Role of GSH in renal uptake of methylmercury

Substantial amounts of GSH are released directly from the liver into plasma as a major source of plasma GSH, and then transported into the kidneys. Therefore, we attempted to clarify the role of hepatic and plasma GSH in the uptake of methylmercury by the kidney which most efficiently accumulates methylmercury.

Several series of experiments were performed to search for a reagent that specifically depressed hepatic GSH, and 1,2-dichloro-4-nitrobenzene (DCNB) was found to

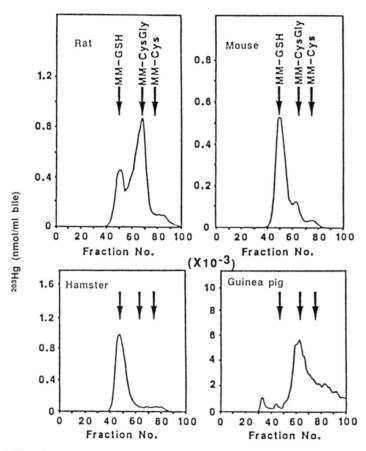

Fig. 1 Gel filtration patterns of mercury in bile of rats, mice, hamsters and guinea pigs after i.v. administration of methylmercuric chloride.

115

significantly decrease GSH concentration in the liver without affecting renal GSH (Naganuma et al., 1988). Pretreatment of ICR male mice with DCNB significantly decreased renal accumulation of methylmercury, suggesting that hepatic GSH rather than renal GSH has an important role in the methylmercury accumulation in the kidney (Naganuma et al., 1988). It is well known that hepatic GSH and its S-conjugates are released from the liver into the plasma. In the kidney, the plasma GSH is hydrolyzed after being filtered through glomeruli by γ-glutamyl transpeptidase (γ-GTP) and dipeptidase into its constituent amino acids, which are rapidly taken up by the kidney cells. Thus, methylmercury may be translocated into the kidney as a complex with GSH, and incorporated into the kidneys through the same route as that of GSH. In fact, renal uptake of methylmercury was increased by intravenous coadministration of GSH with methylmercury (Naganuma et al., 1988).

Pretreatment of mice with acivicin, a potent inhibitor of γ-glutamyl transpeptidase, significantly depressed the renal accumulation of methylmercury, and increased urinary excretion of both GSH and methylmercury. From these results, we concluded that methylmercury was translocated into the kidney as a complex with GSH, and then incorporated into renal cells after degradation of the GSH moiety by γ-glutamyl transpeptidase and dipeptidase (Naganuma et al, 1988).

We subsequently examined the contribution of hepatic GSH and renal γ-glutamyl trans-peptidase to strain and sex differences in mercury accumulation in the mouse kidney (Tanaka et al., 1990). Strain and sex differences in renal methylmercury accumulation were observed in mice 4 hr after administration of methylmercury (Figure 3). The highest concentration of renal methylmercury was observed in male ICR mice. Furthermore, male BALB/c and C57black mice showed higher renal accumulation of methylmercury than female mice of the same strains. Strain and sex-related differences in GSH content were also observed in their livers and kidneys and the GSH content of the liver and kidneys were significantly correlated with the renal accumulation of methylmercury. However, renal mercury accumulation was affected by hepatic GSH rather than renal GSH described above. The effect of pretreatment with various doses of DCNB, a specific hepatic GSH depletor, on tissue GSH levels and renal uptake of methylmercury was examined in ICR, BALB/c or C57BL/6N mice. Figure 4 shows the results in the case of ICR mice. The results show that the renal mercury level is correlated with GSH concentration in the liver and plasma, but not with GSH in the kidney. Since most of the plasma GSH is released from the liver, hepatic GSH, but not renal GSH, may be an important determinant of strain and sex differences in mercury uptake by the kidneys.

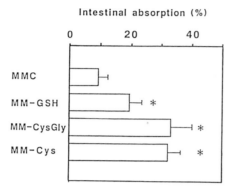

Fig. 2  Intestinal absorption rate of methylmercury compounds from a ligated intestinal
segment of rats 30 min after injection of methylmercuric chloride (MMC) or
methylmercury (MM) premixed with Cys, CysGly or GSH.
*Significantly different from MMC, P<0.005.

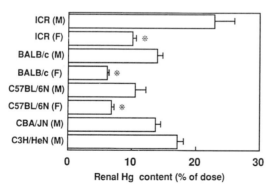

Fig. 3  Strain and sex differences in mercury uptake by mouse kidney 4 hr after
administration of methylmercury chloride.
*Significantly different from the male of the same strain (P<0.005).

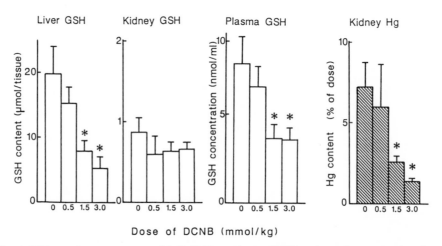

Fig. 4 Effect of pretreatment with DCNB on tissue GSH and renal mercury levels at 10 min after i.v. administration of methylmercuric chloride.

On the other hand, strain and sex differences in the renal activity of γ-glutamyl transpeptidase were also observed. Renal mercury concentration after administration of methylmercury was significantly correlated with the activity of renal γ-glutamyl transpeptidase. From these results, renal γ-glutamyl transpeptidase also appears to be an important factor determining strain and sex differences in mercury accumulation into the kidneys. To further confirm the role of γ-glutamyl transpeptidase on the sex difference in methylmercury uptake by the kidneys, the effects of castration and administration of testosterone on mercury accumulation into the kidneys of male ICR mice were determined. Castration significantly decreased mercury uptake by the kidneys, and the decreased renal mercury accumulation was returned to the normal level by administration of testosterone. Renal activity of γ-glutamyl transpeptidase was also similarly changed by castration and testosterone administration. Administration of testosterone to female mice increased both renal γ-glutamyl transpeptidase activity and mercury accumulation. These results suggest that the sex difference in renal accumulation of mercury may be regulated at least partly by the sex hormone through its action on the renal γ-glutamyl transpeptidase.

These findings indicate that hepatic GSH and renal γ-glutamyl transpeptidase are regulatory factors for renal mercury accumulation and are acting important determinants of strain and sex differences in methylmercury disposition.


## REFERENCES

Abbott, W. and Meister, A., 1986, Intrahepatic transport and utilization of biliary glutathione and its metabolites. Proc. Natl. Acad. Sci. USA. 83:1246.

Ballatori, N. and Clarkson, T.W., 1983, Biliary transport of glutathione and methylmercury. Am. J. Physiol. 244:G435.

Clarkson, T.W., Small, H. and Norseth, T., 1973, Excretion and absorption of methylmercury after polythiol resin treatment. Arch. Environ. Health. 26:173.

Naganuma, A. and Imura, N., 1979, Methylmercury binds to a low molecular weight substance in rabbit and human erythrocytes. Toxicol. Appl. Pharmacol. 47:613.

Naganuma, A. and Imura, N., 1984, Species difference in biliary excretion of methylmercury. Biochem. Pharmacol. 33:679.

Naganuma, A., Koyama, Y. and Imura, N., 1980, Behavior of methylmercury in mammalian erythrocytes. Toxicol. Appl. Pharmacol. 54:405.

Naganuma, A., Oda-Urano, N., Tanaka, T. and Imura, N., 1988, Possible role of hepatic glutathione in transport of methylmercury into mouse kidney. Biochem. Pharmacol. 37:291-296.

Norseth, T., 1973, Biliary excretion and intestinal reabsorption of mercury in the rat after injection of methyl mercuric chloride. Acta Pharmacol. Toxicol. 33:280.

Omata, S., Sakimura, K., Isii, T. and Sugano, H., 1978, Chemical nature of a methylmercury complex with a low molecular weight in the liver cytosol of rats exposed to methymercury chloride. Biochem. Pharmacol. 27:1700.

Tanaka, T., Kobayashi, K., Naganuma, A. and Imura, N., 1991, An explanation for strain and sex differences in renal uptake of methylmercury in mice. Toxicology, in press.

Thomas, D.J. and Smith, J.C., 1979, Partial characterization of a low-molecular weight methylmercury complex in rat cerebrum. Toxicol. Appl. Pharmacol. 47:547.

Urano, T., Iwasaki, A., Himeno, S., Naganuma, A. and Imura, N., 1990, Absorption of methylmercury compounds from rat intestine. Toxicol. Lett. 50:159.

Urano, T., Naganuma, A. and Imura, N., 1988a, Methylmercury-cysteinylglycine constitutes the main form of methylmercury in rat bile. Res. Commun. Chem. Pathol. Pharmacol. 60:197.

Urano, T., Naganuma, A. and Imura, N., 1988b, Species differences in biliary excretion of methylmercury : Role of non-protein sulfhydryls in bile. Res. Commun. Chem. Pathol. Pharmacol. 62:339.

# MECHANISM FOR RENAL HANDLING OF METHYLMERCURY

Kimiko Hirayama, Akira Yasutake* and Tatsumi Adachi*

Kumamoto University College of Medical Science
Kumamoto, Japan

*National Institute for Minamata Disease
Minamata, Japan

## ABSTRACT

Since methylmercury (MeHg) has a high affinity for thiol compounds, the metabolism and transport of glutathione (GSH) and its derivatives are important for determining tissue distribution and elimination of MeHg in organisms which are challenged with this toxic compound.

Our studies revealed that the rate of MeHg secretion from tissues seems to correlate with the rate of tissue efflux of GSH. The transfer of MeHg from extrarenal tissues to the kidney is important for its urinary excretion. Although intracellular MeHg seems to be secreted into the circulation as its GS-conjugate, a significant part of plasma MeHg was identified as its albumin-conjugate. Since rapid exchange reaction of an SH-MeHg conjugate with other thiols occurs, the molecular form of a MeHg-conjugate would be changed by the presence of other thiol compounds.

There may be two mechanisms for renal uptake of MeHg; the one via glomerular filtration followed by its luminal absorption and the other by non-filtrating peritubular uptake. It has been suggested that a reaction catalyzed by $\gamma$-glutamyltransferase ($\gamma$-GTP) and probenecid sensitive transport system may be involved in renal uptake of MeHg. Since urinary excretion of MeHg showed a marked lag period, some non-filtrating mechanism may play an important role in renal uptake of MeHg.

*Advances in Mercury Toxicology*, Edited by T. Suzuki *et al.*
Plenum Press, New York, 1991

Renal brush border membranes have a secretory transport system for GSH and its S-conjugates. Since the rate of urinary elimination of MeHg changed significantly with concomitant changes in the secretory rate of renal GSH, MeHg may be transported to the luminal space via GSH transport system as MeHg-GS. The major part of urinary MeHg was accounted for by its cysteine (CySH) conjugate. In the luminal space, MeHg-GS may be transformed into its CySH conjugate according to degradation of GSH and then excreted in urine. The mechanism of renal uptake and urinary excretion of MeHg will be discussed in detail.

## INTRODUCTION

A great individual variation in the organ distribution, biological half-time, and toxic dose of methylmercury (MeHg) has been reported in animals and humans (Tsubaki, 1968; Al-Sharistani and Shihab, 1974; Tagashira et al., 1980; Magos et al., 1981; Doi et al., 1983; Hirayama and Yasutake, 1986; Yasutake and Hirayama, 1986). Elimination rate of toxic compounds from the body may play a critical role in toxic susceptibility. In various species, strains, ages and sex of animals, greater variations in Hg levels were found in urine than in feces 24 hr after MeHg administration (Figure 1). Thus, the eliminatory pathway of urine may be an important determinant for MeHg toxicity, and it is necessary to understand the mechanism of renal handling of MeHg.

### Transfer of MeHg to the kidney

The transfer of xenobiotics from extra-renal tissues to the kidney is a prerequisite for their urinary excretion. Recent studies revealed that the metabolism and transport of glutathione (GSH) play an important role in the fate of MeHg (Hirayama et al., 1987; Naganuma et al., 1988). Since many cells and tissues have secretory transport systems for GSH and its adducts, and the affinity of MeHg for GSH is high, intracellular MeHg may also be secreted as its GSH-conjugate. The rate of transfer of MeHg to the kidney, including urine, was closely correlated to hepatic efflux of GSH to plasma (Figure 2). Thus, the sum total of Hg levels in kidney and urine was higher in male mice having the faster hepatic efflux of GSH than in females having its slower efflux. In addition, aged mice showed decreased hepatic efflux rate of GSH and MeHg levels in kidney plus urine. These results support the above idea that MeHg may be secreted from the liver into the circulation through the GSH secretion system as a MeHg-GS complex. Although MeHg-GS conjugate is expected to be a predominant form in the circulation, MeHg-albumin conjugate was found to be the major form in plasma (Yasutake et al., 1989), indicating that rapid conversion of GS-MeHg to albumin-conjugate would have occurred in the circulation. It is known that a mercurial compound reversibly forms complexes with various thiols, and hence a rapid exchange reaction between free thiols and MeHg S-conjugates occurs under physiological conditions (Carty and Malone, 1979).

Our recent studies suggest that the SH-group at Cys-34 is not the only site of albumin for the interaction with MeHg compounds. Albumin also interacts with various MeHg compounds even if its Cys-34 residue was masked as shown in Figure 3-A (Yasutake et al., 1990). Furthermore, it has been revealed that not only MeHg but also its conjugates of GSH, cysteinylglycine (CySGly) and of L-cysteine (CySH) bound tightly to both mercapt- and nonmercaptalbumin (Figure 3-B, -C and -D). Binding of MeHg-S-conjugate to albumin increased with concomitant decrease in ligand concentrations. Mercaptalbumin showed much higher affinity for the MeHg mercaptides than did SH-modified albumin. The least binding was observed with S-

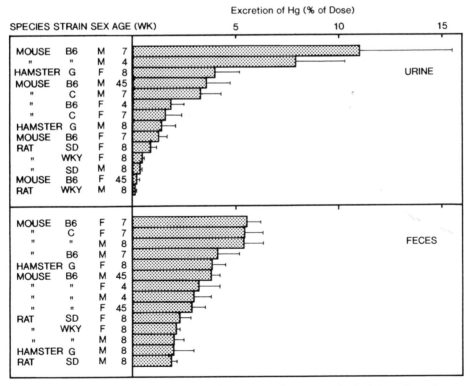

Fig. 1  Urinary and fecal excretion of Hg 24 hr after per oral administration of MeHgCl (20 μmol/kg) in different species, strain, sex and age of animals.

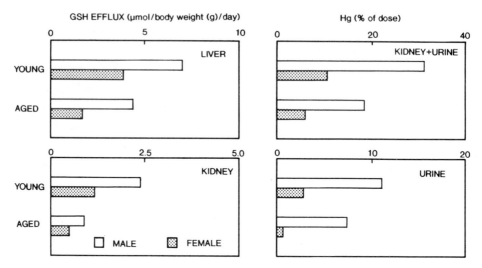

Fig. 2    Hg levels in kidney plus urine or urine 24 hr after per oral administration of MeHgCl (20 µmol/kg), and hepato-renal efflux of glutathione in young (7 wk) and aged (about 90 wk) C57BL/6N mice. Glutathione efflux was estimated from half-time of glutathione determined by using buthionine sulfoximine.

carbamidmethylalbumin (Alb-SCAM) and cysteine-mixed disulfide of albumin (Alb-SSCy). Binding of MeHg ligands to Cys-34 and other sites on albumin (Figure 4) might constitute an important step in the tissue distribution and elimination of the metabolites predominantly due to its reversible nature.

Renal uptake of MeHg

Theoretically, there are two mechanisms for renal uptake of a low molecular weight compound; the one via glomerular filtration followed by its luminal reabsorption, and the other by non-filtrating peritubular uptake. It is considered that binding of substances to the circulating albumin *in vivo* might be important in directing the bound ligand to the kidney, where a peritubular transport system is operating. Since MeHg-albumin conjugate was found to be the major form in plasma, it is speculated that MeHg accumulated in the kidney predominantly via some non-filtrating peritubular mechanism. Alternatively, it is possible that trace amounts of low molecular weight MeHg may be filtered through the glomeruli and reabsorbed from the luminal surface by the proximal tubule cells.

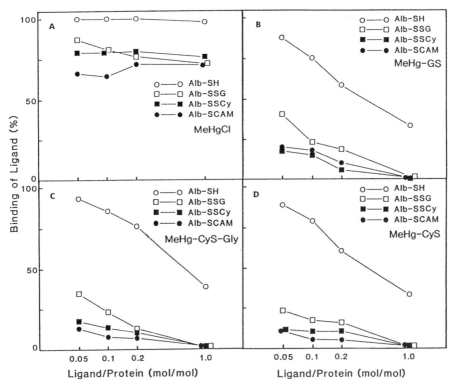

Fig. 3 Binding of MeHg compounds to bovine serum albumin (BSA) species. After mixing each BSA species (0.4 mM) with MeHg ligands (0.02, 0.04, 0.08 and 0.4 mM) in PBS (pH 7.4) at room temperature, the mixtures were filtered immediately through ultrafiltration membranes.

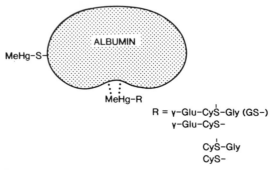

Fig. 4  Complexes of MeHg or MeHg-conjugates with albumin.

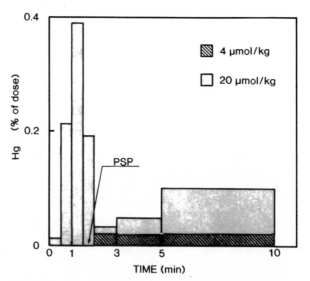

Fig. 5  Time course of Hg in urine after intravenous administration of MeHg-GS conjugate in acivicin (1 mmol/kg) treated C57BL/6N male mice. Urine samples were collected from the tube cannulated into the bladder.

When animals were pretreated with acivicin, an inhibitor of $\gamma$-glutamyltransferase ( $\gamma$-GTP), a significant amount of (about 6% of the dose) of MeHg was excreted into urine within the first 3 hr with concomitant decrease in its renal level (Yasutake et al., 1989). It should be noted that urinary MeHg was scarcely detected by this time in mice which were not treated with acivicin (below 0.1% of the dose). Time course of urinary excretion of MeHg within the first 10 min was examined after intravenous injection of MeHg-GS in acivicin treated mice (Figure 5). When 4 $\mu$mol/kg of MeHg-GS conjugate was injected, MeHg was scarcely found in urine up to 2 min and then its excretion gradually began at the same time that phenolsulfonphtalein (PSP) began to appear. When higher levels of MeHg-GS (20 $\mu$mol/kg) were administered, urinary excretion of MeHg was biphasic: the first stage of Hg excretion appeared as a sharp peak after 1 min and the second stage began after 2 to 3 min. Since it is well known that PSP is taken up at a peritubular site and secreted from the proximal tubule cells, the MeHg excreted in the first phase may be via glomerular filtration while the later one may be via peritubular uptake.

Hg excretion in the first phase accounted for less than 1% of the dose in acivicin treated mice. These results indicate that most of the MeHg in the circulation would have escaped renal filtration and accumulated in the renal cells by some non-filtrating peritubular mechanism.

It has also been reported that the major part of plasma GSH is extracted by the kidney by a non-filtrating peritubular mechanism in which $\gamma$-GTP plays a predominant role (Inoue et al., 1986). A part of MeHg bound to albumin may be transformed into conjugates of GSH or its metabolites, such as CySGly, GluCyS and CySH, at the peritubular site. Since urinary excretion of MeHg is negligible up to 30 sec as shown in Figure 5, the study renal uptake of MeHg after 30 sec is considered to be helpful for understanding the event occurred at peritubular site. Therefore, renal uptake of MeHg was examined 30 sec after administration of MeHgCl or MeHg-conjugates with GSH and its metabolites at a dose of 4 $\mu$mol/kg (Figure 6). Renal uptake of MeHg after MeHgCl administration was inhibited by treatment of probenecid (an inhibitor of organic acid transport) or acivicin. Both probenecid and acivicin treatment caused significantly lower renal uptake of MeHg than acivicin treatment alone. Also, acivicin or probenecid treatment inhibited renal uptake of MeHg after MeHg-GS administration, and both treatment showed much lower uptake of MeHg than after each treatment alone. MeHg-CySGly administration caused the highest renal uptake of MeHg which was inhibited by probenecid but not by acivicin. Moreover, probenecid treatment inhibited renal uptake of MeHg after administration of GluCyS- and CyS-conjugate. Although CySGly and GluCyS are dipeptides, their MeHg- conjugates seem to be transported via the probenecid sensitive transport system. γ-GTP can catalyze hydrolysis and transpeptidation reactions *in vitro*. Since CySH is an excellent acceptor substrate of the enzyme and easily converted to GluCyS, renal uptake of Hg after administration of MeHg-CyS may also be inhibited by probenecid. Thus, γ-GTP-catalyzed degradation of GSH might accelerate the peritubular uptake of MeHg which partly proceeded via the probenecid sensitive transport systems. Although MeHg-GS conjugate may also be taken up by this probenecid-sensitive transport system, the transport of GSH and its adducts by this system is far less efficient than their degradation products. (Inoue et al., 1986).

It has been suggested that MeHg-CyS may share a common transport step with the L-neutral amino acid carrier transport system (Hirayama, 1980, 1985; Aschner, 1989) and a significant percentage of the amino acid accumulation within the renal tubule cells occurs via the basolateral cell membrane. Thus, the amino acid transport system in the basolateral membranes may also play an important role in renal uptake of MeHg. Recently, it has been suggested that the transport system for L-methionine is a candidate for the MeHg-CyS transport system (Aschner and Clarkson, 1989). Our data showed that L-methionine treatment inhibited renal uptake of MeHg 1 hr after intravenous administration of MeHgCl, MeHg-GS or MeHg-CyS, while no change was found in urinary Hg levels in mice at this time (Figure 7). These results suggest that a part of MeHg was transported via the L-neutral amino acid transport system in the basolateral membranes. Thus, the amino acid transport system in the basolateral membranes may also have an important role in renal uptake of MeHg.

## Renal excretion of MeHg

It should be noted that MeHg levels in the kidney were at a maximum 3-6 hr after administration while urinary excretion of MeHg occurred only after 9-12 hr with concomitant decrease in the renal levels (Figure 8) (Yasutake et al., 1989). The reason for the lag time in urinary excretion of MeHg is still unclear. It is possible that the secretion of MeHg taken up by cell organelles, such as mitochondria, lysozomes, etc., may be delayed.

Recent studies by Bergeron et al. (1980) have demonstrated that many mitochondria, especially those in the basal part of the proximal tubule, have a more

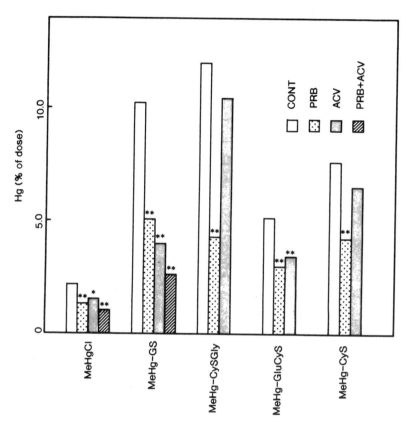

Fig. 6  Effect of probenecid (PRB, 200 μmol/kg) and acivicin (ACV, 1 mmol/kg) on renal accumulation of Hg 30 sec after intravenous administration of 4 μmol/kg of MeHgCl, MeHgGS-cysteinylglycine (-CySGly), -glutamylcysteine(-GlCyS) or cysteine (-CyS) conjugate in C57BL/6N male mice.
  *Significantly different from control, p<0.05
** Significantly different from control, p<0.01

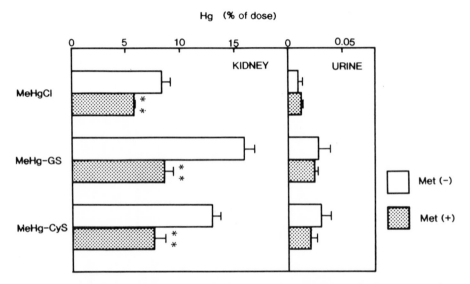

Fig. 7  Effect of methionine (Met, 1 mmol/kg) on renal uptake and urinary excretion
of Hg 1 hr after intravenous administration of 4 μmol/kg of MeHgCl, MeHg-
glutathione (-GS) or MeHg-cysteine (-CyS) conjugate.
** Significantly different from Met (-), p<0.01

extensive, branched or anastomosing x configuration than suspected prevously in rat kidney. In contrast, the mitochondria in the apical region of the cell are smaller, less branched and fewer in number. Such localization of organelles in the proximal tubule cells may be involved in the lag time in urinary excretion of MeHg. Since the rate of GSH efflux from the kidney was found to be correlated with the rate of urinary excretion of MeHg (Figure 2), MeHg-GS conjugate may be secreted via the GSH secretion system from the proximal tubule cells. In luminal space, GSH is hydrolyzed into constituent amino acids by γ-GTP and dipeptidases. Hence, MeHg-CyS conjugate may be formed and reabsorbed from the proximal tubule cells via the neutral amino acid transport system. When L-methionine was treated 15 hr after MeHgCl administration in mice (after appearance of urinary excretion of MeHg), urinary excretion of MeHg was accelerated (data not shown). This result supports the hypothesis that the MeHg-CyS complex may be reabsorbed from the proximal tubule cell membranes via the neutral amino acid transport system. Since most of the urinary MeHg was accounted for by MeHg-CyS (Figure 9), the MeHg-CyS which escaped by the reabsorption from proximal tubule cells may be excreted into the urine (Yasutake et al., 1989).

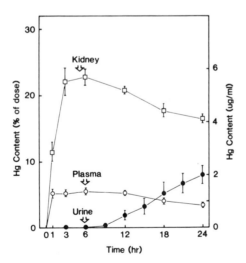

Fig. 8   Time-dependent changes of Hg levels in plasma, urine and kidney. At indicated times after oral admnistration of MeHgCl (20 μmol/kg), animals were exsanguinated by bleeding. Then Hg levels in plasma, urine and kidney were determined. Urinary Hg contents represent cumulative values. Hg contents in kidney and urine were shown as a percent of dose, and plasma level was shown as concentrations. Values represent mean ± SD obtained from five mice.

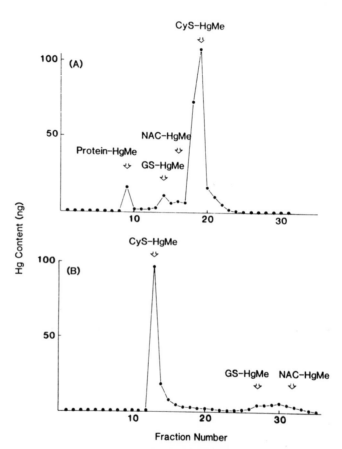

Fig. 9 Chromatographies of urine samples of MeHg-treated mice. Twenty four hours after oral administration of MeHg Cl (20 μmol/kg), freshly obtained urine samples were immediately diluted with an equal volume of 50 mm serine-borate mixture, and chromatographed using column of Cellulofine GCL-25-sf (1.6 x 54 cm) (A). Low molecular weight fraction of the samples obtained by ultrafiltration methods was determined for MeHg distribution in a DEAE-Sephadex a-25 column (1.5 x 34 cm) (B).

CONCLUSION

Figure 10 shows a schematic presentation of renal handling of MeHg. Our data suggested that renal uptake of MeHg from the circulation proceeds predominantly via some non-filtrating peritubular mechanism, such as the probenecid sensitive transport system and the neutral amino acid transport system. γ-GTP-catalyzed degradation of GSH might accelerate the renal uptake of MeHg. Also, a small amount of MeHg filtratedthrough glomerulifiltration is taken up by its luminal absorption.

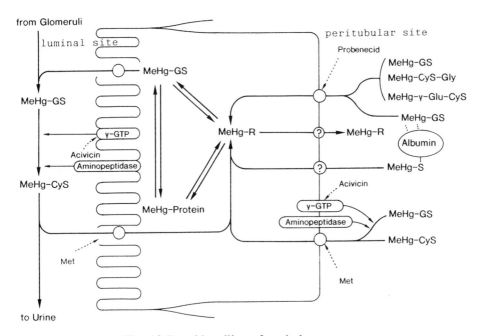

Fig. 10 Renal handling of methylmercury

After some lag period, MeHg accumulated in renal cells may be excreted as GS-conjugate in the luminal space via a secretory transport system for GSH and S-conjugates. In the luminar space, GSH is rapidly degraded into constituent amino acids by γ-GTP and peptidases. Therefore, MeHg-Cys may be formed and reabsorbed from luminal surfaces and the metabolite escaping luminal absorption may be excreted into urine.

133

## REFERENCES

Al-Sharistani, H. and Shihab, K.M.,1974. Variation of biological half-life of methylmercury in man. Arch. Environ. Health 28:342-344.

Aschner, M., 1989. Brain, kidney and liver [203]Hg-methylmercury uptake in the rat: relationship to the neutral amino acid carrier. Pharmacol. Toxicol. 65:17-20.

Aschner, M. and Clarkson, T.W., 1989. Methylmercury uptake across bovine brain capillary endothelial cells *in vitro*: The role of amino acids. Pharmacol. Toxicol. 64:293-299.

Bergeron, M., Guerette, D., Forget, J. and Thiery, G., 1980. Three dimensional characteristics of the mitochondria of the rat nephron. Kidney Int. 17:175-185.

Carty, A.J. and Malone, S.F. (1979). The chemistry of mercury in biological systems. In: The biologeochemistry of mercury in the environment. J.O. Nriagu (ed.). Elsevier, Amsterdam, pp. 437-439.

Doi, R., Tagawa, M., Tanaka, H., and Nakaya, K., 1983. Hereditary analysis of the strain difference of methylmercury distribution in mice. Toxicol. Appl. Pharmacol. 69:400-406.

Hirayama, K., 1980. Effects of amino acids on brain uptake of methylmercury. Toxicol. Appl. Pharmacol. 55:318-323.

Hirayama, K., 1985. Effects of combined administration of thiol compounds and methylmercury chloride on mercury distribution in rats. Biochem. Pharmacol. 34:2030-2032.

Hirayama, K., Yasutake, A., 1986. Sex and age differences in mercury distribution and excretion in methylmercury-administered mice. J. Toxicol. Environ. Health 18:49-60.

Hirayama, K., Yasutake, A. and Inoue, M. 1987. Effect of sex hormones on the fate of methylmercury and glutathione metabolism in mice. Biochem. Pharmacol. 36:1919-1924.

Inoue, M., Shinozuka, S. and Morino, Y., 1986. Mechanism for renal peritubular extraction of plasma glutathione. The catalytic activity of contraluminal γ-glutamyl- transferase is prerequisite to the apparent peritubular extraction of plasma glutathione. Eur. J. Biochem. 157:605-609.

Magos, L., Peristianis, G.C., Clarkson, T.W., Brown, A., Preston, S. and Snowden R.T., 1981. Comparative study of the sensitivity of male and female rats to methylmercury. Arch. Toxicol. 48:11-20.

Naganuma, A., Oda-Urano, N., Tanaka, T. and Imura, N., 1988. Possible role of hepatic glutathione in transport of methylmercury into mouse kidney. Biochem. Pharmacol. 37:291-296.

Tagashira, E., Urano, T. and Yanaura, S., 1980. Methylmercury toxicosis. I. Relationship between the onset of motor incoordination and mercury contents in the brain. Folia. Pharmacol. Japon. 76:169-177.

Tsubaki, T., 1968. Organic mercury intoxication in the Agano River area. Studied by Niigata University Research Group. Clin. Neurol. 8:511-520.

Yasutake, A. and Hirayama, K., 1986. Strain difference in mercury excretion in methylmercury-treated mice. Arch. Toxicol. 59:99-102.

Yasutake, A., Hirayama, K. and Inoue, M., 1989. Mechanism of urinary excretion of methylmercury in mice. Arch. Toxicol. 63:479-483.

Yasutake, A., Hirayama, K. and Inoue, M., 1990. Interaction of methylmercury compounds with albumina. Arch. Toxicol. 64:639-643.

# METHODS FOR DECREASING $^{203}$Hg RETENTION IN RELATION TO AGE AND ROUTE OF EXPOSURE

B. Kargacin and K. Kostial

Institute for Medical Research and Occupational Health
University of Zagreb, M. Pijade 158
41000 Zagreb, Yugoslavia

## ABSTRACT

Different chelating agents are being used for decreasing the body burden of mercury. Data on the efficiency of such treatments in the very young are scarce and almost not available. However, young are known to be at a higher risk than adults at the same level of environmental mercury exposure. Therefore more data on the effect of chelation therapy in the very young are necessary.

We, therefore, performed experiments on albino rats of different ages by using $^{203}$Hg (as chloride from the Radiochemical Centre Amersham, England) orally or intraperitoneally. Different chelating agents - 2,3-dimercaptopropane-sulfonate-(1) (DMPS), dimercapto- succinic acid (DMSA), zinc diethylenetriaminepentaacetate (ZnDTPA) and sodium N-(4-methoxybenzyl)-D-glucamine dithiocarbamate monohydrate (MeOBDCG) were administered as early or late treatment, orally or parenterally and their effect on the toxicokinetics of mercury in relation to age was studied. Radioactivity was determined in the whole body and organs at different time intervals after $^{203}$Hg administration.

The efficacy of the chelating agents was found to be age dependent. After parenteral administration, mercury was more easily removed from the body of older rats than of young rats. After ingestion of $^{203}$Hg, oral administration of chelating agents was found to be very efficient in reducing high gut retention in suckling rats where most of the body burden of mercury is located at this age. This treatment was efficient even after late administration.

*Advances in Mercury Toxicology*, Edited by T. Suzuki *et al.*
Plenum Press, New York, 1991

The efficiency of these chelating agents in relation to age, route, dose and timing of administration will also be presented.

Our results indicate that all these factors should be taken into consideration to improve present methods of treatment especially in the very young.

## INTRODUCTION

Chelation therapy is the treatment of choice in case of mercury intoxication (Catsch and Harmuth-Hoene, 1979). A number of chelating agents have been tested and proposed with the purpose to reduce mercury body burden. The first of these, 2,3-dimercapto-propanol (BAL), was developed more than 40 years ago (Braun et al., 1946) and has been in clinical use about 20 years in spite of its many adverse side effects. The practical usefulness of BAL is limited by a low therapeutic index (Chisolm, 1970) and increased uptake of mercury in the brain (Berlin and Ullberg, 1963). D-penicillamine is another detoxicating agent (Aposhian, 1958). By using derivatives of these agents - 2,3-dimercaptopropane-1-sulfonate, DMPS (Belonozhko, 1958), 2,3-dimercaptosuccinic acid, DMSA (Friedheim and Corvi, 1975), and N-acetyl-DL-penicillamine, NAPA (Aposhian and Aposhian, 1959) better efficacy and lower toxicity was obtained. Two water soluble derivatives of BAL, DMPS, and DMSA showed high efficacy in reducing the body burden of inorganic mercury (Gabard, 1976a; Planas-Bohne, 1981a; Wannag and Aaseth, 1980; Buchet and Lauwerys, 1989) as well as organic mercurials (Gabard, 1976b; Gabard, 1976c; Magos, 1976; Planas-Bohne, 1981b). Therapeutically, ethylenediaminetetraacetate (EDTA) is ineffective and even potentiates the toxicity of mercury (Glömme and Gustavson, 1959) while diethylenetriamine-pentaacetate (DTPA) has poor efficacy (Nigrovic, 1963).

The usual route of mercury and chelating agent administration in most of the studies is the parenteral route. In cases of environmental exposure, the total population, including infants and young children, the oral route of mercury exposure is the expected route of its entry into the body. Also oral administration of chelation therapy would be the convenient method of treatment. Very few data are available on the efficiency of chelation treatment on oral mercury exposure.

Our earlier studies emphasized the importance of age in evaluating the efficacy of chelation therapy for several toxic metals in case of its parenteral (Pb - Jugo et al., 1975; Cd - Kostial et al., 1984a; Ce - Kargacin et al., 1983; Hg - Kostial et al., 1984b) as well as oral administration (Kostial et al., 1987a, 1987b; Kostial et al.,

Experimental Scheme [203]Hg and chelating agent administration

| Experiment | Age (weeks) | [203]Hg route | Chelating agent administration | | | |
|---|---|---|---|---|---|---|
| | | | compound | route | dose mmol/kg | time (hr) |
| 1 | 1 | ip. | DMPS | ip. | 0.21 | 0, 24 |
| | 6 | | DMSA | | 0.20 | |
| | | | MeOBDCG | | 1.0 | |
| 2 | 1 | po. | DMPS | po. | 0.21 | 0, 24 |
| | 6 | | DMSA | | 0.20 | 24, 48 |
| | | | ZnDTPA | | 3.6 | 48, 72 |
| | | | MeOBDCG | | 1.0 | |
| 3 | 1 | ip. | DMPS | po. | 0.37 | 0, 24 |
| | 6 | | DMSA | | 0.20 | 24, 48 |
| | | | | | | 48, 72 |
| 4 | 1 | po. | DMPS | ip. | 0.21 | 0, 24 |
| | 6 | | | | | 24, 48 |
| | | | | | | 48, 72 |
| 5 | 1 | po. | DMPS | po. | 0.10 | 0, 24 |
| | | | | | 0.21 | 24, 48 |
| | | | | | 0.37 | |
| 6 | 1 | ip. | DMPS | ip. po. | 0.21 | 2 weeks |
| | | | MeOBDCG | ip. po. | 1.0 | |
| | 6 | ip. | DMPS | ip. po. | 0.21 | |
| | | | DMSA | ip. po. | 0.20 | |
| | | | MeOBDCG | ip. po. | 1.0 | |

1988). The purpose of the present study was to obtain new data on the effect of age, route, dose and time of metal and chelating agent administration and on the efficiency of chelation therapy with DMPS, DMSA, ZnDTPA and N-(4-methoxybenzyl)-D-glucamine dithiocarbamate monohydrate (MeOBDCG).

Our results indicate that all the variables estimated, i.e, age, dose, time and route of mercury and chelating agent administration influence the efficiency of the treatment and also that the results depend on the specific chelating agent used. By using chelation therapy without estimating these relevant factors the opposite effect might occur, i.e. increased instead of decreased body burden of mercury.

MATERIALS AND METHODS

Animals: The experiments were performed on albino rats from the Institute's breeding farm. The animals were 1- or 6-8 weeks old at the beginning of the experiment with average body weights of 12 and 140g respectively. One week old rats of both sexes were kept in litters of six (reduced to this number one day after birth) in individual cages with their mothers. Three sucklings from each litter were used as controls (received 0.9% saline) and the other three as experimentals (received chelating agent). The 6-8 week old rats were females kept in groups of 6-12 animals per cage.

Radionuclide administration: Animals received $^{203}$Hg (as chloride almost carrier free; Radiochemical Centre, Amersham, England) intraperitoneally or orally according to the data presented on the Experimental scheme. Suckling rats received radionuclide orally by the method of artificial feeding (Kostial et al., 1971) during 8 hr by means of a dropper, and older rats received it by stomach tube.

Chelation therapy: Chelating agents 2,3-dimercaptopropane-1-sulfonate (DMPS; SIGMA Chemical Company, USA), 2,3-dimercaptosuccinic acid (DMSA; SIGMA Chemical Company, USA), zinc diethylenetriaminepentaacetate (ZnDTPA; Heyl and Co., Berlin, Germany) and N-(4-methoxybenzyl)-D-glucamine dithiocarbamate monohydrate (MeOBDCG; by courtesy of Prof. M. M. Jones, Vanderbilt University , Nashville, Tennessee, USA) were administered intraperitoneally or orally, according to the Experimental scheme, as early (immediately and 24 hr), delayed (24 and 48 or 48 and 72 hr), or very late treatment (2 weeks after $^{203}$Hg administration). Two doses of the chelant were always given. Suckling rats received chelating agents orally by the method of artificial feeding by means of a dropper while older rats received them by stomach tube. Sucklings were returned to their dams each day after the end of the artificial feeding or chelating agent administration.

Radioactivity determinination:   Whole body retention was determined after radionuclide administration and six days later when the animals were killed. Only in the experiment with very late chelant administration the animals were killed three weeks after radionuclide administration. The radioactivity of the whole body, gut and carcass (whole body after removal of the gastrointestinal tract) was determined by the use of a twin crystal scintillation counter (Tobor, Nuclear Chicago). The retention in organs - liver, kidney and brain - was determined in an automatic gamma counter (Nuclear Chicago). The results were corrected for radioactive decay and geometry of the samples. They were expressed as percentage of the administered dose and presented as arithmetic means and standard error of the means.

## RESULTS

The effect of age on [203]Hg retention after intravenous, intraperitoneal and oral administration (untreated controls) is presented on Table 1. For all three routes of mercury entry into the body significantly higher retention values were observed in younger than older rats. After intravenous application whole body, liver and brain retentions were 1.4, 13 and 17 times higher while kidney retention was half of the value in older rats. Similar factors were observed after [203]Hg intraperitoneal administration. The greatest differences were observed. however, after oral mercury application since whole body, gut, liver, kidney and brain retentions were 80, 450, 140, 11 and 67 times higher in younger than older rats.

The effect of age on the efficiency of parenteral treatment after intraperitoneal [203]Hg application is presented on Table 2 (Experiment 1). The order of effectiveness for whole body mercury reduction was DMPS, DMSA, MeOBDCG. For all three agents and both their early and delayed administration the treatment was 1.2-2 times more effective in reducing whole body and organ retention in older than in suckling rats. This age related difference was more pronounced for chelating agents with the higher (DMPS) than lower (DMSA, MeOBDCG) efficacy.

In the case of oral [203]Hg and chelating agent administration (Experiment - 2 - Table 3a and b) for all three agents and both early and delayed administration the treatment was always more effective in suckling than in older rats. Again DMPS was the most effective agent in reducing whole body and kidney retention followed by

Table 1. Influence of AGE on 203 retention and distribuition in rats (% dose 6 days after administration)

| Administration route | IV* | | | IP | | | PO | | |
|---|---|---|---|---|---|---|---|---|---|
| Age (weeks) | 2 | 21 | | 1 | 6 | | 1 | 6 | |
| | S | O | S/O | S | O | S/O | S | O | S/O |
| WB | 77.8±0.6 | 56.1±1.3 | 1.4 | 87.1±1.7 | 58.4±1.1 | 1.5 | 79.2±1.1 | 0.97±0.10 | 81 |
| G | - | - | - | 7.7±0.3 | 3.7±0.2 | 2.1 | 54.8±1.0 | 0.12±0.01 | 457 |
| L | 9.2±0.4 | 0.7±0.03 | 13.1 | 16.6±0.6 | 1.4±0.1 | 11.9 | 5.5±0.2 | 0.04±0.003 | 138 |
| K | 27.7±0.3 | 51.2±0.8 | 0.5 | 20.8±0.8 | 43.9±1.4 | 0.5 | 6.0±0.3 | 0.54±0.04 | 11 |
| B | 0.5±0.02 | 0.03±0.001 | 16.7 | 0.8±0.04 | 0.1±0.06 | 8.0 | 0.2±0.01 | 0.003±0.0002 | 67 |

Retention determined in the whole body (WB), gut (G), liver (L), kidneys (K) and brain (B) after intravenous (IV), intraperitoneal (IP) and oral (PO) 203Hg administration in suckling (S) and older (O) rats. 6-30 animals in each group.
*Data from Jugo, 1976.

Table 2. ²⁰³Hg retention in the whole body and organs (% dose 6 days after intraperitoneal administration of ²⁰³Hg and chelants)

| Chelating agent | Dose mmol/kg | | Suckling rats (1-week-old) | | | | | Older rats (6-week-old) | | | | |
|---|---|---|---|---|---|---|---|---|---|---|---|---|
| | | | Controls | Treated Early (0÷24 hr) | C/T | Delayed (24÷48 hr) | C/T | Controls | Treated Early (0÷24 hr) | C/T | Delayed (24÷48 hr) | C/T |
| DMPS | 0.21 | WB | | 63.1±1.1 | 1.4 | | | | 28.0±1.1 | 2.1 | 48.3±1.7 | 1.2 |
| | | C | | 55.3±0.8 | 1.4 | | | | 26.3±1.1 | 2.1 | 44.5±1.6 | 1.3 |
| | | G | | 7.7±0.6 | 1.0 | | | | 1.7±0.1 | 2.2 | 3.8±0.2 | 1.0 |
| | | L | | 16.0±1.0 | 1.0 | | | | 0.92±0.09 | 1.5 | 1.2±0.03 | 1.2 |
| | | K | | 14.8±0.8 | 1.4 | | | | 17.9±0.7 | 2.5 | 29.5±1.5 | 1.5 |
| | | B | | 0.71±0.02 | 1.2 | | | | 0.06±0.002 | 1.8 | 0.11±0.01 | 1.0 |
| DMSA | 0.20 | WB | 87.1±1.7 | 72.0±2.3 | 1.2 | 78.2±2.3 | 1.1 | 58.4±1.1 | 43.9±1.4 | 1.3 | | |
| | | C | 79.4±1.5 | 64.0±1.9 | 1.2 | 70.7±1.8 | 1.1 | 56.1±1.4 | 40.4±1.2 | 1.4 | | |
| | | G | 7.7±0.3 | 8.1±0.5 | 1.0 | 7.5±0.6 | 1.0 | 3.7±0.2 | 3.5±0.2 | 1.1 | | |
| | | L | 16.6±0.6 | 12.7±0.5 | 1.3 | 14.7±0.5 | 1.1 | 1.4±0.1 | 1.0±0.03 | 1.4 | | |
| | | K | 20.8±0.8 | 17.1±0.8 | 1.2 | 19.2±0.7 | 1.1 | 43.9±1.4 | 28.1±1.1 | 1.6 | | |
| | | B | 0.84±0.04 | 0.67±0.02 | 1.3 | 0.66±0.03 | 1.3 | 0.11±0.06 | 0.10±0.003 | 1.1 | | |
| MeOBDCG | 1.0 | WB | | 81.4±1.2 | 1.1 | 83.4±3.2 | 1.0 | | 43.7±1.1 | 1.3 | 50.7±0.9 | 1.2 |
| | | C | | 70.6±1.1 | 1.1 | 73.0±2.9 | 1.1 | | 40.8±1.1 | 1.4 | 47.6±1.0 | 1.2 |
| | | G | | 11.2±0.5 | 0.7 | 10.3±0.4 | 0.7 | | 2.9±0.1 | 1.3 | 3.1±0.1 | 1.2 |
| | | L | | 17.8±0.4 | 0.9 | 15.8±0.7 | 1.1 | | 1.13±0.05 | 1.2 | 1.3±0.1 | 1.1 |
| | | K | | 13.7±0.7 | 1.5 | 14.6±0.7 | 1.4 | | 32.3±1.7 | 1.4 | 39.9±1.2 | 1.1 |
| | | B | | 0.79±0.05 | 1.1 | 0.94±0.05 | 0.9 | | 0.08±0.002 | 1.4 | 0.10±0.003 | 1.1 |

Retention determined in the whole body (WB), carcass (C), gut (G), liver (L), kidneys (K) and brain (B). 16 rats in control and 7-12 in each treated group.

Table 3a. $^{203}$Hg retention in the whole body and organs (% dose 6 days after oral administration of $^{203}$Hg and chelants)

| Chelating agent | Dose mmol/kg b.w. | | Controls | Treated | | | | | |
|---|---|---|---|---|---|---|---|---|---|
| | | | | Early (0+24 hr) | C/T | Delayed (24+48 hr) | C/T | Delayed (48+72 hr) | C/T |
| | | | | Suckling rats (1-week-old) | | | | | |
| DMPS | 0.21 | WB | | 33.1±1.3 | 2.4 | 38.8±2.0 | 2.0 | 53.1±2.7 | 1.5 |
| | | C | | 10.0±0.5 | 2.5 | 16.8±0.7 | 1.5 | 22.4±0.8 | 1.1 |
| | | G | | 21.9±0.8 | 2.5 | 18.3±0.8 | 3.0 | 30.7±2.1 | 1.8 |
| | | L | | 2.6±0.2 | 2.1 | 4.3±0.3 | 1.3 | 5.3±0.5 | 1.0 |
| | | K | | 2.2±0.2 | 2.7 | 4.0±0.3 | 1.5 | 6.0±0.3 | 1.0 |
| | | B | | 0.13±0.001 | 1.8 | 0.23±0.02 | 1.0 | 0.22±0.01 | 1.0 |
| DMSA | 0.20 | WB | 79.2±1.1 | 31.5±1.5 | 2.5 | 30.7±1.0 | 2.6 | 51.0±3.4 | 1.6 |
| | | C | 24.6±0.4 | 13.3±0.7 | 1.8 | 22.7±0.6 | 1.1 | 22.9±0.8 | 1.1 |
| | | G | 54.8±1.0 | 18.2±1.2 | 3.0 | 8.0±0.8 | 6.9 | 26.9±3.1 | 2.0 |
| | | L | 5.5±0.2 | 3.5±0.2 | 1.6 | 5.7±0.2 | 1.0 | 5.3±0.2 | 1.0 |
| | | K | 6.0±0.3 | 3.5±0.2 | 1.7 | 6.3±0.2 | 1.0 | 6.2±0.4 | 1.0 |
| | | B | 0.23±0.01 | 0.14±0.01 | 1.6 | 0.27±0.01 | 0.9 | 0.21±0.01 | 1.1 |
| ZnDTPA | 3.6 | WB | | 32.3±1.0 | 2.5 | 34.0±1.0 | 2.3 | 31.6±1.0 | 2.5 |
| | | C | | 23.3±1.0 | 1.1 | 26.4±0.8 | 0.9 | 26.3±0.8 | 0.9 |
| | | G | | 9.7±0.6 | 5.6 | 7.6±0.5 | 7.2 | 5.4±0.3 | 10.1 |
| | | L | | 5.6±0.2 | 1.0 | 5.4±0.2 | 1.0 | 5.4±0.1 | 1.0 |
| | | K | | 7.4±0.4 | 0.8 | 7.5±0.4 | 0.8 | 7.3±0.2 | 0.8 |
| | | B | | 0.20±0.01 | 1.2 | 0.23±0.01 | 1.0 | 0.23±0.01 | 1.0 |

30 rats in control and 10 in each treated group

Table 3b. $^{203}$Hg retention in the whole body and organs (% dose 6 days after oral administration of $^{203}$Hg and chelants)

| Chelating agent | Dose mmol/kg b.w. | | Controls | Treated | | | | | |
|---|---|---|---|---|---|---|---|---|---|
| | | | | Early (0÷24 hr) | C/T | Delayed (24÷48 hr) | C/T | Delayed (48÷72 hr) | C/T |
| | | | | Older rats (6-week-old) | | | | | |
| DMPS | 0.21 | WB | | 1.51±0.21 | 0.6 | 0.49±0.08 | 2.0 | | |
| | | C | | 1.34±0.19 | 0.6 | 0.40±0.07 | 2.1 | | |
| | | G | | 0.18±0.02 | 0.7 | 0.06±0.01 | 2.0 | | |
| | | L | | 0.06±0.01 | 0.7 | 0.02±0.003 | 2.0 | | |
| | | K | | 0.61±0.08 | 0.9 | 0.20±0.04 | 2.7 | | |
| | | B | | 0.004±0.001 | 0.8 | 0.001±0.0003 | 3.0 | | |
| DMSA | 0.20 | WB | 0.96±0.06 | 0.55±0.04 | 1.7 | 0.71±0.13 | 1.4 | | |
| | | C | 0.84±0.05 | 0.47±0.03 | 1.8 | 0.62±0.12 | 1.4 | | |
| | | G | 0.12±0.01 | 0.08±0.01 | 1.5 | 0.09±0.01 | 1.3 | | |
| | | L | 0.04±0.003 | 0.02±0.001 | 2.0 | 0.03±0.004 | 1.3 | | |
| | | K | 0.54±0.04 | 0.24±0.02 | 2.3 | 0.34±0.08 | 1.6 | | |
| | | B | 0.003±0.0002 | 0.002±0.0001 | 1.5 | 0.002±0.0004 | 1.5 | | |
| ZnDTPA | 3.6 | WB | | 0.73±0.07 | 1.3 | 0.64±0.12 | 1.5 | | |
| | | C | | 0.67±0.08 | 1.3 | 0.55±0.11 | 1.5 | | |
| | | G | | 0.14±0.02 | 0.9 | 0.08±0.01 | 1.5 | | |
| | | L | | 0.03±0.004 | 1.3 | 0.02±0.005 | 2.0 | | |
| | | K | | 0.39±0.03 | 1.4 | 0.31±0.07 | 1.7 | | |
| | | B | | 0.002±0.0002 | 1.5 | 0.001±0.0003 | 3.0 | | |

23 rats in control and 7-19 in each treated group

143

DMSA and ZnDTPA. The decrease in mercury retention in sucklings was mostly due to reduced gut retention (which represents 70% of the whole body retention) and this was especially obvious with delayed treatment which reduced practically only gut retention and had no effect on organ retention. Delayed DMPS treatment reduced not only gut but also organ retention. ZnDTPA decreased only gut retention, caused slightly higher kidney values (20%) and delayed treatment was even more efficient in reducing gut retention than the early one.

In older rats opposite results were obtained with DMPS since with early oral treatment whole body and organ retentions were increased 1.3 - 1.6 times while delayed treatment decreased $^{203}$Hg retention. DMSA and ZnDTPA, opposite to DMPS, decreased mercury retention by both early and delayed treatment about 1.3 - 1.7 times, ZnDTPA having lower efficacy than DMSA.

In the case of intraperitoneal $^{203}$Hg and oral DMPS and DMSA administration (Experiment 3 - Table 4a) both early and delayed treatments had similar efficacy in suckling and older rats. The whole body retention was reduced about 1.4 times while gut retention was unchanged or even increased.

After $^{203}$Hg ingestion intraperitoneal early and delayed DMPS treatment (Experiment 4 - Table 4b) had no effect in sucklings. In older rats early treatment increased whole body (18%), liver (25%, kidney (10%) and brain (100%) retention while delayed treatment caused decreased values. The efficacy of the treatment increased with the time period between $^{203}$Hg and DMPS administration with the exception of mercury retention in the brain.

After oral $^{203}$Hg and DMPS administration in suckling rats (Experiment 5 - Table 5) increasing the dose about 4 times increased the efficiency of the treatment (especially in reducing kidney and brain retention) about 2 times. With delayed treatment dose dependency was more obvious since the lowest dose (0.1 mmol/kg body weight) caused increased organ retention (20-30%) while higher doses reduced mercury retention in the whole body and organs.

When chelating agents (DMPS, DMSA, MeOBDCG) were administered intraperitoneally or orally as late treatment - two weeks after intraperitoneal $^{203}$Hg application (Experiment 6 - Table 6) best results both in suckling and older rats were obtained with intraperitoneal DMPS application. Kidney retention was reduced about 40%. All the other treatments had lower efficacy. The decrease in efficacy with time is more obvious for older than younger rats and in sucklings this late treatment was as effective as the early one in reducing kidney retention.

Table 4a. $^{203}$Hg retention in the whole body and organs (% dose 6 days after ip. $^{203}$Hg and po. chelants administration)

| Chelating agent | Dose mmol/kg b.w. | | Controls | Treated | | | | | |
|---|---|---|---|---|---|---|---|---|---|
| | | | | Early (0+24 hr) | C/T | Delayed (24+48 hr) | C/T | Delayed (48+72 hr) | C/T |
| **Older rats (6-week-old)** | | | | | | | | | |
| DMPS | 0.37 | WB | 57.5±0.9 | 36.0±0.7 | 1.6 | 41.3±1.0 | 1.4 | 42.5±1.4 | 1.4 |
| | | C | 54.0±0.9 | 33.2±0.5 | 1.6 | 36.2±0.9 | 1.5 | 40.0±1.4 | 1.4 |
| | | G | 3.7±0.1 | 2.6±0.1 | 1.4 | 5.2±0.2 | 0.7 | 2.7±0.1 | 1.4 |
| | | L | 1.4±0.1 | 0.69±0.04 | 2.0 | 0.97±0.06 | 1.4 | 1.1±0.06 | 1.3 |
| | | K | 36.3±1.4 | 15.1±0.4 | 2.4 | 21.8±0.8 | 1.7 | 24.3±1.0 | 1.5 |
| | | B | 0.12±0.003 | 0.10±0.003 | 1.2 | 0.12±0.003 | 1.0 | 0.13±0.003 | 0.9 |
| DMSA | 0.20 | WB | | 42.5±1.5 | 1.4 | 46.3±1.3 | 1.2 | | |
| | | C | | 39.0±1.4 | 1.4 | 42.5±1.9 | 1.3 | | |
| | | G | | 3.6±0.1 | 1.0 | 3.8±0.1 | 1.0 | | |
| | | L | | 1.0±0.05 | 1.4 | 1.1±0.08 | 1.3 | | |
| | | K | | 25.5±0.9 | 1.4 | 28.0±1.2 | 1.3 | | |
| | | B | | 0.11±0.004 | 1.1 | 0.12±0.01 | 1.0 | | |
| **Suckling rats (1-week-old)** | | | | | | | | | |
| DMPS | 0.37 | WB | 87.6±1.1 | 66.2±1.2 | 1.3 | 69.6±1.5 | 1.3 | 66.2±1.6 | 1.3 |
| | | C | 78.5±1.0 | 58.1±1.1 | 1.4 | 61.0±1.3 | 1.3 | 55.6±1.5 | 1.4 |
| | | G | 8.0±0.2 | 8.1±0.3 | 1.0 | 8.6±0.4 | 1.0 | 10.7±0.5 | 0.7 |
| | | L | 16.6±0.3 | 14.0±0.5 | 1.2 | 13.7±0.6 | 1.2 | 10.5±0.5 | 1.6 |
| | | K | 18.9±0.6 | 15.4±0.4 | 1.2 | 14.6±0.4 | 1.3 | 12.5±0.5 | 1.5 |
| | | B | 0.85±0.02 | 0.70±0.02 | 1.2 | 0.78±0.02 | 1.1 | 0.70±0.02 | 1.2 |
| DMSA | 0.20 | WB | | 67.7±2.8 | 1.3 | 72.3±1.4 | 1.2 | | |
| | | C | | 62.6±2.6 | 1.3 | 66.1±1.7 | 1.2 | | |
| | | G | | 5.1±0.3 | 1.6 | 6.2±0.4 | 1.3 | | |
| | | L | | 10.4±0.6 | 1.6 | 12.6±0.9 | 1.3 | | |
| | | K | | 15.4±0.9 | 1.2 | 18.4±0.4 | 1.0 | | |
| | | B | | 0.65±0.03 | 1.3 | 0.68±0.02 | 1.3 | | |

35 rats in control and 8-12 in each treated group.

145

Table 4b. $^{203}$Hg retention in the whole body (% dose 6 days after po. $^{203}$Hg and ip. chelants administration)

| Chela-ting agent | Dose mmol/kg b.w. | | Controls | Treated | | | | | |
|---|---|---|---|---|---|---|---|---|---|
| | | | | Early (0÷24 hr) | C/T | Delayed (24÷48 hr) | C/T | Delayed (48÷72 hr) | C/T |
| | | | | | | Older rats (6-week-old) | | | |
| DMPS | 0.21 | WB | 0.47±0.05 | 0.55±0.06 | 0.9 | 0.27±0.04 | 1.7 | 0.22±0.03 | 2.1 |
| | | C | 0.43±0.05 | 0.51±0.05 | 0.8 | 0.24±0.04 | 1.8 | 0.20±0.02 | 2.2 |
| | | G | 0.05±0.004 | 0.05±0.004 | 1.0 | 0.02±0.003 | 2.5 | 0.02±0.002 | 2.5 |
| | | L | 0.017±0.001 | 0.021±0.002 | 0.8 | 0.010±0.001 | 1.7 | 0.010±0.001 | 1.7 |
| | | K | 0.22±0.03 | 0.24±0.03 | 0.9 | 0.10±0.01 | 2.2 | 0.08±0.01 | 2.8 |
| | | B | 0.001±0.0002 | 0.002±0.0002 | 0.5 | 0.001±0.0001 | 1.0 | 0.001±0.0001 | 1.0 |
| DMPS | 0.21 | WB | 72.4±1.7 | 69.0±1.9 | 1.0 | 67.5±1.7 | 1.1 | 71.8±2.9 | 1.0 |
| | | C | 15.9±0.7 | 13.9±0.8 | 1.1 | 15.0±1.0 | 1.1 | 14.7±1.2 | 1.1 |
| | | G | 56.5±1.5 | 55.1±1.6 | 1.0 | 52.5±1.3 | 1.1 | 57.1±2.0 | 1.0 |
| | | L | 4.1±0.2 | 3.6±0.2 | 1.1 | 3.9±0.3 | 1.1 | 3.7±0.3 | 1.1 |
| | | K | 2.5±0.1 | 2.5±0.2 | 1.0 | 2.9±0.2 | 0.9 | 2.6±0.2 | 1.0 |
| | | B | 0.18±0.02 | 0.13±0.01 | 1.4 | 0.14±0.14 | 1.3 | 0.15±0.01 | 1.2 |

10-12 rats in each group.

146

Table 5. $^{203}$Hg retention in the whole body and organs (% dose 6 days after oral administration of $^{203}$Hg and chelant)

| Chelating agent | Dose mmol/kg | Controls 0 | | Suckling rats (1-week-old) Treated | | | | | |
|---|---|---|---|---|---|---|---|---|---|
| | | | | 0.10 | C/T | 0.21 | C/T | 0.37 | C/T |
| DMPS | | | | | | | | | |
| Early (0+24 hr) | WB | 79.2±1.1 | WB | 41.0±1.5 | 1.9 | 33.1±1.3 | 2.4 | 33.2±1.0 | 2.4 |
| | C | 24.6±0.4 | C | 14.5±0.6 | 1.7 | 10.0±0.5 | 2.5 | 8.8±0.3 | 2.8 |
| | G | 54.8±1.0 | G | 26.5±2.1 | 2.1 | 21.9±0.8 | 2.5 | 24.6±0.9 | 2.2 |
| | L | 5.5±0.2 | L | 4.6±0.2 | 1.2 | 2.6±0.2 | 2.1 | 2.0±0.1 | 2.8 |
| | K | 6.0±0.3 | K | 3.9±0.2 | 1.5 | 2.2±0.2 | 2.7 | 1.6±0.1 | 3.8 |
| | B | 0.23±0.01 | B | 0.20±0.01 | 1.2 | 0.13±0.001 | 1.8 | 0.09±0.01 | 2.6 |
| Delayed (24+48 hr) | | | WB | 53.5±1.7 | 1.5 | 38.8±2.0 | 2.0 | 33.4±1.4 | 2.4 |
| | | | C | 23.6±1.1 | 1.0 | 16.8±0.7 | 1.5 | 13.4±0.5 | 1.8 |
| | | | G | 29.9±1.6 | 1.8 | 18.3±0.8 | 3.0 | 20.0±1.2 | 2.7 |
| | | | L | 7.4±0.3 | 0.7 | 4.3±0.3 | 1.3 | 3.1±0.1 | 1.8 |
| | | | K | 7.0±0.4 | 0.9 | 4.0±0.3 | 1.5 | 2.9±0.1 | 2.1 |
| | | | B | 0.32±0.02 | 0.7 | 0.23±0.02 | 1.0 | 0.14±0.01 | 1.6 |

10-30 rats in each group.

Table 6. $^{203}$Hg retention in the whole body and organs (% dose 20 days after $^{203}$Hg intraperitoneal administration)

| Chelating agent | Dose mmol/kg | Route of administr. | | Older rats (6-week-old) | | | Sucklings (1-week-old) | | |
|---|---|---|---|---|---|---|---|---|---|
| | | | | Controls | Treated | C/T | Controls | Treated | C/T |
| DMPS | 0.21 | ip. | WB | | 24.4±0.5 | 1.4 | | 34.5±0.8 | 1.4 |
| | | | L | | 0.29±0.02 | 1.5 | | 5.3±0.3 | 0.9 |
| | | | K | | 22.0±0.8 | 1.5 | | 12.4±0.4 | 2.0 |
| | | | B | | 0.04±0.001 | 1.3 | | 0.65±0.02 | 1.0 |
| | | po. | WB | | 32.0±0.6 | 1.1 | | 44.3±1.5 | 1.1 |
| | | | L | | 0.35±0.02 | 1.3 | | 6.8±0.5 | 0.7 |
| | | | K | | 30.4±1.1 | 1.1 | | 18.1±1.0 | 1.3 |
| | | | B | | 0.05±0.001 | 1.0 | | 0.65±0.02 | 1.0 |
| DMSA | 0.20 | ip. | WB | 35.3±1.6 | 30.4±1.0 | 1.2 | 46.9±0.7 | | |
| | | | L | 0.44±0.05 | 0.35±0.03 | 1.3 | 4.9±0.5 | | |
| | | | K | 33.7±1.7 | 28.3±1.2 | 1.2 | 24.4±0.6 | | |
| | | | B | 0.05±0.002 | 0.05±0.002 | 1.0 | 0.68±0.04 | | |
| | | po. | WB | | 33.1±1.4 | 1.1 | | | |
| | | | L | | 0.41±0.08 | 1.1 | | | |
| | | | K | | 28.5±1.3 | 1.2 | | | |
| | | | B | | 0.05±0.005 | 1.0 | | | |
| MeOBDCG | 1.0 | ip. | WB | | 34.6±0.8 | 1.0 | | 42.2±0.9 | 1.1 |
| | | | L | | 0.35±0.02 | 1.3 | | 5.2±0.3 | 0.9 |
| | | | K | | 32.6±1.0 | 1.0 | | 19.5±0.6 | 1.3 |
| | | | B | | 0.05±0.003 | 1.0 | | 0.62±0.01 | 1.1 |
| | | po. | WB | | 36.5±1.5 | 1.0 | | 46.7±1.0 | 1.0 |
| | | | L | | 0.44±0.04 | 1.0 | | 5.6±0.3 | 0.9 |
| | | | K | | 34.2±1.1 | 1.0 | | 24.6±0.5 | 1.0 |
| | | | B | | 0.05±0.002 | 1.0 | | 0.63±0.01 | 1.1 |

Chelating agents administered intraperitoneally (ip.) or orally (po.) 2 weeks after $^{203}$Hg. 9 rats in each group.

## DISCUSSION

In the present study, the efficiency of four chelating agents for reducing mercury body burden was evaluated in relation to age, route, dose and time of mercury and chelating agent administration . Two of them, DMPS and DMSA are known as efficient chelators for mercury (Gabard, 1976a, b; Planas-Bohne, 1981a,b; Graziano, 1986), while DTPA showed poor efficacy as parenteral treatment (Nigrovic, 1963) and conflicting results were obtained after dithiocarbamate administration (Aaseth et al., 1981; Gale et al., 1985).

Parenteral chelation therapy after 203-Hg intraperitoneal application had lower efficacy in suckling than in older rats both in cases of early and delayed treatment but this was more pronounced for the chelant with higher efficiency - DMPS. This is in agreement with the results of previous studies obtained with lead (EDTA - Jugo et al., 1975), cerium (DTPA - Kargacin et al., 1983) and cadmium (Kostial et al., 1984a). Oral chelation treatment after [203]Hg oral application was, however, more effective in suckling rats and high efficacy was also obtained in cases of delayed treatment, the effect being due almost exclusively to reduced gut retention. Opposite results were obtained in older rats since early oral administration of DMPS, the most effective chelator after parenteral treatment, greatly increased mercury retention. The same results were obtained with early intraperitoneal DMPS treatment after mercury ingestion. This indicates that early oral and intraperitoneal DMPS treatment are contraindicated after mercury ingestion, i.e., while it is still in the gastrointestinal tract. The conclusions made for DMPS, however, do not apply to DMSA and ZnDTPA. DMSA, a chelating agent of similar chemical structure and efficacy had the opposite effect - decreased [203]Hg retention both after early and delayed treatment. DMSA has also other advantages compared to other chelating agents. It is conveniently administered (orally), it is 35 times less toxic than BAL and 3 times less toxic than DMPS, 4 times more efficient than D-penicillamine and in clinical trials lower doses of DMSA may be used (Magos, 1976). ZnDTPA which is the treatment of choice for several radionuclides (actinides - Volf, 1978) and metals (cadmium - Klaassen, 1985) has poor efficacy for parenteral treatment for mercury. Orally applied, it reduced 203-Hg retention in suckling and also in older rats. Parenteral DMPS treatment given as early or delayed therapy after mercury ingestion had no effect in sucklings. Most of the retained [203]Hg is in the gut (70% of the whole body retention) and thus seems to be unavailable for chelation with this chelating agent.

The effect of age on the efficiency of chelating agents is due to specific features of metal metabolism in the young. Much higher absorption and retention, unfavourable distribution and higher toxicity of mercury have been shown in younger experimental animals (Jugo, 1977; Kostial et al., 1978). Factors responsible for these specific features, e.g. immaturity of the kidney and biliary transport, differences in binding affinities and/or contents of metal carrier proteins or differences in disposition and/or biotransformation of chelating agent are discussed in detail elsewhere (Jugo et al., 1975; Kostial et al., 1989).

There is usually a decrease in efficacy with time elapsed after mercury entry into the body with the exception of oral chelation therapy in sucklings. This, however, depends greatly on the particular chelating agent, route of mercury and chelant administered and age. In case of late administration the best treatment both in younger and older animals was parenteral DMPS administration. It was also observed that the rate of removal is greater after DMPS administration than after DMSA but repeated administration of either agent eventually leads to the same total amount of mercury mobilized from the kidney (Buchet and Lauwerys, 1989). The importance of dose should be emphasized especially in case of delayed treatment since higher doses were more effective particularly in reducing mercury retention in critical organs - kidney and brain - while lower doses even caused increased 203-Hg retention in organs.

Treatment with MeOBDCG had poor efficacy, compared to DMPS and DMSA, since five times higher dose had lower efficacy. Our results are in agreement with the ones obtained by Gale and collaborators (1985) who found that substituted dithiocarbamates had only weak mobilizing effects on inorganic mercury in mice. Also, contradictory results were obtained with dithiocarbamates since it has been reported that sodium diethyldithiocarbamate increases the concentration of mercury in brain, liver and kidney in rats after exposure to inorganic mercury (Aaseth et al., 1981; Norseth and Nordhagen, 1977) and in brain and liver after exposure to methylmercury (Norseth and Nordhagen, 1977; Norseth, 1974). It was also shown, in primary cultures of hepatocytes, that diethyldithiocarbamate increases the transport of mercury across the cellular membrane by complex formation with mercury and thus potentiates its toxicity (Hellström-Lindahl and Oskarsson, 1989). Only in cases of very late intraperitoneal and oral administration of MeOBDCG in sucklings we observed slightly increased 203-Hg retention in the liver.

Significantly higher values of mercury retention in sucklings were observed after intravenous (Jugo, 1976), intraperitoneal and especially oral administration (Kostial et al., 1978). Similar or greater differences were found after administration of chelation therapy. After early DMPS parenteral application [203]Hg retention in the whole body, liver, kidney and brain were 2.5, 18, 1.3 and 13 times respectively higher in suckling than older rats. After mercury ingestion and oral chelation therapy the difference is even greater. In spite of high efficacy of oral treatment the retention in sucklings in the whole body, liver, kidney and brain were 22, 43, 4 and 33 times higher respectively than in older rats after early DMPS treatment and 90, 215, 24 and 164 times higher respectively after delayed treatment. In cases of DMSA and ZnDTPA treatment which had lower efficiency these differences were even higher. These results indicate that the young represent a higher risk group than adults at the same level of environmental exposure to mercury.

## CONCLUSIONS

We found that all the variables estimated, i.e. age, dose, time and route of mercury and chelating agent administration influence the efficiency of the treatment and also that the results depend on the specific chelating agent used. Age is a very important factor in estimating the efficacy of parenteral and oral chelation therapy. Chelation treatment of ingested metals is age, dose and time dependent. By using chelation therapy without estimating these relevant factors the opposite effect might occur, i.e. increased instead of decreased body burden of mercury.

## ACKNOWLEDGEMENT

Our thanks are due to Mr. R. Arezina and Mrs. M. Landeka for their skillful assistance in the experimental part of this study and to Mrs. M. Horvat for preparing the manuscript.

## REFERENCES

Aaseth, J., Alexander, J., and Wannag, A., 1981, Effect of thiocarbamate derivatives on copper, zinc and mercury distribution in rats and mice, Arch. Toxicol., 48:29-39.

Aposhian, H.V., 1958, Protection by D-penicillamine against the lethal effects of mercury chloride, Science, 128:93.

Aposhian, H.V., and Aposhian, M.M., 1959, N-acetyl-DL-penicillamine, a new oral protective agent against the lethal effects of mercuric chloride, J. Pharmacol. Exp. Ther., 126:131-135.

Belonozhko, G.A., 1958, Therapeutic action of Unithiol in poisoning with inorganic mercury compounds, Farmakol. Toksikol., 21:69-73.

Berlin, M., and Ullberg, S., 1963, Increased uptake of mercury in mouse brain caused by 2,3-dimercaptopropanol (BAL), Nature, 197:84.

Braun, H.A., Lusky, L.M., and Calvery, H.O., 1946, The efficacy of 2,3-dimercapto-propanol (BAL) in the therapy of poisoning by compounds of antimony, bismuth, chromium, mercury and nickel, J. Pharmacol. Exp. Ther., 87:Suppl. 119-125.

Buchet, J.P., and Lauwerys, R.R., 1989, Influence of 2,3-dimercaptopropane-1-sulfonate and dimercaptosuccinic acid on the mobilization of mercury from tissues of rats pretreated with mercuric chloride, phenylmercury acetate or mercury vapors, Toxicology, 54:323-333.

Catsch, A., and Harmuth-Hoene, A.E., 1979, The pharmacology and therapeutic applications of agents used in heavy metal poisoning, in: "The Chelation of Heavy Metals,"W.G. Levine, ed., 107, International Encyclopedia of Pharmacology and Therapeutics, Pergamon Press.

Chisolm, J.J., 1970, Poisoning due to heavy metals, Pediat. Clin. N. Amer., 17:591-615.

Friedheim, E., and Corvi, C., 1975, Meso-dimercaptosuccinic acid, a chelating agent for the treatment of mercury poisoning, J. Pharm. Pharmac., 27:624-626.

Gabard, B., 1976a, The excretion and distribution of inorganic mercury in the rat as influenced by several chelating agents, Arch. Toxicol., 35:15-26.

Gabard, B., 1976b, Improvement of oral chelation treatment of methyl mercury poisoning in rats, Acta. Pharmacol. Toxicol., 39:250-255.

Gabard, B., 1976c, Treatment of methylmercury poisoning in the rat with sodium 2,3-dimercaptopropane-1-sulfonate: Influence of dose and mode of administration, Toxicol. Appl. Pharmacol., 38:415-424.

Gale, G.R., Atkins, L.M., Smith, A.B., and Jones, M.M., 1985, Effects of substituted dithiocarbamates on distribution and excretion of inorganic mercury in mice, Res. Commun. Chem. Pathol. Pharmacol., 47:293-296.

Glömme, J., and Gustavson, K.H., 1959, Treatment of experimental acute mercury poisoning by chelating agents BAL and EDTA, Acta Med. Scand., 164:175-182.

Graziano, J.H., 1986, Role of 2,3-Dimercaptosuccinic acid in the treatment of heavy metal poisoning, Med. Toxicol., 1:155-162.

Hellstrom-Lindahl, E., and Oskarsson, A., 1989, Increased availability of mercury in rat hepatocytes by complex formation with diethyldithiocarbamate, Toxicol. Lett., 49:87-98.

Jugo, S., Maljkovic, T., and Kostial, K. 1975, The effect of chelating agents on lead excretion in rats in relation to age, Environ. Res., 10:271-279.

Jugo, S., 1976, Retention and distribution of 203-HgCl in suckling and adult rats, Health Phys.,30:240-241.

Jugo, S., 1977, Metabolism of toxic heavy metals in growing organisms: A review, Environ. Res., 13:36-46.

Kargacin, B., Kostial, K., and Landeka, M., 1983, The influence of age on the effectiveness of DTPA in reducing 141-Ce retention in rats, Int. J. Radiat. Biol., 44:363-366.

Klaassen, C.D., 1985, Heavy metals and heavy metal antagonists, in: "The Pharmacological Basis of Therapeutics," A. G. Goodman et al., eds., 1605-1627, 7th ed., MacMillan Publishing Company, New York.

Kostial, K., Simonovic, I., and Pisonic, M., 1971, Lead absorption from the intestine in newborn rats, Nature, 233:564.

Kostial, K., Kello, D., Jugo, S., Rabar, I., and Maljkovic, T., 1978, Influence of age on metal metabolism and toxicity, Environ. Health Perspect., 25:81-86.

Kostial, K., Kargacin, B., Blanusa, M., and Landeka, M., 1984a, Lower efficiency of DTPA in reducing cadmium retention in suckling rats, Environ. Res., 35:254-269.

Kostial, K., Kargacin, B., Blanusa, M., and Landeka, M., 1984b, The effect of 2,3-dimercaptopropane sodium sulfonate on mercury retention in rats in relation to age, Arch. Toxicol., 55:250-252.

Kostial, K., Kargacin, B., and Landeka, M., 1987a, Oral Zn-DTPA treatment reduces cadmium absorption and retention in rats, Toxicol. Lett., 39:71-75.

Kostial, K. Kargacin, B., and Landeka, M.. 1987b, Oral Zn-DTPA therapy for reducing 141-Ce retention in suckling rats, Int. J. Radiat. Biol., 52:501-504.

Kostial, K., Kargacin, B., and Landeka, M., 1988, 2,3-dimercaptopropane-1-sodium sulfonate for reducing retention of ingested 203-Hg in suckling rats, Bull. Environ. Contam. Toxicol., 41:185-188.

Kostial, K., Kargacin, B., and Landeka, M., 1989, Efficiency of chelation therapy in relation to age, Proceedings of the Third International Congress on Trace Elements in Health and Disease, Adana, Turkey, in print.

Magos, L., 1976, The effects of dimercaptosuccinic acid on the excretion and distribution of mercury in rats and mice treated with mercuric chloride and methylmercury chloride, Br. J. Pharmacol., 56:479-484.

Nigrovic, V., 1963, Der Einfluss von Chelatbildnern auf das Verhalten von Quecksilber im Organismus, Arzneim.-Forsch. 13:787-792.

Norseth, T., and Nordhagen, A.-L., 1977, The influence of an industrial complexing agent on the distribution and excretion of lead and mercury, in: "Clinical Chemistry and Chemical Toxicology of Metals". S.S. Brown, ed., 137-140, Elsevier/North-Holland Publishing Co., Amsterdam.

Norseth, T., 1974, The effect of diethyldithiocarbamate on biliary transport, excretion and organ distribution of mercury in the rat after exposure to methyl mercuric chloride, Acts. Pharmacol. Toxicol., 34:76-87.

Planas-Bohne, F., 1981a, The effect of 2,3-dimercaptopropane -1-sulfonate and dimercaptosuccinic acid on the distribution and excretion of mercuric chloride in rats, Toxicology, 19:275-278.

Planas-Bohne, F, 1981b, The influence of chelating agents on the distribution and biotransformation of methylmercuric chloride in rats, J. Pharmacol. Exp. Ther., 217:500-504.

Volf, V., 1978, Treatment of incorporated transuranium elements, IAEA, Technical Reports Series No. 184, Vienna.

Wannag, A., and Aaseth, J., 1980, The effect of immediate and delayed treatment with 2,3-dimercaptopropane-1-sulfonate on the distribution and toxicity of inorganic mercury in mice and in foetal and adult rats, Acts. Pharmacol. Toxicol., 46:81-88.

# IMMUNOHISTOCHEMICAL LOCALIZATION OF METALLOTHIONEIN IN ORGANS OF RATS TREATED WITH EITHER CADMIUM, INORGANIC OR ORGANIC MERCURIALS

Chiharu Tohyama[1] and Abdul Ghaffar[1], Atsuhiro Nakano[2],
Noriko Nishimura[3] and Hisao Nishimura[3]

[1]Environmental Health Sciences Division, National Institute for
Environmental Studies, Onogawa, Tsukuba, Ibaraki 305, Japan

[2]National Research Center for Minamata Disease, Minamata
Kumamoto 867, Japan

[3]Department of Hygiene, Aichi Medical University, Nagakute
Aichi 480- 11, Japan

ABSTRACT

Metallothionein (MT) is a low, molecular, mass protein inducible by heavy metals such as cadmium (Cd), zinc and copper and having high affinity for these metals. In the present study, we have investigated immunohistological localization of metallothionein in the kidney and brain of rats treated with either Cd, inorganic mercury (Hg) or organic Hg.

Kidneys from rats treated with a single dose of Cd showed the presence of MT mainly in the proximal tubular epithelium whereas those from rats treated with Cd for 6 weeks demonstrated very strong MT immunostaining in the proximal and distal tubular epithelium, and the latter showed much stronger staining than the former. In the kidneys from inorganic Hg-treated rats, MT was detected not only in the proximal tubular epithelium but also in the distal tubular epithelium.

When adult rats were administered with either inorganic Hg or organic Hg, MT was induced in ependymal cells, pia mater, arachnoid, vascular endothelial cells and some glial cells. The present results suggest that MT is induced in the distal tubular epithelial cells and that the protein may take part in the blood-brain and cerebrospinal fluid-brain barriers to prevent toxic heavy metals from penetrating into the parenchyma of the brain.

## INTRODUCTION

Metallothionein (MT) is a group of low-molecular-mass protein unique in properties such as inducibility upon exposure to heavy metals, i.e. cadmium (Cd), inorganic mercury (I-Hg) and zinc (Zn), high cysteine content and heat stability. Since MT reduces toxicity of heavy metals such as Cd and Hg ion it is proposed to have a detoxifying role for these heavy metals. In addition to the detoxification, MT is considered to play physiologic roles in transport and storage of Zn and copper ions since MT exists in various tissues under physiologic conditions and can be induced also by stresses and hormones. MT has also been suggested to act as a radical scavenger against active oxygen species. The earlier observation that some tissues, such as liver and kidney from fetal and newborn animals, contain distinctly greater amounts of MT, suggests that MT may be involved in growth (see reviews, Bremner, 1987; Webb, 1979).

In a previous study (Tohyama et al., 1988), it was found that MT is present mainly in the proximal tubular epithelium of the kidney of rats treated with a single dose of Cd. However, in the kidney from rats administered with cadmium for 6 weeks, MT immunostaining was found to be more intense in distal tubular epithelium than proximal tubular epithelium (Tohyama et al., 1988). These observations raised a question whether the increase in MT levels was due to induction of biosynthesis in the distal tubular epithelium, or transported from elsewhere. In order to answer the above question, in the present study, animals were administered with inorganic Hg, which is known to accumulate in the distal tubules.

Recent studies suggested that brains from rats, mice and monkeys contain MT-like proteins (Ebadi and Babin, 1989; Munos et al., 1989; Gulati et al., 1987). An earlier autoradiographic study revealed that radiolabelled Cd is taken up relatively easily by hypothalamus, pia mater, choroid plexus but not by the parenchymal cells (Berlin and Ullberg, 1963a). In order to elucidate possible physiologic functions of MT, immunohistological studies have been carried out to demonstrate localization of MT in various tissues (see references cited in Nishimura et al., 1990). Since distribution of

MT in the brain has not been reported, in the present study, we have also investigated the immunohistochemical localization of MT in the brain of non-treated and heavy-metal treated rats.

## MATERIALS AND METHODS

### Reagents

Goat normal serum, biotinylated goat anti-rabbit IgG and avidin-biotin peroxidase complex (ABC) kit were purchased from Vector Laboratories (Burlingame, California). Other reagents were of analytical grade.

### Animal Treatment

Male rats (Wistar strain, 8 or 10 weeks old) were used. For Cd treatment, rats were injected with $CdCl_2$ at a single or repeated doses of 0.5 mg Cd/kg body weight, and sacrificed 12 hours or 15 weeks after the final injection (Tohyama et al., 1988). For Hg treatment, rats were injected with either $HgCl_2$ or $CH_3HgCl$ at a daily dose of 0.74 and 8.0 mg Hg/kg body weight respectively for 4 consecutive days, and sacrificed at 48 hrs after the final injection. Control rats received physiologic saline or 50% acetone in saline.

### Immunohistochemical Staining

Kidneys and/or whole brains from rats treated with saline, Cd, I-Hg and M-Hg were fixed in 10% buffered formalin and embedded in paraffin. Deparaffinized 5 μm thick tissue sections were subject to immunohistochemical staining with ABC method as described earlier (Nishimura et al., 1989; Tohyama et al., 1988). The immunohistochemical control test has been carried out by replacement of the rabbit antiserum produced against rat MT-1 with preimmune rabbit serum and by the use of the rabbit antiserum that had been adsorbed with rat MT-2. The immuno-histochemical control test was described in detail in an earlier paper (Tohyama et al., 1988). Counter-staining with hematoxylin was carried out in other tissue sections after immunostaining.

### Concentrations of Metallothionein and Heavy Metals in the Kidneys and Brain

Concentrations of MT in the kidneys and brain were determined by radioimmunoassay as described earlier (Nishimura et al., 1989; Tohyama et al., 1988).

Cd, copper and zinc in the kidney were determined by inductively-coupled plasma emission spectrometry (Tohyama et al., 1988). Total and inorganic Hg concentrations in tissues were determined by flameless atomic absorption spectrometry (Magos and Clarkson, 1972) with modifications. In brief, in order to determine inorganic mercury, inorganic mercury was extracted into an aqueous phase from tissue homogenate using chloroform and hydrochloric acid.

Statistical analysis for differences of mean between groups was carried out by Student's $t$-test modified by Welch.

## RESULTS

### Immunohistochemical Localization of Metallothionein

Immunohistochemical control: the use of preimmune rabbit serum did not result in any specific immunostaining in rat kidneys and brains. When the antiserum was replaced with the one preabsorbed with an appropriate amount of MT-2, almost no immunostaining was detected (data not shown; see Tohyama et al., 1988; Nishimura et al., 1989).

### Metallothionein Localization in the Rat Kidney

Concentrations of MT, zinc, copper and Cd in the kidneys of Cd-treated rats are summarized in Table 1. Even saline-treated control rats contained relatively large amounts of MT in the kidney, and MT concentrations increased approximately 2 fold 12 hours after an injection. In saline-treated control rat kidneys, very weak MT immunostaining was observed in proximal tubular epithelial cells (not shown). A single injection of Cd increased the intensity of MT immunostaining in the proximal tubular epithelium particularly near the surface of the kidney and weak MT staining was observed in the distal tubular epithelium (Figure la). In contrast, repeated injections of Cd gave rise to localization of MT not only in the proximal tubular epithelium and its lumen but also in distal tubular epithelium (Figure 1b) and collecting tubular epithelium (Figure 1c). In the proximal and distal tubular epithelium, nuclear staining was conspicuous. A considerably high MT concentration was found to be parallel with an increase in Cd, copper and zinc accumulation in the kidney (Table 1). In the I-Hg-treated rats, MT immunostaining was observed not only in the proximal tubules but also in the distal (Figure 1d) and collecting tubular epithelia (Figure 1e). I-Hg treatment resulted in concomitant accumulation of MT and mercury in the kidney (Table 2). In both cases of Cd-treated rats as well as Hg-treated rats, MT immunoreactivity was not found in the glomerulus, blood vessels and other connective tissues.

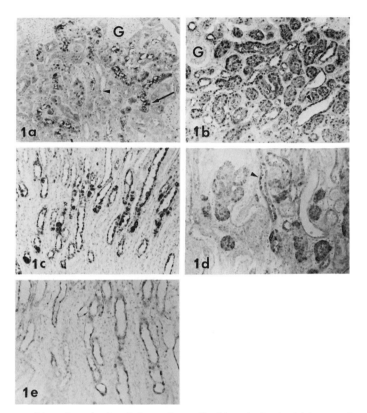

Fig. 1. Immunohistochemical staining of metallothionein in the kidneys of rats treated with cadmium chloride. (a) A single subcutaneous injection of $CdCl_2$ (3.0 mg Cd/kg body weight) increased the intensity of MT immunostaining in the proximal tubular epithelium (arrow) particularly near the surface of the kidney and weak MT staining was observed in the distal tubular epithelium (arrowhead); (b and c) Long-term exposure to $CdCl_2$ resulted in induction of MT not only in the proximal tubular epithelium and its lumen but also in distal tubular epithelium (b, arrowhead) and collecting tubular epithelium (c); (d and e) In the I-Hg-treated rats, MT immunostaining was observed not only in the proximal tubules but also in the distal (d, arrowhead) and collecting tubular epithelia (e). In both cases of Cd-treated rats as well as Hg- treated rats, MT immunoreactivity was not bound in the glomerulus (G), blood vessels and other connective tissues. Original magnification x 50.

Table 1. Metallothionein and heavy metal concentrations in the kidney of rats treated with cadmium chloride[a]

| Treatment | Metallothionein | Cadmium | Copper | Zinc |
|---|---|---|---|---|
| | | ---------------μg/g tissue------------- | | |
| Saline | $128 \pm 1$ | n.d. | $5.13 \pm 0.29$ | $16.8 \pm 0.9$ |
| Single injection | $234 \pm 32^b$ | $3.78 \pm 0.45^b$ | $5.39 \pm 0.29$ | $17.1 \pm 0.6$ |
| Repeated injections | $1507 \pm 78^b$ | $193 \pm 8^b$ | $12.8 \pm 0.5^b$ | $34.9 \pm 0.6^b$ |

a  Values are means and SD for 3 rats injected with saline or $CdCl_2$ at a single dose of 3.0 mg Cd/kg body weight and for 5 rats injected with $CdCl_2$ at a daily dose of 1.5 mg Cd/kg body weight, 4 days a week, for 6 weeks (Tohyama et al., 1988). Rats were sacrificed 12 hours after the single injection and 15 weeks later after a stop of 6-week injections.

b  Significantly different from a saline control value at P<0.05.

## Metallothionein Localization in the Rat Brain

Metallothionein immunostaining was not detected in saline or acetone-saline treated adult rat brains (data not shown). Table 2 shows concentrations of MT, total mercury, and inorganic mercury as well as percentage of inorganic mercury to total mercury in the brain of rats treated with mercuric chloride or methyl mercuric chloride. Particularly, I-Hg treatment induced MT in the brain, but the mean concentration was 34 μg/g wet tissue, which is lower than the concentration often observed in the control rat kidney (Table 1). Control rats treated with acetone in physiologic saline (50% w/w) induced MT in the brain, but no increase in MT was found by immunohistochemical staining as described above. Total mercury concentration in the brain of M-Hg treated rats was approximately 10 times higher than that of I-Hg treated rat brain, but MT concentration of M-Hg treated rats was half that of I-Hg treated rat brain (Table 2).

Compared to minimal change in tissue concentrations of MT, I-Hg treatment resulted in strong MT immunostaining in all the ependymal cells. Staining intensity of the nucleus was less intense than that of the cytoplasm (Figure 2a). In the parenchyma, some glial cells were found to have weak immunostaining in the nucleus and neuroglial processes (Figure 2a). Epithelium of choroid plexus was found negative for the MT staining. It was evident that pia mater and arachnoid (Figure 2b) and vascular endothelial cells contain substantial degrees of MT immunostaining. In M-Hg treated rat brain, observations were very similar to those found in I-Hg treated rats (Figure 2c).

Table 2. Metallothionein, total mercury and inorganic mercury concentrations in the kidney and brain of rats treated with mercuric chloride or methyl mercuric chloride[a]

| Treatment | Metallothionein | Total mercury | Inorganic mercury | % inorganic mercury |
|---|---|---|---|---|
| | | ----------$\mu$g/g tissue---------- | | |
| | | Kidney | | |
| Saline | $67 \pm 15.6$ | b | b | c |
| Mercury chloride | $613 \pm 324^d$ | $66.7 \pm 8.7$ | b | c |
| | | Whole Brain | | |
| Saline | $5.2 \pm 0.7$ | b | b | c |
| Acetone-saline | $11.0 \pm 0.1^d$ | $0.02 \pm 0.00$ | b | c |
| Mercuric chloride | $34.0 \pm 0.7^d$ | $0.86 \pm 0.30^d$ | b | c |
| Methyl mercuric chloride | $16.0 \pm 0.7^e$ | $8.10 \pm 1.50^e$ | $0.55 \pm 0.12$ | 6.8 |

a Values are means and SD for 3 and 4 rats of control and experimental groups of rats, respectively.
b Mercury determination was not carried out.
c Not calculated since amounts of organic mercury was considered to be negligible from there result of control rat tissues.
d Significantly different from a saline control value at P<0.05.
e Significantly different from a acetone-saline control value at P<0.05.

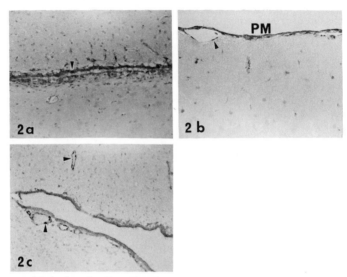

Fig. 2. Immunohistochemical localization of MT in the mercury-treated rat brain.
(a) I-Hg treatment resulted in strong MT immunostaining in all the ependymal cells
(arrowhead). MT staining of the nucleus was less intense than that of the cytoplasm.
In the parenchyma, some glial cells were found to have weak immunostaining in the
nucleus and neuroglial processes. Epithelium of choroid plexus was found negative for
the MT staining. (b) Pia mater (PM) and arachnoid and vascular endothelial cells
(arrowhead) contain substantial degrees of MT immunostaining. (c) M-Hg treatment
gave rise to MT localization very similar to I-Hg-treated rat brain. Strong
immunostaining was observed in the endothelial cells of blood vessels (arrowhead).
Original magnification x 50.

## DISCUSSION

The present study has compared localization of MT in the kidney of Cd-treated-rats
in terms of single and repeated doses of Cd with those of I-Hg-treated rats, as well as
in brains from rats treated with I-Hg and M-Hg.

The comparison of the two dosing schedules of Cd treatment, showed that repeated
administration of Cd resulted in very strong MT immunostaining in the distal tubular
and collecting tubular epithelia than in the proximal tubules (Figures 1b and c), which
is supported by an increase in concentrations of MT, Cd, zinc and copper in the tissue
(Table 1). MT, or Cd- and zinc-containing MT in the case of Cd treatment, induced in
the liver can be released into the blood circulation, followed by filtration through

glomerular basement membrane into proximal tubules, where MT is reabsorbed into the epithelium (Suzuki, 1982). In the epithelial cells, Cd which dissociates from MT exerts toxicity but, at the same time, induces MT in the cells. However, how MT is induced in the distal tubules has not yet been reported. It is conceivable that either Cd, present in the blood circulation, is taken up by the distal tubules and induces MT, or that a portion of the Cd-containing MT, that is not reabsorbed in the proximal tubules, may be reabsorbed by the distal tubular epithelium. Nevertheless, since I-Hg is known to accumulate both in the proximal tubules and in the distal tubular epithelium (Berlin, 1986), MT immunostaining found in the latter from I-Hg- treated and Cd-treated rats suggests the inductions of MT biosynthesis in the distal epithelium.

In the rat brain, biosynthesis of MT was found to occur in the ependymal cells, pia mater and arachnoid upon I-Hg and M-Hg as well as Cd administration. Ependymal cells, a type of glia cells that cover the surface of the ventricles of brain and the central canal of the spinal cord, are very active in cell proliferation in the fetus and newborns (Bruni et al., 1985) and differentiate until a certain period after birth (Walsh et al., 1978; Roessmann et al., 1980; Bruni et al., 1983). Our recent studies have demonstrated that MT is present in the ependymal cells of young rats in an age-dependent manner (from newborn to approximately 3 weeks old) under physiologic conditions whereas MT was found in young and adult mice independent from age (unpublished observation). In the case of mouse brain, administration of I-Hg, M-Hg and Cd failed to induce MT any further probably due to saturated capacity of these cells for MT induction (unpublished observation). The data from rats and mice suggest that ependymal cells furnish a detoxifying potential at a limited extent depending on animal species. However, the present findings of induced MT synthesis in the ependymal cells, pia mater and arachnoid of rats suggest a possible involvement of MT in trapping toxic heavy metals in the barrier between the brain and cerebrospinal fluid.

Administration of Cd, I-Hg and M-Hg resulted in the induction of MT in the endothelium of capillaries, suggesting a possible link of MT with a barrier function. Since the capillaries of the brain are not fenestrated and form well-developed tight occluding junctions between endothelial cells (Bloom and Fawcett, 1986), they are considered to play a major role in the blood-brain barrier which protects the parenchyma against influx of various substances from blood circulation. When Cd was administered to mice, it was found to be relatively smoothly taken up by the choroid plexus, pia mater and hypothalamus, compared with almost no accumulation in the parenchyma (Berlin and Ullberg, 1963). In the brain of monkeys, a significant amount of Cd was reported to be bound to MT-like protein (Gulati et al., 1987). In the endothelium MT is considered to act as a trap for toxic heavy metals. The present results indicate that approximately 7% of total mercury was present as inorganic mercury in the parenchyma 48 hours after the injection of M-Hg. Since MT can be induced by the inorganic form of Hg but not by the organic form, MT staining found in the brain parenchyma was brought up by the I-Hg degraded from M-Hg. In blood-brain and cerebrospinal fluid-brain barriers, MT is considered to reduce toxicity of heavy metals by binding the metal ions in ependymal and vascular endothelial cells, respectively. In conclusion, the finding that MT can be induced in both blood-brain and cerebrospinal fluid-brain barrier systems suggests a new physiologic role of MT in these defense systems in terms of heavy metal handling in the central nervous system.

## ACKNOWLEDGMENT

We thank Dr. M. Murakami (NIES) for encouragement throughout this study.

## REFERENCES

Berlin, M., 1986, Mercury. In Handbook on the Toxicology of Metals, 2nd ed. vol II: Specific Metals, Frigerg, L., Nordberg, G.F., Vouk, V.,eds., 413, Elsevier, Amsterdam.

Berlin, M., and Ullberg, S., 1963, The fate of Cd in the mouse. an autoradiographic study after a single intravenous injection of Cd. Cl. Arch. Environ. Health 17:686.

Bloom, W., and Fawcett, D.W., eds., 1986, A textbook of histology. 10th ed., Saunders Company, 829, Philadelphia.

Bremner, I., 1987, Nutritional and physiological significance of metallothionein. In Metallothionein II. Experientia supplementum vol 52, Kagi JHR, Kojima Y eds., 81, Birkhauser Verlag, Basel.

Bruni, J., Clattenburg, R.E., and Millar, E., 1983, Tanycyte ependymal cells in the third ventricle of young and adult rats; A golgi study. Anat. Anz. Jena. 153:53.

Bruni, J., Delbingo, M.R., and Lin, P.S., 1985, Ependyma: Normal and pathological. A review of the literature. Brain Research Reviews 9:1.

Ebadi, M., and Babin, D., 1989, The amino acid composition of the zinc-induced metallothionein isoforms in rat brain. Neurochem. Res. 14:69.

Gulati, S., Paliwal, V.K., Sharma, M., Gill, K.D., and Nath, R., 1987, Isolation and characterization of a metallothionein-like protein from monkey brain. Toxicology 45:53.

Magos, L., and Clarkson, T.W., 1972, Atomic absorption determination of total, inorganic, and organic mercury in blood. J. AOAC 55:966.

Munos, c., Vormann, J., and Dieter, H.H., 1989, Characterization and development of metallothionein in fetal forelimbs, brain and liver from the mouse. Toxicol. Letters 45:83.

Nishimura, N., Nishimura, H., and Tohyama, C., 1989, Localization of metallothionein in female reproductive organs of rat and guinea pig. J. Histochem. Cytochem. 37:1601.

Nishimura, N., Nishimura, H., and Tohyama, C., 1990, Localization of metallothionein in the genital organs of the male rat. J. Histochem. Cytochem. 39:927.

Roessmann, U., Velasco, M.E., Sindely, S.D., and Gambetti, P., 1980, Glial fibrillary acidic protein (GFAP) in ependyma cells during development. An immunocytochemical study. Brain Res. 200:13.

Suzuki, K.T., 1982, Induction and degradation of metallothionein and their relation to the toxicity of cadmium. In: Biological Roles of Metallothionein, Foulkes, E.C., pp.215-235, Amsterdam, Elsevier.

Tohyama, C., Nishimura, H., and Nishimura, N., 1988, Immunohistochemical localization of metallothionein in the liver and kidney of cadmium- or zinc-treated rats. Acta. Histochem. Cytochem. 21:91.

Walsh, R.J., Brawer, J.R., and Lin, P.S., 1978, Early postnatal development of ependyma in the third ventricle of male and female rats. Amer. J. Anat. 151:377.

Webb, M., ed. 1979, The metallothioneins: The Chemistry, Biochemistry and Biology of Cadmium. ed., Elsevier/North Holland 195, Amsterdam.

# MERCURY VAPOR UPTAKE AND OXIDOREDUCTASES IN ERYTHROCYTES

Stefan Halbach

Institute of Toxicology
GSF-National Center for Environmental Sciences Munich
D-8042-Neuherberg, Germany

## ABSTRACT

The inhibition by ethanol of the absorption of inhaled $Hg^o$ suggested the participation of catalase in this process (Nielsen-Kudsk, 1969). The studies reported there specified the mechanisms by which $H_2O_2$-metabolizing enzymes in erythrocytes enhance or reduce the accumulation of Hg after exposure to $Hg^o$.

When red-blood-cell suspensions were exposed to $Hg_o$ in a closed system, it was found that the mercury was absorbed by the cells only. Thus, a low hematocrit led to a high cellular Hg uptake, alterations of which could therefore be analyzed easier. The uptake was inducible by slow infusion of small amounts of $H_2O_2$ and paralleled the activity of catalase as was shown by selecting red cells from species with largely different $H_2O_2$-metabolizing enzyme activities. The instrumental function of the catalase- $H_2O_2$ intermediate (compound-I) in Hg uptake was further verified by its sensitivity to ethanol or aminotriazole (AT).

The essentiality of $H_2O_2$ as a cofactor strongly suggested a modulation of Hg uptake by glutathione and its peroxidase (GSH/GSH-Px). GSH-dependent $H_2O_2$ removal was inactivated by conjugation of GSH to chlorodinitrobenzole or by t-butylhydroperoxide (t-BOOH) as substrate for GSH-Px. Both treatments significantly enhanced that uptake which could be stimulated by exogenous $H_2O_2$. Yet the presence of t-BOOH strongly augmented the Hg uptake even without added $H_2O_2$. As this augmentation was sensitive to AT, it probably resulted from compound-I formed by endogenous $H_2O_2$ production.

The experiments made evident that elemental Hg is oxidized by a peroxidatic reaction with catalase, a process which is enhanced by the inactivation of the GSH/GSH-Px system.

## INTRODUCTION

The idea of a biological oxidation of atomic mercury vapor (Hg°) can be traced back to 1934 when Stock found a higher uptake of mercury in bovine blood than in aqueous solutions and mentioned the possibility of oxygen transfer from hemoglobin to the metal. Clarkson and coworkers (1961) expanded the role of mercury oxidation by proposing that the binding of the oxidized mercury to intracellular thiol groups constitutes the essential step for cellular fixation of the highly diffusible vapor and for the initiation of the toxic effects.

Later on Nielsen-Kudsk (1965) found incidentally in volunteers inhaling the vapor that the alveolar absorption could be reduced by small amounts of ethanol. He suggested the participation of catalase (E.C. 1.11.1.6) because it is able to oxidize ethanol. These studies, however, when continued with the exposure of human blood *in vitro*, remained inconclusive as to whether catalase or glutathione peroxidase (GSH-Px, E.C. 1.11.1.9) is the oxidizing enzyme. The investigations to be reported here will provide evidence for a direct active role of the catalase-$H_2O_2$ intermediate (compound-I) and for an indirect inhibition by $H_2O_2$ removal through the GSH/GSH-Px system.

## METHODS

The closed-exposure method (Warburg vessels) used here differed from that used by others in the mode of adding hydrogen peroxide. The importance of a slow, continuous and well-controlled generation of the oxidant for the creation of equilibrium concentrations of compound-I had been emphasized by Oshino et al. (1973). Therefore, in our experiments, $H_2O_2$ was delivered to the cell suspensions from a micropump by slow infusion of diluted solutions. Further details of the procedure had been published earlier (Halbach and Clarkson, 1978; Halbach, 1981; Halbach et al., 1988).

## RESULTS AND DISCUSSION

Characteristics of the cellular uptake in a closed-exposure system for Hg-vapor

The presence of $H_2O_2$ accelerated the rate of Hg-uptake by human erythrocytes at least by a factor of 10 (Table 1). In comparison, a $pO_2$ of 95% raised the uptake only two-fold over that in absence of $O_2$ (Clarkson et al., 1961). Hence, $H_2O_2$ appears to be a more efficient cofactor for the uptake than pure oxygen. In this process, the direct oxidation of elemental Hg by $H_2O_2$ is negligible as Hg concentrations in the incubation medium never exceeded 0.5% of the intracellular concentration (Figure 1). This indicated, in addition to the oxidant, the participation of a "cellular factor" in the stimulation of Hg-uptake.

Table 1. Mercury uptake by human erythrocytes in Warburg-vessel exposure

| Reference | Incubation hematocrit % | H$_2$O$_2$ supply | Hg uptake original paper | ng Hg mg Hb x h |
|---|---|---|---|---|
| Nielsen-Kudsk | 45 - 50 | - | $\dfrac{1.1 \, \mu g}{\text{ml cell x 3 h}}$ ** | 1.1 |
| | | + | $\dfrac{12.0 \, \mu g}{\text{ml cell x 3 h}}$ ** | 11.9 |
| Magos et al. | 36* | - | $\dfrac{1.9 \, ng}{\text{mg Hb x 3 h}}$ | 0.63 |
| | | + | $\dfrac{22.0 \, ng}{\text{mg Hb x 3 h}}$ | 7.3 |
| Ogata et al. | 15* | - | $\dfrac{8.6 \, ng}{\text{mg Hb x 3 h}}$ | 2.9 |
| | | + | $\dfrac{110.7 \, ng}{\text{mg Hb x 3 h}}$ | 37.0 |
| Halbach & Clarkson (1978) | 1 | - | $\dfrac{37.9 \, ng}{\text{mg Hb x 0.75 h}}$ | 50.5 |
| | | + | $\dfrac{684.5 \, ng}{\text{mg Hb x 0.75 h}}$ | 912.6 |

*) recalculated from mg Hb/ml suspension, based on 335 mg Hb/ml cell vol.
**) difference in uptake between whole blood and plasma

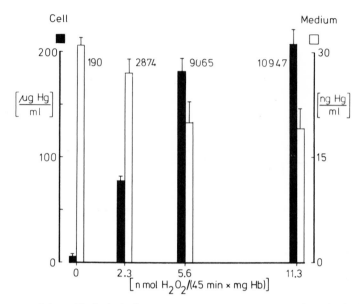

Fig.1 Influence of three $H_2O_2$ infusion rates on mercury concentrations in the incubation medium and in the cells of 1% suspensions of human erythrocytes. Left ordinate and solid bars: μg Hg/ml cell volume; right ordinate and light bars: ng Hg/ml medium. Figures at the top of columns indicate the intra-/extracellular concentration ratio. Means ± SEM of 3 experiments.

The very high (possibly saturated) concentration of Hg vapor and the simplicity of the Warburg-vessel exposure system would predict similar rates of erythrocytic Hg uptake obtained in different laboratories. Yet this assumption is at variance with a comprehensive inspection of published figures as can be seen in Table1 in which the original data were standardized. The extreme variations of the standardized rates of uptake seemingly cast doubt on the reliability of this exposure method, particularly since they are also observed in the absence of $H_2O_2$. These variations could eventually be explained by arranging the cellular Hg uptake in ascending order which was then accompanied by a reversed order of the hematocrit, i.e. cells take up more Hg when there are fewer cells.

It was also observed that the uptake in the most active human cells reached some ceiling at infusion rates above 5.6 nmol $H_2O_2$/(mg Hb x 45 min) which means that the maximum of the uptake was limited by some third factor (Figure 2). Concerning the roughly inverse parallelism between uptake and hematocrit, specific experiments demonstrated a hyperbolic correlation between these parameters (Figure 3); concomitantly, the amount of mercury taken up by the whole suspension remained nearly unchanged. This suggests, in conjunction with the lack of Hg accumulation in the medium (see above), that a given amount of Hg vapor will be completely absorbed by the cells. The magnitude of this amount will be determined by the evaporation rate of the metallic Hg and by the degree of $H_2O_2$ stimulation of the uptake; the first is constant and the second can be varied experimentally. Therefore, the exposure can be operated under different conditions: (1) the uptake can be varied by biochemical interventions as long as sufficient $Hg^0$ is available, i.e. uptake < evaporation, or (2) the uptake is activated to the extent that it equals the evaporation.

If the exposure system is sensitive enough both modes of operation should be observed. Sensitivity, being defined as the span of cellular uptake obtainable by addition of $H_2O_2$, will be greatest if the suspension is made up of few cells only, each containing much of the "cellular factor". This can be seen in the almost linear increase in Hg uptake by human cells (1% hematocrit) in dependence on the infused amount of $H_2O_2$. At high rates of $H_2O_2$ infusion, however, the additional oxidant became less effective (Figure 2); from there on the uptake depended on exposure time (Table 2), i.e. the rate of uptake had attained the maximum determined by the evaporation rate. The concept of an $H_2O_2$-dependent working range of Hg uptake that is capped by the availability of the vapor, and the gain in sensitivity of the uptake obtained by lowering the hematocrit improved the usefulness of the closed-exposure system in investigating the contribution of the red cell enzymes to the uptake process.

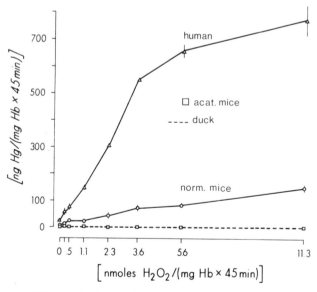

Fig.2 Dependence of Hg uptake by erythrocytes from various species on the rate of $H_2O_2$ infusion. 1% hematocrit; 45 min incubation. Means $\pm$ SEM of 3 to 4 experiments.

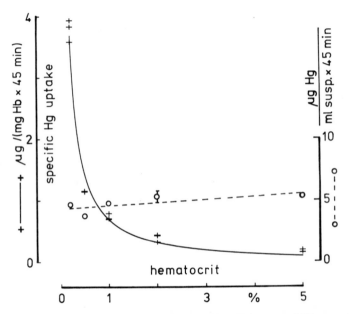

Fig.3   Relationship between hematocrit of the suspensions and Hg uptake by human erythrocytes (left ordinate, +——+) or Hg uptake by whole suspension (right ordinate, 0-----0). Hyperbolic regression:  µg Hg/(mg Hb x 45 min) = 0.75/% hematocrit - 0.06; r = 0.992; n = 15.

Table 2. Uptake of mercury vapor by suspensions (5% hematocrit) of human erythrocytes after different exposure times

| Rate of $H_2O_2$ infusion* | Hg uptake, µg/2 ml suspension | |
|---|---|---|
| nmol/ (mg Hb45 min) | after 45 min | after 90 min |
| ---- | 0.052 | 0.057 |
| 5.63 | 3.651 | 6.957 |
| 11.25 | 3.181 | 6.435 |
| 22.5 | 3.26 | 6.663 |
| 112.5 | 4.061 | 6.883 |

*) same rates for 90 min exposure

Table 3. Specific activities of catalase and glutathione peroxidase in red cells from various species

| Activities | Erythrocytes | | | |
|---|---|---|---|---|
| | Human | Duck | Cs[a] Mouse | Cs[b] Mouse |
| Catalase | 354 ± 13* | 1 ± 0.1 | 174 ± 6.4 | 13 ± 0.3 |
| $s^{-1} g^{-1}$ ml | (18) | (6) | (9) | (6) |
| GSH peroxidase | 22 ± 2.5 | 88 ± 8.5 | 297 ± 11.5 | 305 ± 8.7 |
| µmol NADPH | | | | |
| min x g Hb | (6) | (3) | (3) | (3) |

*) Means ± SEM of (n) determinations

The evidence for the interaction between catalase-compound-I and Hg vapor

Two $H_2O_2$-metabolizing enzymes had been proposed by Nielsen-Kudsk (1969) to be linked to the absorption of Hg° by red blood cells: either catalase or GSH-Px. The distinction with regard to Hg-uptake could be made by using a variety of erythrocytes with different activities of these enzymes or by application of alternative substrates and specific inhibitors. The extreme positions of catalase activities are held by human and avian red cells, the former being very high and the latter virtually devoid of activity. An intermediate activity occurs in mouse cells which, if derived from two particular substrains, gave access to experiments with homologous cells except for the expression of catalase. Likewise, large variations of GSH-Px in these four cell types offered the possibility to assess the role of this enzyme in the uptake process; its activity followed the order man < duck < mouse (Table 3).

When 1% suspensions of these cells were exposed to $Hg^0$ in absence of added peroxide, differences in the rate of Hg absorption were barely discernible. Upon infusion of $H_2O_2$, however, the rate of uptake was increased remarkably in human cells and moderately in those from normal mice, but was insensitive to the oxidant in acatalatic-mouse and duck erythrocytes (Figure 2). The $H_2O_2$-inducible Hg uptake paralleled the activity of catalase but showed no relation to that of GSH-Px. The requirement of $H_2O_2$ and catalase indicated the formation of compound-I which initiated the cellular absorption of Hg. Conceivably, other agents of high affinity to compound-I should be able to interfere with the uptake. The compound-I substrates, formic acid and more so ethanol, demonstrated this interaction by a downward shift of the uptake curve (Figure 4). In contrast, the uptake curve in presence of the inhibitor aminotriazole (AT) displayed an inflection point: no effect at low but maximal inhibition at high peroxide supplementation. The $H_2O_2$-dependent increase in uptake-inhibiting efficiency of AT correlated well with the increasingly higher inactivation of catalase at elevated $H_2O_2$ infusion rates (Figure 5). This resulted probably from the property of AT to bind to compound-I only which is more abundant at elevated $H_2O_2$ concentrations.

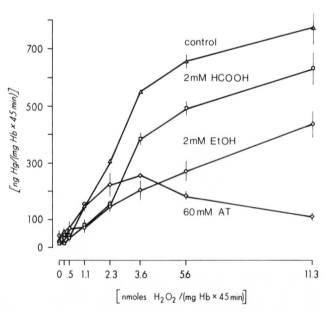

Fig.4  Hg uptake by human erythrocytes in presence of other substrates or inhibitors of catalase. Means ± SEM of 4 experiments (control) and of 3 (inhibitors). Conditions as in Figure 2.

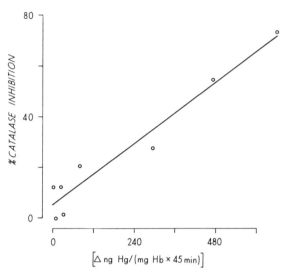

Fig.5   Inhibition of Hg uptake by human erythrocytes and of catalase activity by 60 mM
        aminotriazole (AT).  Ordinate:  % inhibition = 100 - % remaining activity (means of
        3 experiments); abscissa:  difference in uptake between means of control and means
        of AT curve of Figure 4.  Regression:  % inhibition = 0.1 x difference in ng Hg +
        5.3; n = 8; r = 0.97.

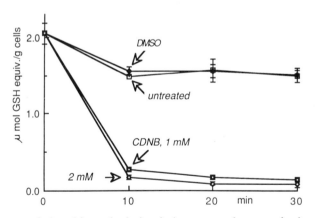

Fig.6   Time course of glutathione depletion in human erythrocytes by incubation with 1-
        chloro-2, 4-dinitrobenzene (CDNB).  10 µl of DMSO containing 0.2 or 0.4 M
        CDNB were added to 2 ml of 25% cell suspensions.  Means ± SEM of 3
        experiments.

## Inhibition of GSH-dependent hydrogen peroxide removal

Although the evidence obtained so far excluded a direct involvement of GSH-Px in the mechanism of Hg uptake, an indirect influence of this enzyme could be assumed because of its concomittant degradation of $H_2O_2$ depriving catalase of some of this cofactor. This could be investigated by inhibiting the $H_2O_2$-removing activity of GSH-Px in two ways: (1) conjugation of GSH to chlorodinitrobenzole (CDNB) eliminates the hydrogen donor for $H_2O_2$ reduction or (2) t-butylhydroperoxide (t-BOOH) replaces $H_2O_2$ in the reaction with GSH-Px whereby GSH is also decreased in favor of GSSG. A short incubation of human erythrocytes in presence of CDNB diminished the content of reduced GSH below 20% (Figure 6). These cells, when resuspended at 1% hematocrit in CDNB-free medium, absorbed significantly more mercury than control cells of normal GSH content (Figure 7). Likewise, the addition of t-BOOH enhanced the $H_2O_2$-inducible uptake remarkably (Figure 8).

Notably, the endpoints of the curves of Figures 7 and 8 were worth a detailed inspection. The curves concurred at the high rates of uptake because the availability of Hg vapor became rate limiting (see above). In the region of low uptake it can be seen that the absorption of Hg could be substantially increased even without exogenous $H_2O_2$, in particular if 100 µmol/1 t-BOOH was added.

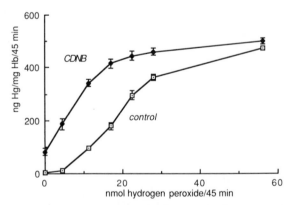

Fig.7  $H_2O_2$-induced Hg uptake by human erythrocytes without or after GSH depletion by preincubation with 1 mM 1-chloro-2,4-dinitrobenzene (CDNB). Means ± SEM of 6 experiments. Conditions as in Figure 2.

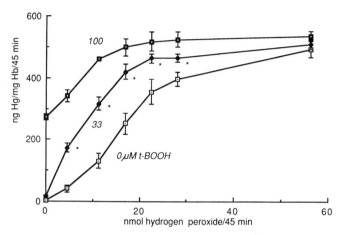

Fig.8   H$_2$O$_2$-induced Hg uptake by human erythrocytes in presence of various concentrations of t-butylhydroperoxide (t-BOOH). Means ± SEM of 4 experiments. Conditions as in Figure 2.

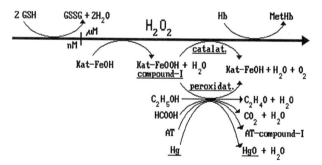

Fig.9   Reaction scheme of the degradation of hydrogen peroxide in the erythrocyte. The bold horizontal arrow visualizes the increasing intracellular concentration of the oxidant. Mercury has been integrated in the series of substrates for peroxidation by compound-I.

If the assumption underlying the data of Figures 7 and 8 is correct, i.e. that the inhibition of the GSH/GSH-Px system increases the proportion of exogenous $H_2O_2$ available to catalase and hence increases Hg uptake, it should also hold for any Hg uptake induced by production of endogenous peroxide. Endogenous $H_2O_2$ production can be probed by the inhibition of compound-I after preincubation of cell suspensions with AT (Cohen and Hochstein, 1963). A parallel series of equally pretreated cells served to test the sensitivity of the t-BOOH-inducible Hg uptake towards AT (without exogenous $H_2O_2$; Table 4). The comparison of t-BOOH-activated Hg uptake (D-C vs. H-G) with catalase activity (G vs. C) demonstrated an inhibition of 40% by AT. The effectiveness of AT could be augmented if it had been administered in the preincubation together with t-BOOH; this lowered the t-BOOH-induced uptake to 24% (K-I vs. F-E) and eliminated catalase activity almost completely (I). In each case the effects on catalase and on t-BOOH-activated Hg uptake could be explained by the presence of endogenous $H_2O_2$, of which only a fraction is supplied to catalase if GSH-Px is not affected, but which becomes much more available for compound-I formation if GSH-Px alternatively degrades the organic peroxide (Table 4, AT-effect I > G). This stepwise inactivation of catalase is necessarily followed by a corresponding suppression of t-BOOH-inducible Hg uptake.

Table 4. Catalase activity and t-BOOH-induced mercury uptake in human erythrocytes (1% hematocrit) after preincubation with t-BOOH and AT

|   | preincubation | | catalase activity | Hg exposure | Hg uptake |
|---|---|---|---|---|---|
|   | AT mM | t-BOOH mM | ml sec x g Hb | t-BOOH mM | ngHg mg Hb/45 min |
| A | no preincubation | | 235 +/-5 | 0 | 2 +/-0 |
| B |   |   |   | 0.1 | 244 +/-8 |
| C | 0 | 0 | 205 +/-4 | 0 | 10 +/-1 |
| D |   |   |   | 0.1 | 401 +/-23 |
| E | 0 | 0.1 | 205 +/-4 | 0 | 38 +/-13 |
| F |   |   |   | 0.1 | 446 +/-32 |
| G | 60 | 0 | 122 +/-5 | 0 | 5 +/-2 |
| H |   |   |   | 0.1 | 231 +/-1 |
| I | 60 | 0.1 | 5 +/-1 | 0 | 16 +/-11 |
| K |   |   |   | 0.1 | 115 +/-25 |

## CONCLUSIONS

The use of cell suspensions of a low hematocrit greatly enlarged the potential range of Hg accumulation in the cell and this facilitated the detailed observation of subtle changes in the uptake. The increase in sensitivity of the closed-exposure method revealed the "working range" of this procedure within which the Hg uptake reacts very sensibly to $H_2O_2$ supplementation as long as the supply of Hg vapor is not exhausted. The response to $H_2O_2$ was also closely linked to the activity of catalase in the cells and was readily affected by other substrates or inhibitors of the enzyme.

The findings assign a key role to catalase-compound-I in the cellular fixation of $Hg^O$. Catalase in its compound-I state appears to be unique among the oxidoreductases because of its ability to transfer two electrons in one single step whereby the hematin iron cycles between +III and +V (Dounce and Sichak, 1988). Thus it is conceivable that compound-I (Fe-V) can be reduced to native catalase in a single step by oxidizing atomic mercury to Hg-II. Figure 9 shows the integration of mercury oxidation in the peroxidatic mode of catalase action.

The essentiality of $H_2O_2$ as a cofactor in this process increases, in turn, the probability of an involvement of other $H_2O_2$ -relevant pathways, i.e. the degradation by GSH-Px. This enzyme, by counteracting the supply of peroxide to catalase (Figure 9), restricts the compound-I induced Hg uptake to submaximal rates. If $H_2O_2$ degradation through GSH-Px becomes inactive the endogenous peroxide supply is sufficient for a remarkable Hg uptake that would otherwise almost be negligible in presence of a normally operating GSH/GSH-Px system. It remains to be investigated whether the dampening effect of intact GSH-Px observed *in vitro* is responsible for the relatively low absorbed proportion of a dose of Hg vapor estimated in the blood of exposed human subjects in comparison to the absorption in other organs (blood < urine < brain < kidney <; Clarkson et al, 1988). Inversely, it would be interesting to measure blood mercury of individuals exposed to $Hg^O$ who have low levels of reduced red-cell GSH due to genetic disorders (glucose-6-phosphate-dehydrogenase deficiency).

## ACKNOWLEDGEMENT

The support of these studies given by the Environmental Health Sciences Center at the University of Rochester, New York, is gratefully acknowledged.

REFERENCES

Clarkson, T.W., Gatzy, J., Dalton, Ch., 1961, Studies on the equilibration of mercury vapor with blood. The University of Rochester, Atomic Energy Project, report nr. UR-582.

Clarkson, T.W., Friberg, L., Hursh, J.B., Nylander, M., 1988, The prediction of intake of mercury vapor from amalgams. In: Biological monitoring of toxic metals, T.W. Clarkson, L. Friberg, G.F. Nordberg, P.R. Sager, eds., 247-264, Plenum Press, New York.

Cohen, G., and Hochstein, P., 1963. Glutathione peroxidase: The primary agent for the elimination of hydrogen peroxide in erythrocytes. Biochemistry 2:1420.

Dounce, A.L., Sichak, S.P., 1988, Hematin iron valence in catalase and peroxidase compound I: relationship to free radical reaction mechanism. Free Radical Biology & Medicine, 5:89.

Halbach, S., 1981, Limitations on the uptake of mercury vapour by human erythrocytes in a closed exposure system. J. Appl. Toxicol. 1:303.

Halbach, S., Ballatori, N., Clarkson, T.W., 1988, Mercury vapor uptake and hydrogen peroxide detoxification in human and mouse red blood cells. Toxicol. Appl. Pharmacol. 96:515.

Halbach, S., Clarkson, T.W., 1978, Enzymatic oxidation of mercury vapor by erythrocytes. Biochem. Biophys. Acta 523:522.

Magos, L., Sugata, Y., Clarkson, T.W., 1974, Effects of aminotriazole on mercury uptake by in vitro human blood samples and by whole rats. Toxicol. Appl. Pharmacol. 28:367.

Nielsen-Kudsk, F., 1965, The influence of ethyl alcohol on the absorption of mercury vapour from the lungs in man. Acta Pharmacol. Toxicol. 23:263.

Nielsen-Kudsk, F., 1969, Factors influencing the in vitro uptake of mercury vapour in blood. Acta Pharmacol. Toxicol. 27:161.

Ogata, M., Ikeda, M., Sugata, Y., 1979, In-vitro mercury uptake by human acatalasemic erythrocytes. Arch. Environ. Health 34, 218-22.1

Oshino, N., Oshino, R., Chance, B., 1973, The characteristics of the "peroxidation" reaction of catalase in ethanol oxidation. Biochem J. 131:555.

Stock, A., 1934, Über Verdampfung. Löslichkeit und Oxydation des metallischen Quecksilbers. Zeitschr. Anorgan. Allgem. Chemie, 217:241.

# DIFFERENTIAL DETERMINATION OF IONIZABLE AND

# UNIONIZABLE (INERT) FORMS OF INORGANIC MERCURY

# IN ANIMAL TISSUES

Hitoshi Takahashi, Katsutoshi Suetomi and Tetsuro Konishi

Kumamoto University Medical School
Kumamoto-City, Japan

ABSTRACT

For our study on biotransformation of organic Hg to inorganic Hg in the animal body, we had to develop a direct determination method for minor inorganic Hg in the presence of fair amounts of organic Hg in tissue. This was achieved by using $H_2O_2$ as a reducing agent in a strong alkaline medium (Konishi and Takahashi, 1983). Inorganic Hg is known to be excreted faster than alkyl-Hg, however, inorganic Hg remains much longer than alkyl-Hg in tissues when alkyl-Hg is administered to animals. This suggests the formation of inorganic Hg in some stable form in tissue. We tried to develop a method which only determines ionizable (toxic) inorganic Hg separate from unionizable (stable) inorganic Hg in tissues. Ionizable Hg is first released from proteins by the addition of NaCl in a medium acidified with sulfuric acid, and reduced with stannous chloride. The Hg vaporized was trapped by gold amalgamation and finally released to atomic absorption spectrometry by quick heating. By subtracting ionizable Hg from total inorganic Hg determined by the first method, we could find the existence and the amount of the unionizable inert form of inorganic Hg. When we applied this method to the various tissues of rats freshly injected with $HgCl_2$, we could hardly detect this inert form of Hg. However, formation of this inert form became evident after two weeks in the liver followed by the kidney of rats similarly treated.

Various organs of Minamata Disease patients were also found to contain considerable amounts of unionizable stable inorganic Hg. Effect of dietary sulfur or selenium compounds on the formation of inert from of inorganic Hg was studied by feeding a basic synthetic diet low in S and Se to the rats given $HgCl_2$. Supplementation of Se in the form of selenite increased the amount of the unionizable form of Hg in the liver much more than that of S in the form of methionine. The rats injected with a large amount of HgS or HgSe (50 mg hg/kg) survived for a long time and very little of Hg migrated to liver and kidney. Thus we propose that HgSe and probably HgS are the chemical forms of inert inorganic Hg formed in animal tissues.

INTRODUCTION

During our studies (Konishi and Takahashi, 1983a; Suda and Takahashi, 1986; Takahashi and Suda, 1986; Suda and Takahashi, 1990) on the biotransformation mechanism of alkyl mercury to inorganic mercury in the animal body, we had to develop a proper and sensitive analytical method which would directly determine very small amounts of inorganic mercury in the presence of fair amounts of alkyl mercury. Such a method (Konishi and Takahashi, 1983b) was successfully established by using hydrogen peroxide as a reducing agent after alkali digestion. By this method we could determine inorganic mercurials including Mercury(II) sulfide without decomposition of alkyl mercury. Addition of potassium cyanide was necessary to convert sugar aldehydes to cyanohydrin, otherwise some decomposition occurred on alkyl mercury during alkali digestion.

When we analyzed the organs of Minamata disease patients who lived many years before death, considerable amounts of inorganic mercury were found besides methyl mercury in the livers, kidneys, and even in the brains. This suggested to us an existence of an inert form of inorganic mercury and some blind space to allow such a form of mercury without giving further damage to other cells.

To study inert forms of inorganic mercury resistant to ionization and non-toxic, a suspension of a large amount (50 mg Hg/kg) of either mercury(II) sulfide or mercury(II) selenide finely ground was injected intramuscularly to rats. These rats survived without developing obvious sick conditions. When these healthy looking rats were killed 15 or 30 days later, relatively small amounts of mercury were found in the livers and the kidneys, though mercury(II) selenide tended to give a little higher mercury value than expected in the kidneys. Thus we thought mercury(II) sulfide and mercury(II) selenide to be the candidates for inert inorganic mercury.

## METHODS

In the past, no method was available to determine inert forms of inorganic mercury separate from toxic ionizable forms of inorganic mercury in the tissue. After many trials, the analytical method was finally established by releasing ionizable inorganic mercury from tissue proteins with a fair amount of sodium chloride in an acidified condition with sulfuric acid, and vaporizing it to gold amalgamation by reduction with Tin(II) chloride and finally releasing it quickly to atomic absorption spectrometry. The amount of ionizable inorganic mercury was subtracted from the total inorganic mercury determined by the first method, for obtaining the amount of inert inorganic mercury.

Table 1 Ionizable Hg determined by the present method from four kinds of biological samples containing $HgCl_2$, HgS, HgSe or MeHgCl

| Biological sample | Ioniz. Hg present (ng) | Hg added (ng Hg) | | Hg found (ng) | Recovery (%) |
|---|---|---|---|---|---|
| Liver (0.3 g) | 0.01 | $HgCl_2$ | 50 | 49.91 | 99.8 |
| Liver (0.3 g) | 0.01 | HgS | 1000 | 13.61 | 1.4 |
| Liver (0.3 g) | 0.01 | HgSe | 1000 | 0.71 | 0.1 |
| Liver (0.3 g) | 0.01 | MeHgCl | 1000 | 5.65 | 0.6 |

When we tested this method to the tissue homogenates containing 50 ng Hg of mercury(II) chloride, 1000 ng Hg of mercury(II) sulfide, 1000 ng Hg of mercury(II) selenide, or 1000 ng Hg of methyl mercury chloride, we could recover almost 100% of Hg from mercury(II) chloride, but negligible amounts from the other three compounds, indicating that ionizable inorganic mercury could be determined separately from the stable inorganic mercury as well as alkyl mercury (Table 1). The procedure we used for determining ionizable inorganic mercury in tissue homogenates is shown in the Figure 1. After a time course study, the time for vaporization was 30 minutes with safety.

RESULTS AND DISCUSSION

When we looked for inert inorganic mercury in the tissues by the present method, not much of this form was found in the organs of the rats that received two injections of 1 mg Hg/kg of mercury(II) chloride and killed on the following day. However, we didfind such an inert fraction of inorganic mercury definitely in the liver and less definitely in the kidney of the rats killed 14 days later. This is shown in the Figure 2. It was surprising for us to see such a large portion of inorganic mercury that became inert after 14 days in the liver and also that it was much higher in the liver than in the kidney.

Thus we next prepared an experimental diet low but sufficient in sulfur and selenium content (control diet, CD) by using Torula yeast according to Nishikido et al. (1987). This diet was fed to the young rats for 7 weeks. Mercury(II) chloride was similarly injected just before the last 2 weeks. In these rats, very small amounts of inert inorganic mercury were detected in the liver, and none in the kidney or in the brain 14 days later, indicating little contribution of sulfur or selenium to the formation of inert inorganic mercury. When the same experimental diet, supplemented with sulfur as methionine (methionine supplemented diet, MSD) or supplemented with selenium as sodium selenite (selenium supplemented diet, SeSD), were fed to the young rats for 7 weeks, the injection of mercury(II) chloride just before the last 2 weeks increased inert inorganic mercury in the liver and in the kidney but not in the brain. The result with three different diets is shown in Figure 3 (brain data are not shown). The composition of three experimental diets are shown in the Table 2A and 2B.

In this experiment, the contribution of dietary sulfur seemed low in the formation of inert inorganic mercury. Thus in the similar experiment with CD, we supplemented sulfur as L-cysteine (250 mg/kg) by ip injections for the last 2 weeks. In this case, contribution of sulfur became evident in the liver 14 days after mercury(II) chloride injection (Figure 4).

Finally, we would like to show the data analyzing autopsy tissue specimens obtained from some of the Minamata disease victims by the present method. Inert portions of inorganic mercury as well as organic mercury are shown in Table 3. In the cerebrum and cerebellum, most of the inorganic mercury was in an inert form. However, in the liver and kidney, both ionizable and inert forms of inorganic mercury were found together in some specimens. The relation between the patients' personal histories and the amounts of inert inorganic mercury should be studied by increasing numbers of cases.

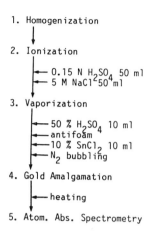

1. Homogenization

2. Ionization

   — 0.15 N $H_2SO_4$  50 ml
   — 5 M NaCl 50 ml

3. Vaporization

   — 50 % $H_2SO_4$  10 ml
   — antifoam
   — 10 % $SnCl_2$  10 ml
   — $N_2$ bubbling

4. Gold Amalgamation

   — heating

5. Atom. Abs. Spectrometry

Fig. 1   Analysis for ionizable inorg. Hg

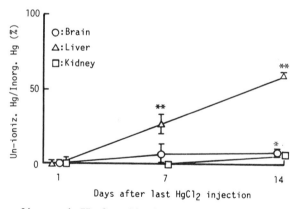

Fig. 2  Inert form of inorganic Hg found in the organs after feeding rats with stock diet. Asterisks indicate significant differences from the values found in the rats killed 1 day after last HgCl2 injection. (*), $p<0.05$; (**), $p<0.01$.

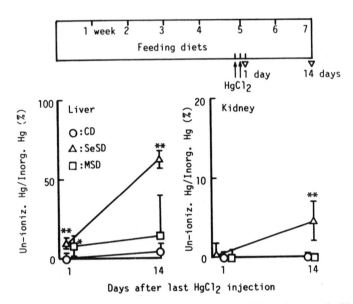

Fig. 3  Effects of three different diets on the formation of inert inorganic Hg in the rat organs. (↑), HgCl2 (1 mg Hg/kg) was s.c. injected for 2 consecutive days. Asterisks indicate significant differences from the values found in the rats fed CD (low S and Se). (*), p<0.05; (**), p<0.01.

Table 2a  Composition of control diet

| Ingredient | % |
|---|---|
| Torula yeast | 30.0 |
| Sucrose (fine granule) | 50.0 |
| Corn starch | 8.2 |
| Lard | 5.0 |
| Cod-liver oil | 2.0 |
| *  Mineral mixture | 3.5 |
| Vitamine mixture | 1.0 |
| ** L-Methionine | 0.3 |

Table 2b Final concentration of selenium and methionine in each 3 diets

| Diet | Se* | L-Methionine** |
|---|---|---|
| CD | 0.04 ppm | 0.6% |
| SeSD | 0.40 ppm | 0.6% |
| MSD | 0.04 ppm | 1.8% |

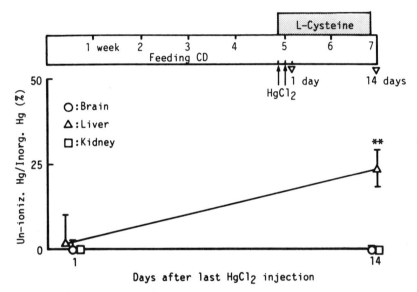

Fig. 4 Effect of L-cysteine supplementation on the formation of inert inorganic Hg in the rats fed with CD (low S and Se). ▨ L-Cysteine (250 mg/kg) was i.p. injected daily for 15 days. † HgCl2 (1 mg/kg) was s.c. injected for 2 consecutive days. Asterisks indicate significant differences from the values found in the rats killed 1 day after last HgCl2 injection. (**), p<0.01.

Table 3   Mercury contents in the autopsy specimens of Minamata disease victims

| | Autopsy case No. | Total Hg ($\mu$g/g) | Inorg. Hg ($\mu$g/g) | Ioniz. Hg ($\mu$g/g) | (%) | Un-ioniz. Hg (%) |
|---|---|---|---|---|---|---|
| Cerebrum | 8918 | 0.22 ± 0.05 | 0.23 | 0.01 | ( 4%) | (96%) |
| | 7903 | 0.44 ± 0.05 | 0.42 | 0.03 | ( 7%) | (93%) |
| | 7494 | 0.37 ± 0.02 | 0.35 | 0.03 | ( 9%) | (91%) |
| | 6820 | 1.01 ± 0.05 | 0.84 | 0.05 | ( 6%) | (94%) |
| Cerebellum | 8918 | 0.56 ± 0.03 | 0.51 ± 0.01 | 0.02 | ( 4%) | (96%) |
| | 7903 | 0.61 | 0.44 | 0.06 | (14%) | (86%) |
| | 7494 | 0.26 ± 0.01 | 0.29 | 0.03 | (10%) | (90%) |
| | 6820 | 0.85 ± 0.02 | 0.86 | 0.06 | ( 7%) | (93%) |
| Liver | 8918 | 0.79 | 0.80 ± 0.07 | 0.48 | (60%) | (40%) |
| | 7903 | 0.77 | 0.74 ± 0.03 | 0.52 ± 0.03 | (70%) | (30%) |
| | 7494 | 0.38 | 0.40 | 0.24 | (60%) | (40%) |
| | 6820 | – | – | – | – | – |
| Kidney | 8918 | 1.71 | 1.28 ± 0.08 | 0.40 ± 0.01 | (31%) | (69%) |
| | 7903 | 2.67 | 2.41 ± 0.27 | 1.65 ± 0.06 | (68%) | (32%) |
| | 7494 | 2.01 | 1.35 ± 0.08 | 0.08 ± 0.00 | ( 6%) | (94%) |
| | 6820 | 4.94 ± 0.27 | 1.98 ± 0.21 | 0.25 ± 0.00 | (13%) | (87%) |

Data are expressed as means ± SD of 3-4 analyses. Others are means of 2 analyses.

## CONCLUSION

In summary, we would like to say that our present method which determines ionizable forms of inorganic mercury in tissues, can be used to analyze coexisting inert or stable forms of inorganic mercury, by combining with our other method which directly determines total inorganic mercury without decomposing organic mercury. Such inert forms of inorganic mercury are not immediately formed after injection of mercury(II) chloride, but tends to increase as time passes.   Depletion and supplementation of sulfur as methionine or cysteine and those of selenium as sodium selenite show clear effects on the amount of inert forms of inorganic mercury in the tissues.   It is difficult to identify at present the real chemical forms of such stable mercury compounds until the pure form has been isolated.   And also it is hard to say that these inert mercury compounds are absolutely stable.   We believe that conversion to the inert form occurs on the inorganic mercury biotransformed from organic mercury in the animal body.   Thus, estimation of the initial amount of mercury in the tissues such as brain is almost impossible by just using a biological half life of Hg or MeHg and the amount of Hg found in the tissues at autopsy.

REFERENCES

Konishi, T. and Takahashi, H., 1983a, Lack of methylmercury biotransformation in fetus and newborn rat, Kumamoto Medical Journal, 36:97-102.
Konishi, T. and Takahashi, H., 1983b, Direct determination of inorganic mercury in biological materials after alkali digestion and amalgamation, Analyst, 108:827-834.
Nishikido, N., Furuyashiki, K., Naganuma, A., Suzuki, T., and Imura, N., 1987, Maternal selenium deficiency enhances the fetolethal toxicity of methyl mercury, Toxicol. Appl. Pharmacol., 88:322-328.
Suda, I. and Takahashi, H., 1986, Enhanced and inhibited biotransformation of methyl mercury in the rat spleen, Toxicol. Appl. Pharmacol, 82:45-52.
Suda, I. and Takahashi, H., 1990, Effect of reticuloendotherial system blockade on the biotransformation of methyl mercury in the rat, Bull. Environ. Contam. Toxicol., 44:609-615.
Takahashi, H. and Suda, I., 1986, Metabolic fate of methyl mercury in animals, in "Recent advances in Minamata Disease studies", Tsubaki, T. and Takahashi, H., ed., 135-150, Kodansha Ltd., Tokyo.

ROLE OF NEURONAL ION CHANNELS

IN MERCURY INTOXICATION

Toshio Narahashi, Osamu Arakawa and
Masanobu Nakahiro

Department of Pharmacology
Northwestern University Medical School
Chicago
Illinois

ABSTRACT

Mercury compounds exert multiple actions on the nervous system. At skeletal neuromuscular junctions, mercury increases spontaneous release of acetylcholine from nerve terminals and suppresses the nerve-evoked synchronized release of acetylcholine. Voltage-activated sodium and potassium channels of neuronal membranes are suppressed by mercury causing conduction block. Our recent patch clamp study with the rat dorsal root ganglion neurons has unveiled a highly potent and efficacious action of mercuric chloride in augmenting the GABA-activated chloride channel current, a prominent effect being observed at 1 µM. Mercuric chloride also induced a slow inward current by itself, which is likely to account for an increase in leakage current, resting membrane conductance, and membrane depolarization. It was concluded that the stimulation of GABA-induced chloride current plays an important role in mercury intoxication.

*Advances in Mercury Toxicology*, Edited by T. Suzuki *et al.*
Plenum Press, New York, 1991

INTRODUCTION

The symptoms of poisoning by mercury compounds in humans are characterized by neurological disorders as observed in Minamata disease in Japan (Takeuchi et al., 1962) and in outbreak of intoxication in Iraq (Bakir et al., 1973). Clinical signs include impaired vision, speech and hearing, paresthesia, weakness of the extremities, and ataxia (Hunter et al., 1940; Takeuchi et al., 1968; Chang, 1977, 1980). A myasthenia gravis-like muscle weakness, which could be treated by neostigmine, was also observed (Rustam et al., 1975). Similar symptoms of mercury poisoning were produced in experimental animals (Miyakawa et al., 1970; Somjen et al., 1973; Berlin et al., 1975; Fehling et al., 1975). Despite the grave environmental concern, the mechanisms by which mercury compounds exert their neurotoxic effects remains largely to be seen. A large number of experimental observations conducted in many laboratories fall far short of drawing the integrated picture of mercury neurotoxicity.

In the present paper, a concise review of our previous studies of the mechanism of action of mercury on the nervous system will be given first. This will be followed by a brief account of our most recent studies of mercury interaction with the γ-aminobutyric acid (GABA) receptor-channel complex, which is regarded as one of the most important target sites of mercury compounds.

HISTORICAL REVIEW

In human patients poisoned with mercury, no change was observed in the properties of peripheral motor and sensory nerves including the conduction velocity(Le Quesne et al., 1974; Von Burg and Rustam, 1974; Snyder and Seelinge, 1976). Only when large doses of mercury were administered, a reduction of conduction velocity and histopathological damages were observed in peripheral nerves (Miyakawa et al., 1970; Cavanaugh and Chen, 1971; Chang and Hartmann, 1972; Herman et al., 1973; Somjen et al., 1973; Fehling et al., 1975; Jacobs et al., 1975; Misumi, 1979). Conduction block caused by methylmercury in squid giant axons was ascribed to block of sodium channels (Shrivastav et al., 1976).

There was general agreement that synapses are the major site of action of mercury compounds. In earlier studies conducted in the 1970s through mid-1980, neuromuscular preparations isolated from frogs and rats were used as materials, partly because of easy applicability of intracellular microelectrode techniques and partly because of the documented case of neuromuscular weakness in mercury intoxication (Rustam et al., 1975). However, since patch clamp techniques were developed in the 1980s, it has now become possible to measure the activity of various ion channels not only in the cholinergic system as in the case of neuromuscular preparations, but also in any other system. Patch clamp experiments have finally been applied to the study of mercury and uncovered striking modulations of the GABA-activated chloride channels by mercury (Arakawa et al., 1990).

## EFFECTS ON NEUROMUSCULAR TRANSMISSION

Mercury compounds exert two major effects on skeletal neuromuscular junctions. Spontaneous quantal release of acetylcholine (ACh) is stimulated by mercury as evidenced by an increase in the frequency of spontaneous miniature end-plate potentials (MEPPs) (Barrett et al., 1974; Manalis and Cooper, 1975; Juang and Yonemura, 1975; Juang, 1976; Atchison and Narahashi, 1982; Miyamoto, 1983; Cooper and Manalis, 1983; Cooper et al., 1984; Atchison, 1986; Traxinger and Atchison, 1987). An example of such an experiment with the rat phrenic nerve-hemidiaphragm preparation is illustrated in Figure 1 (Atchison and Narahashi, 1982). Methylmercury did not change the MEPP frequency at a concentration of 4 μM, but markedly increased the frequency at 20 μM. At 100 μM, the latency for the increase in frequency was greatly shortened, although the maximum level of the frequency did not further increase beyond the level attained in 20 μM methylmercury. Thus the access of methylmercury to the site of action is a rate limiting factor. The effect was not reversed after washing with mercury-free media. However, a prolonged application of methylmercury eventually caused the MEPP frequency to decrease to a very slow level (Figure 1). In the face of drastic changes in MEPP frequency, the mean amplitude of MEPPs remained unchanged by methylmercury at a concentration of 4, 20 or 100 μM.

Mercury causes a second major effect on the nerve-evoked end-plate potential (EPP). the EPP amplitude was suppressed by mercuric chloride or methylmercury (Barrett et al., 1974; Manalis and Cooper, 1975; Juang, 1976; Atchison and Narahashi, 1982; Cooper and Manalis, 1983; Atchison et al., 1986, but in some cases suppression was preceded by a transient increase in the amplitude of EPP (Manalis and Cooper, 1975; Juang, 1976). In order to elucidate the mechanism of action of mercury on EPP, various experiments were performed with the rat muscle preparation (Atchison and Narahashi, 1982).

Figure 2 shows a dose-response relationship for the methylmercury

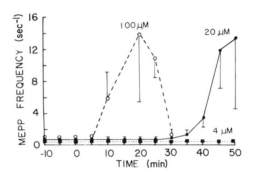

Fig. 1 Increase in MEPP frequency by methylmercury. Rat phrenic nerve-
hemidiaphragm preparation. Methylmercury was not effective at 4 µM, but
increased the MEPP frequency at 20 µM and 100 µM, and the latency was
shortened with increasing concentration. Values are the mean ± S.E.M. of
3-8 determinations. Room temperature. From Atchison and Narahashi
(1982).

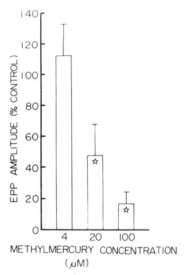

Fig. 2 Effects of methylmercury on nerve-evoked EPP amplitude. Rat phrenic
nerve-hemidiaphragm preparation. Methylmercury was not effective at 4 µM,
but suppressed the EPP amplitude at 20 and 100 µM. The EPP was recorded
in solutions contianing high $Mg^{2+}$ (8 mM) and low $Ca^{2+}$ (1 mM) to block
muscle action potentials. At least 100 EPPs were analyzed for each
preparation . The asterisk indicates a statistically significant decrease
($p < 0.05$). Room temperature. From Atchison and Narahashi (1982).

suppression of EPP amplitude. Although methylmercury had no effect at 4 µM, it drastically decreased the amplitude of EPP at 20 µM or 100 µM. The decrease in EPP amplitude was ascribed to a decrease in the quantal content of EPP. The ACh sensitivity of end-plate membrane was not changed by methylmercury as examined by iontophoretic application of ACh. One possible mechanism underlying the decrease in EPP amplitude and quantal content and the increase in the frequency of spontaneous MEPPs is a depolarization of nerve terminals. If methylmercury selectively opened the sodium channels thereby causing a membrane depolarization, tetrodotoxin (TTX) would antagonize the effect through membrane repolarization. However, 1 µM TTX had no effect on the methylmercury-induced increase in MEPP frequency.

The mechanism of the increase in MEPP frequency caused by methylmercury was further explored by using a variety of techniques. Delay in the increase in MEPP frequency following application of methylmercury was interpreted as being due to accumulation by the cell of either methylmercury or free $Ca^{2+}$ (Atchison, 1986). The route by which mercury enters the nerve terminals has been controversial somewhat. Experiments in which calcium and sodium channels were blocked by $Ca^{2+}$ and TTX, respectively, have led to the conclusions that mercuric chloride reaches an intracellular site via the sodium and calcium channels, and that methylmercury gains access to an intracellular site via membrane lipid (Miyamoto, 1983). On the contrary, experiments using sodium and calcium channel openers and ionophores have led Atchison (1987) to conclude that methylmercury gains entrance via the calcium channels. More recent experiments in which a variety of inhibitors of mitochondrial function (e.g. dinitrophenol, dicumarol and ruthenium red) and a variety of chemical agents known to disrupt intraterminal $Ca^{2+}$ buffering (e.g. caffeine, N,N-dimethylamino-8-octyl-3,4,5-trimethoxybenzoate and dantrolene) have led to the conclusion that methylmercury induces release of $Ca^{2+}$ from nerve terminal mitochondria thereby causing an increase in MEPP frequency (Levesque and Atchison, 1987, 1988).

One important point that merits attention with respect to the mercury-induced increase in MEPP frequency is the high concentrations required for the effect. In both frog and rat preparations, mercuric chloride or methylmercury needs to be applied at 20-100 µM to observe the effect. Although these neuromuscular preparations can serve as an excellent model for the study of mercury, relevance to the clinical situation still remains to be seen.

## EFFECTS ON NERVE MEMBRANES

When applied externally to the squid giant axon, methylmercury at a

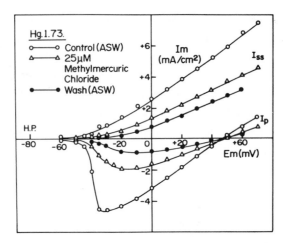

Fig. 3  Current-voltage relationships for peak sodium current ($I_p$) and steady-state potassium current ($I_{ss}$) in a squid giant axon before and during a 15 minute application of 25 μM methylmercury chloride and after washing with normal artificial sea water (ASW) for 90 minutes. Holding potential (H.P.) -80 mV. From Shrivastav et al. (1976).

concentration of 25 μM suppressed both sodium and potassium currents (Figure 3) (Shrivastav et al., 1976). The effects were irreversible after washing with methylmercury-free media, and progressed further during wash-out. The leakage conductance increased to about five times of the control value during a 15 minute application of methylmercury. This effect was also irreversible after washing with drug-free media. The membrane was irreversibly depolarized with a high concentration (500 μM) of methylmercury. The depolarization progressed further after washing with drug-free media. The suppression of sodium channel activity by 25 μM methylmercury is responsible for conduction block.

In frog skeletal muscle, 40 μM inorganic mercury was found to decrease both the resting and action potentials, although methylmercury at the same concentration was devoid of such activity (Juang, 1976). Membrane depolarization was also observed in crab and barnacle muscles following application of inorganic mercury (26-370 μM) or methylmercury (100 μM) (Hift and Schultz, 1976; Marco et al., 1979).

In order to relate the effect of mercury compounds to clinical situations of humans, it is preferable to use mammalian neuronal preparations as experimental materials. Thus we conducted experiments using the mouse neuroblastoma N1E-115 cell line (Quandt et al., 1982). Intracellular microelectrode and suction pipette methods were used to measure membrane potential and ionic currents. Resting potential, resting membrane resistance, and action potential were all decreased by 20-100 μM methylmercury. Even at 1 μM, the action potential was decreased to 57% of the control without change in resting potential. Figure 4 illustrates membrane potential changes in response to a hyperpolarizing current pulse. In control experiments (record A), a hyperpolarizing response is followed by an anode break action potential. Methylmercury at 40 μM depolarizes the membrane by 4 mV and abolishes the anode break action potential (record B). A stronger hyperpolarizing current pulse generates an anode break action potential (record C) which is smaller than that of the control. Voltage clamp experiments revealed that 60 μM methylmercury suppressed the sodium current substantially with a small decrease in the potassium current.

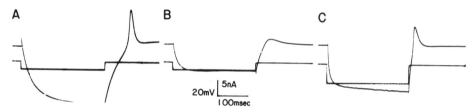

Fig. 4. Effects of methylmercury on the membrane parameters of a neuroblastoma cell, N1E-115 line. Upper trace, potential record; lower trace, current record. A. Control; anode break action potential. B. Six minutes following application of 40 μM methylmercury; the hyperpolarizing response to inward current is decreased, and no anode break action potential is produced. C. Same as B, but the response to a stronger inward current. Temperature 22°C. From Quandt et al. (1982).

## EFFECTS ON NEUROTRANSMITTER-ACTIVATED ION CHANNELS

Whereas methylmercury even at a high concentration of 100 μM did not affect ACh-induced depolarization of the rat end-plate membrane (Atchison and Narahashi, 1982), there was a possibility that the neuronal ACh receptor-channel complex might respond to mercury compounds differently as the latter was known to be pharmacologically different from the muscle end-plate ACh receptors. The mouse neuroblastoma N1E-115 line, which is endowed with at least one nicotinic receptor and two muscarinic receptors (Kato et al., 1983), was used for the experiments (Quandt et al., 1982). The fast nicotinic depolarizing response, and the slow muscarinic depolarizing response were all decreased by 10, 30 and 100 μM

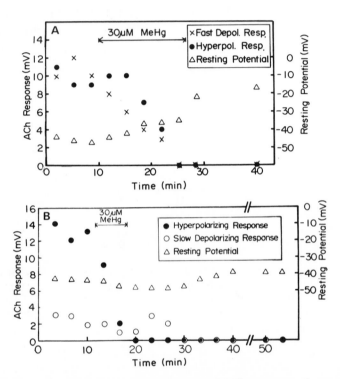

Fig. 5 Effects of methylmercury on three types of ACh responses in neuroblastoma cells, N1E-115 line. A. Decreases in fast nicotinic depolarizing response, muscarinic hyperpolarizing response, and resting potential in 30 μM methylmercury. B. Another cell. Decreases in muscarinic hyperpolarizing response, slow muscarinic depolarizing response, and resting potential in 30 μM methylmercury. Temperature 33°C. From Quandt et al. (1982).

methylmercury before a sizable membrane depolarization took place (Figure 5). These observations are in keeping with those in binding experiments in which methylmercury inhibited binding of agonists to the nicotinic and muscarinic receptors (Eldefrawi et al., 1977; Bondy and Agrawal, 1980; Von Burg et al., 1980; Abd-Elfattah and Shamoo, 1981; Abdallah and Shamoo, 1984).

The effects of methylmercury on ACh responses as described above are not necessarily a non-specific action universal for all transmitter-activated channels. Thus the dopamine-induced current in neuroblastoma cells was not affected by 30 μM methylmercury (Quandt et al., 1982). However, binding of spiroperidol to the dopamine receptor was inhibited by mercuric chloride and methylmercury (Bondy and Agrawal, 1980). The reason for the discrepancy remains to be seen.

Development of patch clamp techniques during the past decade since its first invention by Neher and Sakmann (1976) and its improvement by Hamill et al. (1981) has made it possible to measure any channel activity, in the form of either whole cell currents or single channel currents, from practically any type of cells. Thus it is now possible to examine the effects of mercury compounds on any neuronal channels. During the course of experiments designed to find high affinity target sites of mercury compounds, we have recently found a highly potent and efficacious action of mercuric chloride on the GABA-activated chloride channel (Arakawa et al., 1991).

Whole cell patch clamp techniques (Hamill et al., 1981) were applied to the rat dorsal root ganglion neurons in primary culture to record the GABA-induced current. This current was previously shown to be carried by chloride ions (Ogata et al,. 1988). GABA at a low concentration of 3 μM generated a steady-state inward current at a holding potential of -60 mV. At higher concentrations (30-300 μM), GABA-induced response consisted of a transient current followed by a small steady-state current due to desensitization (Nakahiro et al., 1989).

Mercuric chloride at concentrations of 1-100 μM augmented the transient inward chloride current induced by 30 μM GABA. An example of such an experiment with 10 μM mercuric chloride is illustrated in Figure 6. The transient current was increased to approximately 200% of the control. In addition, mercury generated a slow inward current. Both effects were not reversed after washing with $Hg^{2+}$-free media. The dose-response relationship for $Hg^{2+}$-induced augmentation of GABA-activated chloride current is shown in Figure 7. Mercuric chloride was effective at 1 μM, and a small augmentation was observed even at 0.1 μM.

Methylmercury, even at 100 μM, did not augment the GABA-induced chloride current but rather suppressed it (Figure 8). A slow inward current was produced as in the case of mercuric chloride.

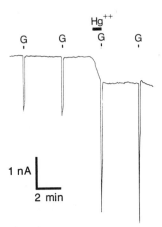

Fig. 6   Mercuric chloride (10 μM) generates a slow inward current and enhances the
         peak current induced by 30 μM GABA (G) in a rat dorsal root ganglion
         neuron. Both effects are irreversible after washing with Hg++-free solution.
         GABA and Hg++ were applied as indicated by bars above the record.
         Temperature 22°C. From Arakawa et al. (1991).

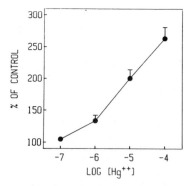

Fig. 7   Dose-response relationship for Hg++-induced enhancement of the peak
         current evoked by 30 μM GABA.  Data are given as mean ± S.D. (n=4~5).
         At  0.1 μM, S.D. is smaller than the size of the symbol. Temperature 22°C.
         From  Arakawa et al. (1990).

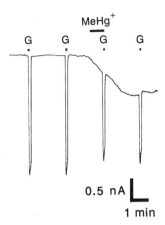

Fig. 8    Methylmercury (100 µM) generates a slow inward current and suppresses the peak current evoked by 30 µM GABA. GABA (G) and MeHg+ were applied as indicated by bars above the record. Temperature 22°C. From Arakawa et al. (1991).

The characteristics of the slow inward current generated by mercuric chloride were studied using various pharmacological agents. Bicuculline and picrotoxin, GABA receptor-channel blockers, blocked the GABA-induced chloride current in the absence and presence of mercuric chloride, but did not abolish the $Hg^{2+}$-induced slow inward current. Thus, the slow inward current is not mediated by the GABA-activated chloride channels. Tetrodotoxin, a sodium channel blocker, did not abolish the $Hg^{2+}$-induced slow inward current, indicating that the current does not go through the sodium channel. Lanthanum, a calcium channel blocker, had no effect on the $Hg^{2+}$-induced slow current. Therefore, the current is not mediated via the calcium channels. When the chloride concentration of the internal solution was reduced from the control level of 142 mM to 20 mM, the reversal potential for the $Hg^{2+}$-induced current was shifted in the direction of depolarization only by 8.4 mV, a magnitude much less than that predicted by the Nernst potential for chloride. Thus, the contribution of chloride ions to the $Hg^{2+}$-induced inward current is small, and an ion or ions other than chloride largely carry the inward current.

Hg$^{2+}$-induced slow inward current resembles inward currents produced by other heavy metals including Pb$^{2+}$, Cd$^{2+}$, Cu$^{2+}$ and Al$^{3+}$ via non-specific cation channels (Weinreich and Wonderlin, 1987; Oortgiesen et al., 1990a,b). Internal calcium ions are also known to generate slow inward currents via non-selective cation channels (Yellen, 1982; Swandulla and Lux, 1985; Partridge and Swandulla, 1988). These slow inward currents might be related to the so-called "leakage current" which is known to be increased by certain heavy metals including mercury compounds (Shrivastav et al., 1976; Juang, 1976; Quandt et al., 1982; Paiement and Joly, 1985).

## DISCUSSION

Whereas mercury compounds exert multiple actions on a variety of receptors and channels, some of the effects are by no means non-specific. The stimulating action of mercuric chloride on the GABA-activated chloride channel is the most potent and efficacious among electrophysiological effects observed in various systems. The current is augmented by as much as 30% by 1 μM mercuric chloride (Fig. 7). Increase in MEPP frequency and suppression of the nerve-evoked EPP are observed in most experiments at 10-100 μM mercuric chloride or methylmercury (Juang and Yonemura, 1975; Manalis and Cooper, 1975; Atchison and Narahashi, 1982; Atchison, 1986, 1987; Traxinger and Atchison, 1987; Levesque and Atchison, 1987, 1988). The sodium and potassium channels are partially blocked by methylmercury at 20-60 μM (Shrivastav et al., 1976; Quandt et al., 1982). Neuronal ACh-activated channels are suppressed by 10-100 μM methylmercury (Quandt et al., 1982). Therefore, the stimulating action on the GABA system is considered to play an important role in generation of neurotoxic symptoms of poisoning with mercury.

Humans chronically intoxicated by methylmercury are known to develop the symptoms of poisoning when the level in the blood and brain reaches 4-80 μM (Skerfving, 1972). More than 99% of methylmercury in the blood is present in red blood cells or bound to plasma protein (Hughes, 1957; Yoshino et al., 1966; Norseth and Clarkson, 1970), and the concentration of free methylmercury in plasma may be a small fraction of the total concentration in blood (Somjen et al., 1973). Furthermore, chronic neurological disorders begin after a delay of several days or longer. Since mercuric chloride exists as a divalent cation and is less hydrophobic than methylmercury, the pharmacodynamics would be considerably different between them.

The GABA receptor-channel complex is inhibitory in function. It remains to be seen exactly what aspect of the symptoms of mercury poisoning is caused by the stimulation of the GABA system. It should be noted that several other classes of drugs are known to augment the GABA-induced response or chloride currents, including the general anesthetics halothane, enflurane and isoflurane (Nakahiro et al., 1989), ethanol and longer-chain alcohols (Nishio and Narahashi, 1990; Nakahiro et al., 1990), barbiturates (Barker and Ransom, 1978; Brown and Constanti, 1978; Macdonald and Barker, 1979), and benzodiazepines (Macdonald and Barker, 1979; Choi et al., 1981; Skerritt and Macdonald, 1984; Bormann and Clapham, 1985; Chan and Farb, 1985).

Methylmercury is known to be converted into $Hg^{2+}$ *in vivo* (Omata et al., 1980). The percentage of conversion varied greatly in different organs; 80% in kidney cytosol at day 48 after single injection, 40% in liver mitochondria at day 26, and less than 3% in brain mitochondria, microsomes and cytrosol at day 15. It appears that mercuric ions play an important role in producing the symptoms of poisoning in methylmercury intoxication.

## ACKNOWLEDGEMENTS

Author's workers quoted in this article were supported by NIH grant ES02330. We thank Vicky James-Houff for secretarial assistance.

## REFERENCES

Abdallah, E.A.M. and Shamoo, A.E., 1984. Protective effect of dimercaptosuccinic acid on methylmercury and mercuric chloride inhibition of rat brain muscarinic acetylcholine receptors. Pesticide Biochem. Physiol. 21:385-393.

Abd-Elfattah, A.S.A., and Shamoo, A.E., 1981, Regeneration of a functionally active rat brain muscarinic receptor by d-penicillamine after inhibition with methylmercury and mercuric chloride. Evidence for essential sulfhydryl groups in muscarinic receptor binding sites, Mol. Pharmacol., 20:492-497.

Arakawa, O., Nakahiro, M., and Narahashi, T., 1991, Mercury modulation of GABA-activated chloride channels and non-specific cation channels in rat dorsal root ganglion neurons, Brain Res.551:58-63.

Atchison, W.D., 1986, Extracellular calcium-dependent and -independent effects of methylmercury on spontaneous and potassium-evoked release of acetylcholine at the neuromuscular junction, J. Pharmacol. Exp. Ther., 237:672-680.

Atchison, W.D., 1987, Effects of activation of sodium and calcium entry on spontaneous release of acetylcholine induced by methylmercury, J. Pharmacol. Exp. Ther., 241: 131-139.

Atchison, W.D., and Narahashi, T., 1982, Methylmercury-induced depression of neuromuscular transmission in the rat, NeuroToxicoloty, 3:37-50.

Bakir, F., Damluji, S.F., Amin-Zaki, L., Murtadha, M., Khalidi, A., Al-Rawi, N.Y., Tikriti, S., Dhahir, H.I., Clarkson, T.W., Smith, J.C., and Doherty, R.A., 1973, Methylmercury poisoning in Iraq, Science, 181:230-240.

Barker, J.L., and Ransom, B. R., 1978, Pentobarbitone pharmacology of mammalian central neurones grown in tissue culture, J. Physiol. (London), 280:355-372.

Barrett, J., Botz, D., and Chang, D.B., 1974, Block of neuromuscular transmission by methylmercury, in: "Behavioral Toxicology, Early Detection of Occupational Hazards," C. Xintaras, B.L. Johnson and I. de Groot, eds., Vol. 5, 277-287, U.S. Dept. of Health, Education and Welfare, Washington.

Berlin, M., Carlson, J., and Norseth, T., 1975, Dose-dependence of methylmercury metabolism, Arch. Environ. Health, 30:307-317.

Bondy, S.C., and Agrawal, A.K., 1980, The inhibition of cerebral high affinity receptor sites by lead and mercury compounds, Arch. Toxicol., 46:249-256.

Bormann, J., and Clapham, D.E., 1985, $\gamma$-Aminobutyric acid receptor channels in adrenal chromaffin cells: A patch-clamps study, Proc. Natl. Acad. Sci. USA, 82:2168-2172.

Brown, D.A., and Constanti, A., 1978, Interaction of pentobarbitone and $\gamma$-aminobutyric acid on mammalian sympathetic ganglion cells, Brit. J. Pharmacol., 63:217-224.

Cavanaugh, J.B., and Chen, F.C.K., 1971, The effects of methyl-mercury-dicyandiamide on the peripheral nerves and spinal cord of rats, Acta Neuropathol., 19:208-215.

Chan, C.Y., and Farb, D.H. 1985, Modulation of neurotransmitter action: Control of the $\gamma$-aminobutyric acid response through the benzodiazepine receptor, J. Neuroscience, 5:2365-2373.

Chang, L.W., 1977, Neurotoxic effects of mercury--a review, Environ. Res., 14:329-373.

Chang, L.W., 1980, Mercury in: "Environmental and Clinical Neurotoxicology," P.S. Spencer and H.H. Schaumburg, eds., 508-526, The Williams and Wilkins, Baltimore.

Chang, L.W., and Hartmann, H.A., 1972, Ultrastructural studies of the nervous system after mercury intoxication. II. Pathological changes in the nerve fibers, Acta Neuropathol., 20:316-331.

Choi, D.W., Farb, D.H., and Fischbach, G.D., 1981, Chlordiazepoxide selectively potentiates GABA conductance of spinal cord and sensory neurons in cell culture, J. Neurophysiol., 45:621-631.

Cooper, G.P., and Manalis, R.S. 1983, Influence of heavy metals on synaptic transmission, NeuroToxicology, 4:69-84.

Cooper, G.P., Suszkiw, J.B., and Manalis, R.S., 1984, Heavy metals: Effects on synaptic transmission, NeuroToxicology, 5:247-266.

Eldefrawi, M.E., Monsour, N.A., and Eldefrawi, A.T., 1977, Interactions of acetylcholine receptors with organic mercury compounds, in: "Membrane Toxicity," M.W. Miller and A.E. Shamoo, eds., 449-463, Plenum, New York.

Fehling, C., Abdulla, M., Brun, A., Dictor, M., Schütz, A., and Skerfving, S., 1975, Methylmercury poisoning in the rat: A combined neurological, chemical and histopathological study, Toxicol. Appl. Pharmacol., 33:27-37.

Hamill, O.P., Marty, A., Neher, E., Sakmann, B., and Sigworth, F.J., 1981, Improved patch-clamp techniques for high-resolution current recording from cells and cell -free membrane patches, Pflügers Arch., 391:85-100.

Herman, S.P., Klein, R., Talley, F.A., and Krigman, M.R., 1973, An ultrastructural study of methylmercury-induced primary sensory neuropathy in the rat, Lab. Invest., 28:104-118.

Hift, H., and Schultz, R., 1976, Methylmercury induced injury of single barnacle muscle fibers, Environmental Res., 11:367-385.

Hughes, W.L., 1957, A physicochemical rationale for the biological activity of mercury and its compounds, Ann. New York Acad. Sci., 65:454-460.

Hunter, D., Bomford, R., and Russell, D.S., 1940, Poisoning by methylmercury compounds, Quart. J. Med., 9:193-241.

Jacobs, J.M., Carmichael, N., and Cavanaugh, J.B., 1975, Ultrastructural changes in the dorsal root and trigeminal ganglia of rats poisoned with methylmercury, Neuropathol. Appl. Neurobiol., 1:1-9.

Juang, M.S., 1976, An electrophysiological study of the action of methylmercuric chloride and mercuric chloride on the sciatic nerve-sartorius muscle preparation of the frog, Toxicol. Appl. Pharmacol., 37:339-348.

Juang, M.S., and Yonemura, K., 1975, Increased spontaneous transmitter release from presynaptic nerve terminal by methylmercuric chloride, Nature (London), 256:211-213.

Kato, E., Anwyl, R., Quandt, F.N., and Narahashi, T., 1983, Acetylcholine-induced electrical responses in neuroblastoma cells, Neuroscience, 8:643-651.

Le Quesne, P.M., Damaluji, S.F., and Rustam, H., 1974, Electrophysiological studies of peripheral nerves in patients with organic mercury poisoning, J. Neurol. Neurosurg. Psychol., 37:333-338.

Levesque, P.C., and Atchison, W.D., 1987, Interactions of mitochondrial inhibitors with methyl mercury on spontaneous quantal release of acetycholine, Toxicol. Appl. Pharmacol., 87:315-324.

Levesque, P.C., and Atchison, W.D., 1988, Effect of alteration of nerve terminal $Ca^{2+}$ regulation on increased spontaneous quantal release of acetylcholine by methylmercury, Toxicol. Appl. Pharmacol., 94:55-65.

Macdonald, R.L., and Barker, J.L., 1979, Enhancement of GABA-mediated postsynaptic inhibition in cultured mammalian spinal cord neurons: A common mode of anticonvulsant action, Brain Res., 167:323-336.

Manalis, R.S., and Cooper, G.P., 1975, Evoked transmitter release increased by inorganic mercury at frog neuromuscular junction, Nature (London), 257:690-691.

Marco, L.A., Isaacson, L., and Torri, J.C., 1979, Effects of mercuric chloride on the resting membrane potentials of blue crab (*Callinectes sapidus*) muscle fibers, Toxicology, 12:41-46.

Misumi, J., 1979, Electrophysiological studies *in vivo* on peripheral nerve function and their application to peripheral neuropathy produced by organic mercury in rats. III. Effects of methylmercuric chloride on compound action potentials in the sciatic and tail nerve in rats, Kumamoto Med. J., 32:15-22.

Miyakawa, T., Deshimaru, M., Sumiyoshi, S., Teraoka, A., Udo, N., Hattori, E., and Tatetsu, S., 1970, Experimental organic mercury poisoning--Pathological changes in peripheral nerves, Acta Neuropathol., 15:45-55.

Miyamoto, M.D., 1983, $Hg^{2+}$ causes neurotoxicity at an intracellular site following entry through Na and Ca channels, Brain Res., 267:375-379.

Nakahiro, M., Yeh, J.Z., Brunner, E., and Narahashi, T., 1989, General anesthetics modulate GABA receptor channel complex in rat dorsal root ganglion neurons, FASEB J., 3:1850-1854.

Nakahiro, M., Arakwara, O., and Narahashi, T., 1991. Modulation of GABA receptor-channel complex by alcohols. J. Pharmacol. Exp. Ther., in press.

Neher, E., and Sakmann, B., 1976, Single-channel currents recorded from membrane of denervated frog muscle fibres, Nature (London), 260:779-802.

Nishio, M., and Narahashi, T. 1990, Ethanol enhancement of GABA-activated chloride current in rat dorsal root ganglion neurons, Brain Res., 518:283-286.

Norseth, T., and Clarkson, T., 1970, Studies on the biotransformation of $^{203}$Hg-labeled methylmercury chloride in rats, Arch. Environ, Health, 21:717-727.

Ogata, N., Vogel, S.M., and Narahashi, T., 1988, Lindane but not deltamethrin blocks a component of GABA-activated chloride channels, FASEB J., 2:2895-2900.

Omata, S., Sato, M., Sakimura, K., and Sugano, H., 1980, Time-dependent accumulation of inorganic mercury in subcellular fractions of kidney, liver, and brain of rats exposed to methylmercury, Arch. Toxicol., 44:231-241.

Oortgiesen, M., van Kleef, R.G.D.M., and Vijverberg, H.P.M., 1990a, Novel type of ion channel activated by $Pb^{2+}$, $Cd^{2+}$ and $Al^{3+}$ in cultured mouse neuroblastoma cells, J. Membrane Biol., 113:261-268.

Oortgiesen, M., van Kleef, R.G.D.M., Bajnath, R.B., and Vijverberg, H.P.M., 1990b, Nanomolar concentrations of lead selectively block neuronal nicotinic acetylcholine responses in mouse neuroblastoma cells, Toxicol. Appl. Pharmacol. 103:165-174.

Paiement, J., and Joly, L. P., 1985, Effect of organic mercury on the electrical resistance of phosphatidylserine bilayers, Biochim. Biophys. Acta, 816:179-181.

Partridge, L.D., and Swandula, D., 1988, Calcium-activated non-specific cation channels, Trends in Neurosciences, 11:69-72.

Quandt, F.N., Kato, E., and Narahashi, T., 1982, Effects of methylmercury on electrical responses of neuroblastoma cells, NeuroToxicology, 3:205-220.

Rustam, H., Von Burg, R., Amin-Zaki, L., and El Hassani, S., 1975, Evidence for a neuromuscular disorder in methylmercury poisoning, Arch. Environ. Health, 30:190-195.

Shrivastav, B.B., Brodwick, M.S., and Narahashi, T., 1976, Methylmercury: Effects on electrical properties of squid axon membranes, Life Sci., 18:1077-1082.

Skerfving, S., 1972, Organic mercury compounds. Relation between exposure and effects, in "Mercury in the Environment," L. Friberg and J. Vostal, eds., 141-168, CRC Press, Cleveland.

Skerritt, J.H., and Macdonald, R.L., 1984, Diazepam enhances the action but not the binding of the GABA analog THIP, Brain Res., 297:181-186.

Snyder, R.D., and Seelinge, D.F., 1976, Methylmercury poisoning clinical follow up and sensory nerve conduction studies, J. Neurol. Neurosurg. Psychol., 39:701-704.

Somjen, G.G., Herman, S.P., and Klein, R., 1973, Electrophysiology of methyl mercury poisoning, J. Pharmacol. Exp. Ther., 186:579-592.

Swandulla, D., and Lux, H.D., 1985, Activation of a nonspecific cation conductance by intracellular $Ca^{2+}$ elevation in bursting pacemaker neurons of *Helix pomatia*, J. Neurophysiol., 54:1430-1444.

Takeuchi, T., Morikawa, N., Matsumoto, H., and Shiraishi, Y., 1962, A pathological study of Minamata disease in Japan, Acta Neuropathol., 2:40-57.

Takeuchi, T., Matsumoto, H., Sasaki, M., Kambara, T., Shiraishi, Y., Hirata, Y., Nobuhiro, M.,and Ito, H., 1968, Pathology of Minamata disease, Kumamato Med. J., 34:521-524.

Traxinger, D.L., and Atchison, W.D., 1987, Comparative effects of divalent cations on the methylmercury-induced alterations of acetycholine release, J. Pharmacol. Exp. Ther., 240:451-459.

Von Burg, R., and Rustam, H., 1974, Conduction velocities in methylmercury poisoned patients, Bull. Environ. Contam. Toxicol., 12:81-85.

Von Burg, R., Northington, F.K., and Shamoo, A., 1980, Methylmercury inhibition of rat brain muscarinic receptors, Toxicol. Appl. Pharmacol., 53:285-292.

Weinreich, D., and Wonderlin, W.F., 1987, Copper activates a unique inward current in molluscan neurones, J. Physiol. (London), 394:429-443.

Yellen, G., 1982, Single $Ca_2$-activated nonselective cation channels in neuroblastoma, Nature (London) 296:357-359.

Yoshino, Y., Mozai, T., and Nakao, K., 1966, Distribution of mercury in the brain and its subcellular units in experimental organic mercury poisonings, J. Neurochem., 13:397-406.

# ROLE OF OXIDATIVE INJURY IN THE PATHOGENESIS OF METHYLMERCURY NEUROTOXICITY

M. Anthony Verity and Ted Sarafian

Department of Pathology (Neuropathology)
  and Brain Research Institute
UCLA School of Medicine
Los Angeles, California

## ABSTRACT

We have previously demonstrated (J. Neuropath. Exp. Neurol. 48:1-10, 1989) that the pathogenesis of methylmercury (MeHg) induced cytotoxicity in suspensions of cerebellar granule neurons was not strictly coupled to the reduction of ATP or combined inhibition of ATP or macromolecule synthesis, but suggested a component of free-radical injury. The present studies reveal that MeHg initiates a dose- and time-dependent lipoperoxidation measured as malonaldehyde generation or induction of a 2',7'-dichlorofluorescein (DCFA) signal representing oxygen radical species generation. A simultaneous decline in GSH occurred. Partial protection was given by EGTA (a $Ca^{2+}$ chelator) or desferoxamine ($Fe^{2+}$ chelation with inhibition of the Fenton reaction and OH radical production). However, no cytoprotection was found with alpha-tocopherol succinate although significant inhibition of lipoperoxidation was observed. Analogous experiments in cerebellar granule cell culture revealed a dose-dependent (0.5-5µM) increase in the specific activity of GSH accompanied by increased lipoperoxidation and neuronal cell injury. Such paradoxical induction of GSH occurred in glial cells, whose endogenous content was higher than that of neuron culture. Inhibition of gamma-glutamyl cysteine synthetase by buthionine sulfoximine (BSO) lowered cellular GSH and strongly potentiated

MeHg-induced lipoperoxidation and cell death in neuron culture but had minimal effect in glial culture.

It is likely that activated oxygen species and lipoperoxidation significantly contribute to the pathogenesis of alkylmercury induced injury but are not singly causal to the final cytotoxic event. Other processes, especially intracellular protein degradation of -SH sensitive proteins or permeases, modification of cytoskeletal proteins, activation of phospholipase $A_2$ and activation of protein kinase C contribute to the final lethal event.

## INTRODUCTION

Since Ganther et al. (1972) revealed a protective effect of selenium (Se) against methylmercury (MeHg) toxicity there has been interest in the mechanism of Se protection, the role of Se in fish known to contain high levels of MeHg (Nishigaki et al., 1974) and experimental confirmation of such Se protection both *in vivo* (Chang and Suber, 1982; Ohi et al., 1976; Iwata et al., 1973) and in *in vitro* preparations, e.g. tissue culture (Potter and Matrone, 1977; Kasuya, 1976). These studies, culminating in the Ganther hypothesis (1978, 1980) suggested that organomercury compounds may be converted to free radicals, in turn causing hydrogen (H) abstraction or alkylation of specific target molecules. Further evidence was forthcoming for the role of oxidant or free radical injury in the pathogenesis of MeHg neurotoxicity. For instance, alpha-tocopherol administration protected against MeHg toxicity both *in vivo* (Welsh and Soares, 1976; Chang et al., 1978) and in tissue culture (Kasuya, 1975; Prasad and Ramanujam, 1980) likely due to an antioxidant effect on lipoperoxidation (inhibition of the propagative process) or via the maintenance of protein - SH potential even in the face of GSH reduction (Pascoe and Reed, 1989). Moreover, MeHg toxicity was associated with a reduction in glutathione peroxidase activity (Hirota et al., 1980; Chang and Suber, 1982) a selenium containing enzyme central to cellular antioxidant activities and the GSH-redox cycle.

However, direct tests of the Ganther hypothesis were not made as the majority of studies utilized *in vivo* preparations with the difficulty of identifying causal steps in the oxidant injury pathway; the lack of direct evidence of oxidant mediated cell injury; failure to demonstrate free radical species; and the failure to measure lipoperoxidation of brain membrane lipids considered causal to the cytotoxic event. However, considerable advantage may be gained by the use of a neuronal cell suspension or a neuronal tissue culture. The dissociated neonatal cerebellar granule cell system fulfills many advantages by providing a nearly homogeneous population of neurons (Messer, 1977; Sarafian and Verity, 1986) for acute *in vitro* experiments in suspension or as starting material for cerebellar granule cell culture.

Studies of MeHg induced neurotoxicity using cerebellar granule cells in suspension

We have utilized an *in vitro* model of cerebellar granule cell suspensions to investigate the early cytotoxic events leading to MeHg-induced neuronal cell death. This model has allowed for assessment of the detailed relationship between MeHg impairment of cellular energetics, macromolecular synthesis (Sarafian et al., 1984; Sarafian and Verity, 1985; Sarafian and Verity, 1986) and cytotoxicity (Sarafian et al., 1989). Using this *in vitro* preparation, we have observed that MeHg induced a concentration dependent increase in membrane lipoperoxidation assessed by the spectrofluorometric assay of Yagi (1976) for TBA-reactive substances (Figure 1). The onset of neuronal cytotoxicity was assayed using trypan blue permeability change which closely followed the generation of lipoperoxides. In addition to the increased lipoperoxidation, 5-20 μM MeHg also produced a substantial

Table 1

EFFECT OF DEFEROXAMINE AND CHELATING AGENTS

ON MeHg INDUCED LIPOPEROXIDATION AND GSH

Assays performed after 3 hr incubation, 37° C.
values expressed as % change from control

|  | VIABILITY | GSH | LIPOPEROXIDATION |
|---|---|---|---|
| MeHg, 20 μM (9) | $14 \pm 5$ | $20 \pm 4$ | $176 \pm 13$ |
| MeHg, 20 μM + deferoxamine, 2.5 mM (5) | $64 \pm 12$ | $34 \pm 8$ | $64 \pm 6$ |
| MeHg, 20 μM + EGTA, 1 mM (3) | $37 \pm 3$ | $32 \pm 8$ | $60 \pm 12$ |

decline in the intracellular level of reduced glutathione (GSH) which was detected immediately upon addition of MeHg to the neuronal suspension (Table 1). These studies confirmed the dose-dependent association between MeHg-induced cytotoxicity, GSH depletion and the genesis of oxidant injury as manifested by membrane lipoperoxidation.

Recently, the intracellular generation of oxygen radicals has been demonstrated by the use of 2',7'-dichlorofluorescein diacetate (DCFA). This compound is oxidized by reactive oxygen species to a fluorescent product. We have used this principle to confirm the lipoperoxidation data. Cells were loaded with DCFA, washed and suspended in various concentrations of MeHg. Figure 2 reveals a dose-dependent increase in fluorescent signal generation in the presence of MeHg coupled to cytotoxicity. Of note is the apparent latent period when significant cytotoxicity, especially at 20 μM MeHg, occurs in the absence of signal generation above control. This experiment reveals that increased signal does not precede the cytotoxic event but likely accompanies or occurs later in the genesis of oxidant injury.

Critical to the Ganther hypothesis is the establishment of lipoperoxidation as a causal event in the evolution of cytotoxicity. Could the accelerated lipoperoxidation signal reflect an epiphenomenon associated with cell death *per se*? This is unlikely as control studies of neurons subjected to hypo-osmolar stress producing an equivalent degree of cytotoxicity were not associated with the production of thiobarbituric acid-reactive substance. In view of the *in vivo* studies revealing alpha-tocopherol protection of MeHg toxicity, alpha-tocopherol succinate was added to the neuronal suspension and found to significantly inhibit lipoperoxidation in both control and MeHg stressed cells (Table 2). However, only minor, non-statistical significant protection of cytotoxicity, as measured by trypan blue, was observed after a 3 hr. exposure period. Similarly, EDTA significantly blocked lipoperoxidation without change in cytotoxicity. However, exogenous glutathione and the $Fe^{2+}$-chelator desferoxamine proved strongly effective in both maintaining viability and blocking MeHg-induced lipoperoxidation (Table 1). However, mannitol, catalase and/or superoxide dismutase were ineffective confirming the need for an intracellular compartment in the generation of free radicals, ·OH and lipoperoxidation versus non-specific exogenous events mediated by contiguous cell necrosis. In summary, these studies confirm the generation of a lipoperoxide signal but reveal a dissociation of such signal from cytotoxicity. Similar observations have been made by Stacey and Klaassen (1981) and Stacey and Kappus (1982) who revealed inhibtion of lipid peroxidation without prevention of cellular injury in isolated rat hepatocytes and confirmed the generation of lipoperoxides by mercury. Also Ichikawa et al. (1987) revealed that the peroxidation of erythrocyte membrane lipids was not the cause of hemolysis induced by either mercury or MeHg.

As noted above, the $Ca^{2+}$ chelator EGTA, was partially protective against MeHg-induced cytotoxicity and blocked lipoperoxidation. This observation suggests that $Ca^{2+}$ flux and/or intracellular $[Ca^{2+}]$ may play a role in MeHg neurotoxicity. For instance, Kauppinen et al. (1989) revealed an increase in cytosolic $[Ca^{2+}]$ induced by 30 μM MeHg in cerebrocortical synaptosomes. Such increase was associated with mitochondrial

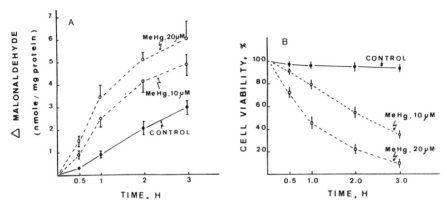

Fig. 1 A. Increase in lipoperoxidation, measured as malonaldehyde equivalent occuring in freshly prepared cerebellar granule cell suspensions incubated in control, 10 μM or 20 μM MeHg. B. Correlation of decline in cell viability as measured by trypan blue exclusion in suspension in the presence of 10 or 20 μM MeHg.

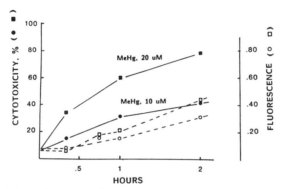

Fig. 2 Correlation of cytotoxicity measured by trypan blue exclusion in cerebellar granule cell suspensions with generation of fluorescent signal from dichlorofluorescein acetate representing intracellular production of activated oxygen species. Note: Dose-response of fluorescent signal and 30' latent period during which early, measurable cytotoxicity was determined.

depolarization suggesting that the rise in $[Ca^{2+}]$ was due to mitochondrial energy failure and uncoupling (Verity et al., 1975) as opposed to increased influx, whch was noted to occur at 100 μM [MeHg]. Chavez and Holguin (1988) also showed that mercury can cause release of mitochondrial $Ca^{2+}$. These observations assume importance in view of the studies of Braughler et al. (1985) who demonstrated $Ca^{2+}$ enhancement of free radical induced damage in synaptosomes and cultured spinal cord neurons. Moreover, Jones et al.

Table 2

EFFECT OF α-TOCOPHEROL SUCCINATE ON MeHg-INDUCED CELL INJURY

|  | MALONALDEHYDE nmol/mg protein | VIABILITY (%) |
|---|---|---|
| CONTROL | 2.8 ± 0.3 (6) | 94.0 ± 2 (8) |
| CONTROL + α-TOCOPHEROL, 25 μM | 1.3 ± 0.4 (6) | 96.1 ± 2 (8) |
| MeHg, 10 μM | 4.8 ± 0.5 (7) | 37.1 ± 8 (7) |
| MeHg (10 μM) + α-TOCOPHEROL, 25 μM | 2.6 ± 0.3 (8) | 48.2 ± 5 (8) |

Values represent means + SEM (number of experiments). Values for lipoperoxidation are expressed as the increase in malonaldehyde content during 3 hr incubation, 37°C. Viability was determined using the method of trypan blue exclusion.

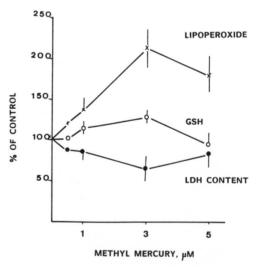

Fig. 3  Effect of MeHg, 0.5-5 μM on lipoperoxide generation (malonaldehyde equivalent,) reduced gluathione (GSH) and LDH content of cerebellar granule cells in culture.  Assays were performed on individual, 12-well cultures following 24h incubation with MEHg.  Values represent mean±SEM (7-12).  Note the significant elevation of GSH at 1 and 3μM MeHg in the face of significant lipoperoxide generation.

(1983) clearly demonstrated that microsomal ATP-dependent $Ca^{2+}$ sequestration was highly sensitive to oxidative damage and that GSH protected against such damage. These authors contended that inactivation of $Ca^{2+}$ sequestration may be due to a direct effect of a free radical product on the $Ca^{2+}$-pump protein and not mediated through a generalized peroxidation of membrane lipids. These studies attest to the likely potentiation of minor oxidative injury by elevated cytoplasmic $Ca^{2+}$. Hence, MeHg-induced cytotoxicity and lethal injury results from the combination and synergetic interaction of changes in intracellular $Ca^{2+}$ homeostasis and disruption of intracellular -SH status culminating in lipoperoxidation, activation of $Ca^{2+}$-dependent proteolysis, endonuclease activation and phospholipid hydrolysis.


Studies in cerebellar granule cell culture and glial culture

While cerebellar granule neurons in suspension provided significant data following acute MeHg exposure times, we wished to evaluate a subacute or chronic exposure model at lower [MeHg] with respect to lipoperoxide formation, glutathione loss and cytotoxicity. We utilized analogous cerebellar granule cultures prepared from 6-day old Sprague-Dawley rats (Sarafian and Verity, 1986; Weir et al., 1990). Both trypan blue exclusion assay and lactate dehydrogenase release were used for quantitation of neuronal cell survival. Addition of 0.5-5 µM MeHg to cerebellar granule cell cultures induced dose-dependent cytotoxicity at 24h assessed by the trypan blue exclusion assay or residual lactate dehydrogenase and increased lipoperoxidation in agreement with the neuronal suspension experiments (Figure 3).

Paradoxically, the GSH content of the culture or the specific activity of GSH (nmole/mg protein) showed a significant increase especially at 2-3 µM MeHg. This observation suggested that neuronal induction of GSH by MeHg was simultaneous with ongoing lipoperoxidation and cytotoxicity which appeared unlikely and other hypotheses were suggested to account for the data. For instance, a selective loss of neurons with low GSH specific activity (nmole/mg protein) would result in a residual, "protected" population of neurons with high GSH specific activity. Alternatively, the preferential loss of cytoplasmic protein, e.g. LDH, with retention of cytoplasmic GSH would provide similar data. This hypothesis appears unlikely in view of the low molecular weight of GSH, its predominant cytoplasmic location and ease of transmembrane flux with early membrane damage. Finally, MeHg may have selectively induced GSH in a "second" cell population

Table 3

COMPARISON OF ENDOGENOUS GSH CONTENT AND MeHg-INDUCED

GSH IN CEREBELLAR NEURONAL AND GLIAL DERIVED CULTURE

Cultures incubated with 1 µM MeHg, 24 hrs. Washed,
GSH and protein content assayed. Means ± SEM (n).

| GSH | | |
|---|---|---|
| | (n.mole/mg protein) | |
| CULTURE | ENDOGENOUS | 1 µM MeHg |
| NEURONAL | 18.8 ± 1.3 (11) | 19.6 ± 1.7 (12) |
| GLIA | 24.1 ± 4.3 (11) | 31.7 ± 4 (12)* |

* Student t-test, p <.01.

Table 4

EFFECT OF BUTHIONINE SULFOXIMINE (BSO)

ON GSH CONTENT IN CEREBELLAR NEURONE AND GLIAL CULTURE

| | GSH | |
|---|---|---|
| | (n.moles.mg protein $^{-1}$) | |
| | NEURONE | GLIA |
| CONTROL | 23.6 ± 5.6 | 30.4 ± 7.6 |
| BSO, 100-300 µM | 14.3 ± 2.4* | 17.2 ± 2.9* |
| MeHg + BSO | 7.3 ± 2.2* | 9.6 ± 2.6* |

*Significant Student t-test from appropriate control, p <.01.

Table 5

BSO-INDUCED GSH DEPLETION POTENTIATES MeHg NEUROTOXICITY

| | NEURONE | | GLIA | |
|---|---|---|---|---|
| ADDITION | GSH | CYTOTOXICITY | GSH | CYTOTOXICITY |
| BSO | 61 | 25 ± 7 | 57 | 6 ± 2 |
| MeHg | 101 | 35 ± 12 | 133 | 5 ± 4 |
| BSO + MeHg | 31 | 83 ± 9 | 32 | 19 ± 9 |

Cultures treated as previously. Cytotoxicity assayed by trypan blue exclusion or LDH
release. GSH expressed as percent of control (neuron 23.6; glia 30.4 n.moles/mg protein).
Cytotoxicity represents percent cell death (control cultures 3-7%).

not involved in lipoperoxide generation. This hypothesis appeared attractive as the cerebellar granule cell culture routinely contains 5-8% glial cells. To examine this hypothesis, we prepared a glial culture from the cerebellum and compared the effect of 1 μM MeHg on the GSH content in both neuronal and glial culture following treatment for 24h (Table 3).

This experiment confirmed the higher endogenous specific activity of GSH in the glial culture (24.1 ± 4.3 nmoles/mg protein) and confirmed our hypothesis that the mercurial induction of GSH in the glial culture was higher than the neuronal culture. Hence, MeHg-induction of GSH in the neuronal culture can be accounted for by selective induction in the 5-8% glial cells. Further, the specific activity of GSH in glial cells is significantly greater than neurons.

This latter observation suggests the interesting hypothesis that the apparent preferential neuronal sensitivity to MeHg is a function of the low endogenous GSH content and/or the lack of ability of the neuron to synthesize GSH in the presence of mercurial stress. We have examined this question further. Glutathione is known to play an important role in protecting cells against the destructive effects of reactive oxygen intermediates and free radicals (Meister, 1983) and our hypothesis couples free radical production, membrane lipoperoxidation and elevated cytoplasmic $[Ca^{2+}]$ as cooperative events in the pathogenesis of mercurial neurocytotoxicity. Selective depletion of GSH in a variety of cell systems increases their sensitivity to oxidant damage (Dethmers and Meister, 1981; Tsan et al., 1985). Treatment of the neuronal or glial culture with 100-300 μM buthionine sulfoximine (BSO), an irreversible inhibitor of gamma-glutamyl cysteine synthetase (Griffith and Meister, 1979) depleted the glutathione content of both cultures approximately 50% (Table 4). However, the depletion of intracellular GSH by buthionine sulfoximine markedly sensitized the neuronal culture to lethal cytotoxicity by MeHg with a further decline in GSH (Table 5).

In contrast, only minor cytotoxicity was revealed in the glial culture under similar conditions of dose and time. While the BSO induced reduction in GSH in the neuronal culture was associated with increased lipoperoxidation and cytotoxicity even in the absence of MeHg, no increased cytotoxicity or lipoperoxidation was evident in the glial culture under similar conditions, although the percent reduction in GSH content was similar in both cultures. A further observation is of interest. While the mechanism of MeHg-induced stimulation of GSH in the glial cell is unknown, it is noted that such induction was inhibited by preincubation with BSO in the presence of MeHg. It is likely, therefore, that the apparent increase of intracellular GSH induced by MeHg utilizes a system in which gamma-glutamylcysteine synthetase provides a key step. It is reasonable to state also that MeHg itself does not inhibit gamma-glutamylcysteine synthetase on the basis of these observations.

We have indicated that glutathione functions in the protection of mammalian cells against oxidative damage. Studies have also shown that cellular protection from oxidant or irradiation injury may be enhanced by increasing the level of intracellular glutathione (Williamson and Meister, 1981; Anderson and Meister, 1987). However, examination of MeHg and/or BSO induced glutathione depletion reveals a significant difference in cytotoxicity at apparent equivalent levels of intracellular glutathione. Expressed differently, the present studies of *in vitro* culture do not reveal a correlation between GSH content and cytoprotection or cytotoxicity. Presumably, a decreased level of cellular GSH turns on glutathione synthesis during the evolution of oxidant injury ultimately limited by the available amino acid substrates, ATP, NADPH and the enzymes involved. Thus, the capacity of mammalian cells to synthesize glutathione in the presence of challenge (e.g. MeHg) may be of key importance and of greater significance than the level of glutathione within the cell at the time of challenge. Of interest in this respect are the experiments of Moore et al. (1989) who revealed an increased capacity for glutathione synthesis enhanced the resistance to irradiation in *E. coli*. Although the gene-enriched strain had a higher glutathione content than the wild strain, the observed radioresistance was associated with an increased capacity of the gene-enriched strain to synthesize glutathione when irradiated rather than the cellular level of glutathione *per se*. Such resistance was abolished in the presence of BSO, very analogous to the MeHg observations.

The neuronal suspension studies revealed an apparent dissociation between cytotoxicity and membrane lipoperoxidation measured by TBA reactive substances, or the intracellular generation of oxygen reactive species as measured by the signal generated from cells labeled with 2',7'-dichlorofluorescein diacetate. For instance, the inhibition of MeHg-induced lipoperoxidation by alpha-tocopherol was without effect on cytotoxicity. However, cellular lipoperoxidation injury is often associated with and follows the well recognized stimulation of intracellular proteolysis (Davies and Goldberg, 1987; Davies and Delsignore, 1987). Various oxygen radical generating systems promote early protein degradation (or denaturation) in red cells at a time when lipoperoxidation had not been initiated (Reglinski et al., 1988). Moreover, the addition of various antioxidants decreased lipid peroxidation without affecting proteolysis. These studies emphasize that cellular proteins, especially plasmalemmel or organelle proteins may be especially important targets of radical damage and specific sites initiating lethal ion fluxes.

CONCLUSION

Numerous potentially toxic mechanisms are activated when cells are exposed to agents which initiate oxidative stress commonly measured as lipoperoxidation. There is a strong association between MeHg induced neuronal cytotoxicity and lipoperoxidation also associated with oxygen radical formation as measured by the 2',7'-dichlorofluorescein diacetate technique. While partial protection against mercurial injury was found with desferoxamine, known to block ·OH formation through the chelation of iron and inhibition of the Fenton reaction, alpha-tocopherol was ineffective likely due to its inhibition of lipoperoxide propagation within the membrane with generation of intracellular thiol redox change and/or selective sulfhydryl protein denaturation. Certainly our studies support the view that MeHg induced cytotoxic injury is relatively neuron specific due to low endogenous glutathione content compared to the highly inducible system in astrocytes. More specifically, the thiol-$Ca^{2+}$ hypothesis of MeHg induced oxidative neuronal cytotoxicity strongly suggests a primary membrane protein degradation step as a likely proximate target in the evolution of the cytotoxic state.

The contribution of the mercurial *per se*, the oxidant radical species or the sustained rise in cytosolic $Ca^{2+}$ in mediating the intracellular events leading to the final cytotoxic pathway are unclear. Evidence has been presented elsewhere that $Hg^{2+}$ may behave as a $Ca^{2+}$-mimetic agent (Shier and DuBourdieu, 1983). in this respect mimicking the effects of elevated cytosolic $Ca^{2+}$. It is not difficult to envisage synergism between the lipophilic alkyl mercurial, direct-SH group inactivation and cytoplasmic $Ca^{2+}$, all contributing to activation of degradative processes previously considered to underly the mechanism of $Ca^{2+}$-mediated cytotoxicity. Such processes may included intracellular protein degradation especially of thiol-sensitive proteases, modification of cytoskeletal proteins, activation of protein kinase C via phospholipase/protease involvement or the activation of an endogenous endonuclease (Duke et al., 1983; Orrenius et al., 1988).

In a study of the pathogenesis of rapid cell death induced by MeHg in suspension of cerebellar granule neurons (Sarafian et al., 1989)., we demonstrated that the genesis of such cytotoxicity was not strictly coupled to MeHg reduction of cytoplasmic ATP and/or combined inhibition of macromolecule synthesis by cyclohexamide or actinomycin D. However, comparable rates of cytotoxicity were obtained in the presence of free-radical generating systems suggesting that oxidative or free radical generation was capable of reproducing the temporal pattern of neuronal cell destruction manifested by MeHg.

The present studies confirm that MeHg does initiate lipoperoxidation (TBA) or 2'-7'dichlorofluorescin signal associated with the onset and magnitude of cytotoxicity. However, the lipoperoxide signal *per se* does not appear causally related to cellular injury but it is likely that under the conditions of oxidative stress, a distruption of intracellular $Ca^{2+}$ homeostasis, -SH depletion and perturbation of glutathione cycling will occur. Such phenomena will rapidly lead to the activation of phospholipases, proteases and endonucleases leading to irreversible neuronal injury. It appears that the perceived preferential sensitivity of the neuron to such mercurial injury may be related to the lack of efficient GSH redox activity and/or the absolute glutathione content or GSH synthesis. Glutathione cycling in the astrocyte appears poised for more efficient handling of MeHg and providing appropriate protection to the neuron.

## REFERENCES

Anderson, M.E. and Meister, A., 1987, Intracellular delivery of cysteine. Methods Enzymol. 143:313.

Braughler, J.M. Duncan, L.A. and Goodman, T., 1985, Calcium enhances *in vitro* free radical-induced damge to brain synaptosomes, mitochondria and cultured spinal cord neurons. J. Neurochem. 45:1288.

Chang, L.W., Gilbert, M. and Sprecher, J., 1978, Modificaiton of methyl mercury neurotoxicity by vitamin E. Environ. Res. 17:356.

Chang, L.W. and Suber, R., 1982, Protective effect of selenium on methyl mercury toxicity: A possible mechanism. Bull. Environ. Contam. Toxciol. 29:285.

Chavez, E. and Holguin, J.A., 1988, Mitochondrial calcium release as induced by mercuric ion. J. Biol. Chem. 263:3582.

Davies, K.J.A. and Goldberg, A.L., 1987. Oxygen radicals stimulate intracellular proteolysis and lipid peroxidation by independent mechanism in erythrocytes. J. Biol. Chem. 262:8220.

Davies, K.J.A. and Delsignore, M.E., 1987. Protein damage and degradation by oxygen radicals. III. Modification of secondary and tertiary structure. J. Biol. Chem. 262:9908.

Dethmers, J.K. and Meister, A., 1981. Glutathione export by human lymphoid cells: Depletion of glutathione by inhibition of its synthesis decreases export and increases sensitivity to irradiation. Proc. Natl. Acad. Sci. U.S.A. 78:7492.

Duke, R.C., Chervenak, R. and Cohen, J.J., 1983. Endogenous endonuclease-induced DNA fragmentation: An early event in cell-mediated cytolysis. Proc. Natl. Acad. Sci. U.S.A. 80:6361.

Ganther, H.E., Goudie, C., Sunde, M.L., Kopecky, M.J., Wagner, R., Oh, S-H., and Hoekstra, W.G., 1972. Selenium: Relation to decreased toxicity of methyl mercury added to diets containing tuna. Science 75:1122.

Ganther, H.E., 1978. Methylmercury toxicity and metabolism by selenium and vitamin E: Possible mechanism. Environ. Health Perspect. 25:71.

Ganther, H.E., 1980. Interactions of vitamin E and selenium with mercury and silver. Ann. N.Y. Acad. Sci. 355:212.

Griffith, O.W. and Meister, A., 1979. Potent and specific inhibition of glutathione synthesis by buthionine sulfoximine (S-n-butyl homocysteine sulfoximine). J. Biol. chem. 254:7558.

Hirota, Y., Yamguchi, S., Shimojoh, N. and Sano, K-I, 1980. Inhibitory effect of methylmercury on the activity of glutathione peroxidase. Toxicol. Appl. Phrmacol. 53:174.

Ichikawa, H., Ronowicz, K., Hicks, M. and Gebicki, J.M., 1987. Lipid peroxidation is not the cause of lysis of human erythrocytes exposed to inorganic or methylmercury. Arch. Biochem. Biophys.259:45.

Iwata, H., Okamoto, H. and Ohsawa, Y., 1973. Effect of selenium on methyl mercury poisoning. Res. Commun. Pathol. Pharmacol. 5:673.

Jones, D.P., Thor, H., Smith, M.T., Jewell, S.A. and Orrenius, S., 1983. Inhibition of ATP-dependent microsomal calcium ion sequestration during oxidative stress and its prevention by glutathione. J. Biol Chem. 258:6390.

Kasuya, M., 1975. The effect of vitamin E on the toxicity of alkyl mercurials on nervous tissue in culture. Toxicol. Appl. Pharmacol. 32:347.

Kasuya, M., 1976. Effect of selenium on the toxicity of methylmercury on nervous tissue in culture. Toxicol. Appl. Pharmacol. 35:11.

Kauppinen, R.A., Komulainen, H. and Taipale, H., 1989. Cellular mechanisms underlying the increase in cytosolic free calcium concentration induced by methyl-mercury in cerebrocortical synaptosomes from guinea pig. J. Pharmacol. Exp. Ther. 248:1248.

Meister, A., 1983. Selective modification of glutathione metabolism. Science 220:4761.

Messer, A., 1977. The maintenance and identification of mouse cerebellar granule cells in monolayer culture. Brain Res. 130:1.

Moore, W.R., Anderson, M.E., Meister, A., Murata, K., and Kimura, A., 1989. Increased capacity for glutathione synthesis enhances resistance to radiation in *Escherichia coli:* A possible model for mammalian cell protection. Proc. Natl. Acad. Sci. U.S.A. 86:1461.

Nishigaki, S., Kamura, Y., Maki, T., Yamada, H., Shimamura, Y., Ochiai, S. and Kimura, Y, 1974. Mercury-selenium correlations in connection with body weight in muscle of sea fish. Ann. Rep. Tokyo Mat. Res. Lab. Pub.Health 25:235.

Ohi, G., Nishigaki, S., Seki, H., Kamura, Y., Maki, T., Konno, H., Ochiai, S., Yamada, H., Shimamura, Y., Mizoguchi, I. and Yagyu, H., 1976. Efficacy of selenium in tuna and selenite in modifying methylmercury intoxicaiton. Environ. Res. 12:49.

Orrenius, S., McConkey, D.J. and Nicotera, P., 1988. Mechanisms of oxidant-induced cell damage. In: Oxy-radicals in Molecular Biology and Pathology. Alan R. Liss, Inc., pp. 327-339.

Pascoe, G.A. and Reed, D.J., 1989. Cell calcium, vitamin E and the thiol redox system in cytotoxicity. In: Free Radical Biology and Medicine 6:209.

Potter, S.D. and Matrone, G., 1977. A tissue culture model for mercury-selenium interactions. Toxicol. Appl. Pharmacol. 40:201.

Prasad, K.N. and Ramanujam, S., 1980. Vitamin E and vitamin C alter the effect of methylmercuric chloride on neuroblastoma and glioma cells in culture. Environ. Res 21:343.

Reglinski, J., Hoey, S., Smith, W.E. and Sturrock, R.D., 1988. Cellular reponse to oxidative stress at sulfydryl group receptor sites on the erythrocyte membrane. J. Biol. Chem. 263:12360.

Sarafian, T.A., Cheung, M.K. and Verity, M.A., 1984. *In vitro* methyl mercury inhibition of protein synthesis in neonatal cerebellar perikarya. Neuropathol. Appl. Neurobiol. 10:85.

Sarafian, T. and Verity, M.A., 1985. Inhibition of RNA and protein synthesis in isolated cerebellar cells by *in vitro* and *in vivo* methyl mercury. Neurochem. Pathol. 3:27.

Sarafian, T. and Verity, M.A., 1986. Influence of thyroid hormones on rat cerebellar cell aggregation and survival in culture. Devel. Brain Res. 26:261.

Sarafian, T. and Verity, M.A., 1986. Mechanism of apparent transcription inhibition by methylmercury in cerebellar neurons. J. Neurochem. 47:625.

Sarafian, T., Hagler, J., Vartavarian, L. and Verity, M.A., 1989. Rapid cell death induced by methylmercury in suspension of cerebellar granule neurons. J. Neuropathol. Exp. Neurol. 48:1.

Shier, W.T. and DuBourdieu, D.J., 1983. Stimulation of phospholipid hydrolysis and cell death by mercuric chloride: Evidence for mercuric ion acting as a calcium-mimetic agent. Biochem. Biophys. Res. Commun. 110:758.

Stacey, N.H. and Klaassen, K., 1981. Inhibition of lipid peroxidation with prevention of cellular injury in isolated rat hepatocytes. Toxicol. Appl. Pharmacol. 58:8, 1981.

Stacey, N.H. and Kappus, H., 1982. Cellular toxicity and lipid peroxidation in response to mercury. Toxicol. Appl. Pharmacol. 63:29.

Tsan, M-F, Danis, E.H., DelVecchio, P.J. and Rosano, C.L., 1985, Enhancement of intracellular glutathione protects endothelial cells against oxidant damage. Biochem. Biophys. Res. Commun. 127:270.

Verity, M.A., Brown, W.J. and Cheung, M., 1975. Organic mercurial encephalopathy: *In vivo* and *in vitro* effects of methyl mercury on synaptosomal respiration. J. Neurochem. 25:759.

Weir, K., Sarafian, T., and Verity, M.A., 1990. Methylmercury induces paradoxical increase in reduced glutathione (GSH) in cerebellar granule cell culture. Toxicol. 10:25.

Welsh, S.O. and Soares, J.H. Jr., 1976. The protective effect of vitamin E and selenium against methyl mercury toxicity in the Japanese quail. Nutrition Rep. Int. 13:43.

Williamson, J.M. and Meister, A., 1981. Stimulation of hepatic glutathione formation by administration of L-2-oxothiazolidine-4-carboxylate, a 5-oxo-L-prolinase substrate. Proc. Natl. Acad. Sci. U.S.A. 78:936.

Yagi, K., 1976. A simple fluorometric assay for lipoperoxide in blood plasma. Biochem. Med. 15:212.

# ALTERATIONS IN GENE EXPRESSION DUE TO METHYLMERCURY IN

## CENTRAL AND PERIPHERAL NERVOUS TISSUES OF THE RAT

Saburo Omata, Yasuo Terui, Hidetaka Kasama, Tohru Ichimura,
Tsuneyoshi Horigome, and Hiroshi Sugano

Department of Biochemistry
Faculty of Science,   Niigata University
Niigata 950-21
Japan

## ABSTRACT

Polyadenylated messenger RNAs (poly (A) + mRNA) obtained from the brain of control and methylmercury (MeHg) -treated rats were translated in reticulocyte lysate system in the presence of $^{35}$S-methionine or $^{3}$H-leucine. The $^{3}$H-labelled translation products of the control poly (A)$^{+}$ mRNA were added to each of the $^{35}$S-labelled translation products as an internal standard, and the mixture was analyzed by two-dimensional electrophoresis. The results showed that the effect of MeHg on the synthesis of proteins in the brain was not uniform for individual protein species in the early, latent and symptomatic periods of MeHg intoxication. Among 120 protein species examined, the numbers of protein species the synthetic activities of which were significantly reduced/increased were 25/34 on the early period, 5/67 on the latent period and 4/99 on the symptomatic period.

These results suggest that MeHg affect differently the synthesis of brain proteins and the unusual reduction or elevation of certain protein species caused by perturbation of the synthetic rates by MeHg may be responsible for the impairment of normal nerve functions.

*Advances in Mercury Toxicology*, Edited by T. Suzuki *et al.*
Plenum Press, New York, 1991

## INTRODUCTION

Although the biochemical mechanism underlying the neurotoxicity of methylmercury (MeHg) has not yet been clarified, results obtained in several laboratories supported the notion that a defect in protein synthetic activity in the nervous system is closely related with the neurotoxicity of MeHg (for reviews, see Omata and Sugano, 1985; Miura and Imura, 1987). Previous studies in our laboratory indicated that the decrease in protein synthesis due to MeHg in the brain and peripheral nerve tissues of the rat occurred through the direct action of this compound on the protein synthesizing system and not the indirect effects such as malnutrition, changes in cellular amino acid uptake, and changes in amino acid pool size and/or their metabolism in the tissues (Omata et al., 1978, 1980, 1982). However, the results only reflected MeHg-induced quantitative changes in the whole tissues. The present study was performed to determine whether the rates of synthesis of various protein species in nervous tissues are affected uniformly or differently by MeHg.

## MATERIALS AND METHODS

*Chemicals:* Methylmercury chloride was obtained from Wako Pure Chemical Ind. Ltd. (Osaka, Japan). Acrylamide, N,N'-methylenebisacrylamide was purchased from Seikagaku Kogyo Co. (Tokyo, Japan), and Ampholine from LKB (Bromma, Sweden). Disodium ATP, trisodium GTP, tripotassium phosphoenol pyruvate, pyruvate kinase, Antifoam A and molecular weight markers were obtained from Sigma Chemical Co. (St. Louis, USA). L-[$^{35}$S]methionine ( > 1000 Ci/mmol), L-[$^3$H] leucine. (131 Ci/mmol), reticulocyte lysate and amino acid mixtures were products of the Radiochemical Center (Amersham, England). Oligo (dT) cellulose was obtained from Collaborative Research (Waltham, USA).

*Animal treatment:* Administration of MeHg was carried out according to the dosage schedule of Klein et al. (1972) to produce acute MeHg intoxication. Female Wistar rats weighing 200-230 g received daily sc. injections of MeHg chloride (10 mg/kg) in 10 mM $NaHCO_3$-$Na_2CO_3$, pH 9.2 for 7 consecutive days and then were killed at specified times; Day 4 (early period), Day 10 (latent period) and Day 15 (symptomatic period). Day 1 was the day on which the first injection was given. Control rats received injections of vehicle alone.

**Preparation and translation of poly (A)+ mRNA:** Brains of 5 rats pooled from each experimental group were subjected to repeated extraction/precipitation cycles with guanidine salts/ethanol essentially according to Chirgwin et al. (1979), followed with chloroform/phenol extraction, then precipitation with 4 M NaCl and 2.5 vol. of ethanol. The total RNA fraction obtained was fractionated further by oligo (dT) cellulose column chromatography (Aviv and Leder, 1972) to separate the poly(A)$^+$ messenger RNA. The poly(A)$^+$ mRNA was stored at -70° C in sterilized double distilled water until use.

The poly(A)$^+$ mRNA was translated in a reticulocyte lysate system (Pelham and Jackson, 1976) with modification as described previously (Namba et al., 1984) except that the volume of the reaction mixture was 25 ul. The poly(A)$^+$ mRNA from each experimental group (control, Day 4, Day 10 and Day 15) was translated in the presence of 50 uCi of $^{35}$S-methionine at 30° C for 60 min. In addition, the poly(A)$^+$ mRNA from the control group was translated in the presence of 150 uCi of $^3$H-leucine. Aliquots of the translation mixtures were processed to assess "total" protein synthesis as described previously (Kasama et al., 1989).

*Sample preparation for electrophoresis:* An aliquot equivalent to 3.60 x 10$^5$ dpm protein radioactivity was taken from the control translation mixture labeled with $^{35}$S-methionine and mixed with an aliquot equivalent to 7.20 x 10$^5$ dpm protein radioactivity from the control translation mixture labeled with $^3$H-leucine to produce $^{35}$S-control/$^3$H-control sample. In order to prepare the $^{35}$S-MeHg/$^3$H-control samples, an aliquot of the translation mixture (3.60 x 10$^5$ dpm protein radioactivity) from each MeHg-treated group labeled with $^{35}$S-methionine was mixed with an aliquot (7.20 x 10$^5$ dpm protein radioactivity) of the control translation mixture labelled with $^3$H-leucine. The $^3$H-leucine labeled control protein fraction was added as an internal standard to correct for difference in the recoveries of the applied proteins from gel to gel.

Each sample was mixed with 4 vol. of acetone (-80°C) and processed similarly as described by Kasama et al. (1989).

*Two-dimensional electrophoresis and fluorography:* The two-dimensional electrophoretic analyses were carried out essentially according to O'Farrell (1975).

Details of the electrophoretic and fluorographic procedures were as described previously (Kasama et al., 1989).

*Extraction of protein spots and counting*: Protein spots on the gels were cut out and homogenized in 0.5 ml of a 1% SDS/0.5 N NaOH mixture, and then the homogenate was transferred to a scintillation vial. The homogenizer and pestle were washed with two portions of 0.5 ml of the SDS/NaOH mixture. The combined mixture was incubated at 60°C for 18 hr to extract proteins. To the vial was added 15 ml Scintisol and a few drops of formic acid, and then the radioactivity was determined with an Aloka LSC-900 liquid scintillation counter.

RESULTS

Evolution of MeHg intoxication in the rats was very similar to that reported previously (Omata et al., 1978). On Day 4 (early phase), no sign of intoxication was found in the animals. On Day 10 (latent phase), weakness, diarrhea, and dragging of hind limbs appeared and by Day 15 (symptomatic phase), crossing of hind limbs, a typical sign of MeHg poisoning (Klein et al., 1972), was observed in the animals. Rats in the symptomatic phase which were sacrificed on Day 14 or Day 15 are referred to simply as rats on Day 15.

Yield of total RNA fraction by the guanidine/ethanol method was 380-500 ug/g brain in the control and MeHg-treated groups, whereas that of polyadenylated mRNA was in the range of 15-21 μg/g brain. The amount of poly(A)$^+$ mRNA was similar to that reported by Darmon and Paulin (1985) for the adult rat brains. The optimal concentrations of MgCl2 and K-acetate in the reaction mixture were found to be 3.0 mM and 130 mM, respectively. The optimal mRNA concentration was about 1.0 μg/25 μl reaction mixture. Table 1 shows the total incorporation of $^{35}$S-methionine into the protein fraction as directed by poly(A)$^+$ mRNAs from control and MeHg-treated brains. Incorporation of $^3$H-leucine was also shown. The translation activities of the mRNA fractions obtained from the brains of MeHg-treated rats were slightly lower than the control value. The translation of the control mRNA was reduced to 75% of the control value in the presence of 0.4 mM MeHg in the reaction mixture.

*Analysis by two-dimensional electrophoresis*: After the translation, the $^{35}$S-labeled protein fractions were mixed with the $^3$H-labeled protein fractions as described in Materials and Methods, and each mixture, $^{35}$S-control/$^{3}$H-control and $^{35}$S-MeHg (Day 4, 10 and 15) /3H-control, was subjected to separate analysis on a two-dimensional gel.

Figure1(a) shows the fluorogram of the $^{35}$S-control/$^{3}$H-control mixture, and Figure 1(b) is a tracing of the major 120 spots in Figure. 1(a). The spots were numbered sequentially in decreasing order, as to Mr, and according to decreasing pI when Mr of the spots were similar. The fluorograms of $^{35}$S-MeHg/$^{3}$H-control mixtu res (results not shown) were very similar to the pattern of the control, and no newly appearing or missing protein spots after MeHg treatment was detected on visual inspection.

*Calculations*: After fluorography, each protein spot on the gels were located, cut out, extracted and then radioactivity was determined. The radioactivities of protein spots ranged from about 70 to 5200 dpm for $^{35}$S and from about 150 to 9800 dpm for $^{3}$H. The $^{35}$S/$^{3}$H ratio was calculated for each protein spot in $^{35}$S-control/$^{3}$H-control and $^{35}$S-MeHg/$^{3}$H-control gels. The ratio in the case of the $^{35}$S-control/$^{3}$H-control gel

Table 1 Translation of poly (A) + mRNAs from brians of control and MeHg-treated rats in reticulocyte lysate system

| Amino acids | mRNAs | Protein synthesis (dpm x $10^{-4}$/$\mu$l) | % |
|---|---|---|---|
| $^{35}$S−Met | Control | 4. 51 | 100 |
| | Day 4 | 4. 26 | 94. 5 |
| | Day 10 | 4. 33 | 96. 0 |
| | Day 15 | 4. 19 | 92. 9 |
| | Control* | 3. 37 | 74. 8 |
| $^{3}$H−Leu | Control | 9. 82 | −− |

Translation of mRNA (1. 0 $\mu$g/25 $\mu$l reaction mixture) was performed in the presence of $^{35}$S−methionine (50 $\mu$Ci) or $^{3}$H−leucine (150 $\mu$Ci).
* Control mRNA was translated in the presence of 0. 4 mM MeHgOH.

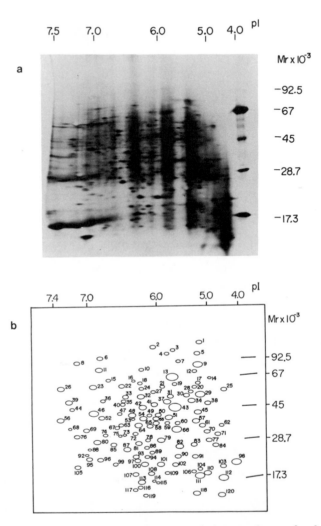

Fig.1.Two-dimensional gel separation of the translation products of poly(A)+ mRNA from the brain of the control rats. The mixture containing [35]S-control/[3]H-control was processed as described under *Materials and Methods*. Fluorogram obtained was shown in (a). The molecular weights of [14]C-labelled marker proteins, and pIs of the first dimension gel slice are indicated. A tracing of the protein spots in (a) with spot nos. is shown in (b).

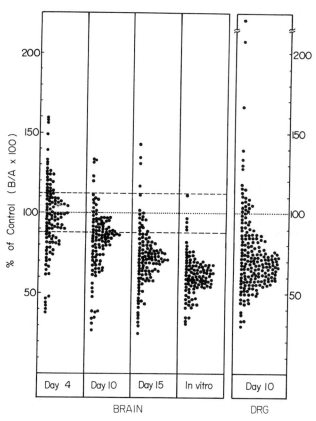

Fig. 2. Changes in the synthesis of individual protein species after MeHg treatment., Days 4, 10 and 15 in the brain shows the *in vivo* effect of MeHg, while "*in vitro*" represents the results in which control poly (A)+ mRNA was translated in the presence of 0.4 mM MeHg in the reaction mixture. See text for details.

was termed A and that in the case of $^{35}S$-MeHg/$^3H$-control gel was termed B. Thus, the B/A ratio for each protein spot represents the effect of MeHg on the synthesis of a given protein species. On the other hand, the confidence limits for variation in B/A was obtained as follows. Aliquots of $^{35}S$-control/$^3H$-control mixture were separately electrophoresed and the $^{35}S/^3H$ ratio in a gel ($A_1$) was compared with that in the second gel ($A_2$). The mean value $\pm$ SD of the calculated $A_2/A_1$ of all protein spots was 1.022 $\pm$ 0.124, thus providing $\pm$ 12.4% as the confidence limit for deviation of B/A ratio. In the subsequent experiments, only the protein spots with the B/A ratio higher than 113% or lower than 87% of the control value were regarded as showing significant changes, and this range is shown as two broken lines in Figure 2.

*Changes in the synthesis of individual protein species*: Figure 2 shows effect of MeHg on the synthesis of individual protein species. Each closed circles represents each protein species and the ordinate is B/A (see above) expressed as %. The graph for DRG on Day 10 was redrawn from the results reported by us (Kasama et al., 1989) in which the methods similar to the present procedures were applied to the mixtures of the extracts of DRG slices of control and MeHg-treated rats labelled with $^{35}S$-methionine or $^3H$-leucine.

On Day 4, changes in the synthetic activities of individual protein species ranged from about 40 to 160%. With the progress of MeHg intoxication, the protein species the synthetic activities of which were lowered increased, while those the synthetic activities of which were elevated decreased (see Days 10 and 15). Most of the protein species in which the control mRNA was translated in the presence of 0.4 mM MeHg, exhibited decreased activities though the extent was not uniform from protein to protein. In DRG on Day 10, depression of protein synthesis seemed to be larger than that in the brain, although some protein species showed the synthetic activity much higher than those in the brain.

The protein species were classified according to their patterns of changes in the synthetic activities during the intoxication periods (Table 2). The class I protein species showed no appreciable change throughout the experimental period. About half the number of protein species exhibited decreased synthetic activities on Day 15 or on Days 10 and 15 and these groups were designated as the Class II and III, respectively. The synthetic activities of the Class IV protein species were lowered throughout the experimental period and the time-dependent change was shown in Figure 3. Most of the protein species showed the decrease towards Day 15 with several exception of nos. 71, 87, 88, 90, 91, 98, 101, 106 and 115. The synthetic activities of the protein species that were elevated on Day 4 were classified to the Class V.

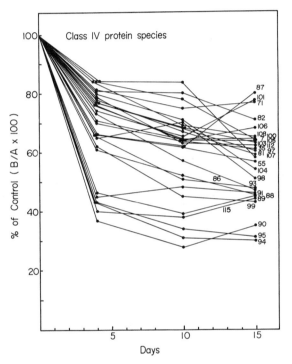

Fig. 3. Changes in the synthetic activities of the Class IV protein species during MeHg intoxication. The changes of the protein species the synthetic activities of which were reduced throughout the experimental period were shown. For the classification of the protein species, see text and Table 2.

Table 2   Changes in synthetic activities of protein species in the brains of rats during MeHg intoxication

| Class | Patterns of Changes | | | | Numbers of spots (%) | Spot No. |
|---|---|---|---|---|---|---|
| | Day 4 | Day 10 | Day 15 | | | |
| I | 0 | 0 | 0 | | 8 (6.7) | 2, 8, <u>37</u>, 38, 42, 48, 69, 83 |
| II | 0 | 0 | — | | 30 (25.0) | <u>9</u>, 10, 11, 12, <u>13</u>, 15, 17, 18, <u>20</u>, 24 25, <u>26</u>, 27, 30, 33, 39, 40, 44, 46, 52 <u>54</u>, 56, 59, <u>64</u>, 70, 74, 76, 116, <u>120</u> |
| III | 0 | — | — | | 27 (22.5) | 3, 14, 16, 19, <u>28</u>, 32, 41, <u>43</u>, 45, 47 58, 60, 61, 62, 63, 65, <u>66</u>, <u>73</u>, <u>77</u>, 85 92, 109, 110, 111, 113, <u>118</u>, 119 |
| IV | — | — | — | | 28 (23.3) | 7, 35, 55, <u>71</u>, 81, <u>82</u>, 86, 87, 88, 89 90, 91, 93, 94, 95, 97, <u>98</u>, 99, 100, <u>101</u> <u>103</u>, 104, 105, 106, 107, 108, <u>112</u>, 115, |
| V | + | 0 | 0 | (5) | 17 (14.2) | 1, 36, 67, 72, 117 |
| | + | + | 0 | (1) | | <u>80</u> |
| | + | + | + | (4) | | 4, <u>31</u>, 75, 114 |
| | + | 0 | — | (7) | | 5, 6, 21, 23, <u>29</u>, <u>34</u>, <u>84</u> |
| VI | 0 | — | 0 | (4) | 10 (8.3) | 50, 51, 79, 102 |
| | ( ·· x ·· ) | | | (6) | | 22, 49, 53, 68, 78, 96 |
| Total | | | | | 120 (100) | |

0, unchanged;   +, increased;   —, decreased;   x, difficult to identify on the gels (see detail for text).

From the results of Figure 4, it is seen that about a third of the protein species in the group V retained their elevated activities throughout the experimental period, while the rest showed gradual or sharp decrease towards the symptomatic period. The Class VI group consists of miscellaneous protein species; four protein species showed decreased synthetic activities only on Day 10; among other six protein species, identification of the spots on the fluorograms was very difficult in one (spot nos. 22, 68, and 78), two (spot nos. 49 and 96) and three (spot no. 53) phases of intoxication.

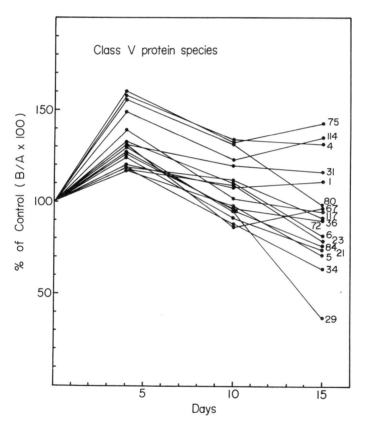

Fig. 4. Changes in the synthetic activities of the Class V protein species during MeHg intoxication. The changes of the protein species the synthetic activities of which were elevated on Day 4 were shown. For the classification of the protein species, see text and Table 2.

*Assignment of the protein spots to the known protein species*: The protein spots detected on the fluorograms were assigned to the known protein species in the brain and DRG on the basis of their pI and Mr according to Hall (1982), Hall et al. (1984) and Fujio et al. (1987). Table 3 summarizes the results. Heat shock protein 90, tubulin β subunit, neuron-specific enolase (14-3-2) and a member of 14-3-3 protein family (spot no. 84) seems to belong to the Class V protein species, while changes in the synthetic activity was small in creatine kinase. Although all protein species listed showed significant decrease on Day 15, notable changes were seen in tubulin (α and β subunits) and calmodulin. In DRG on Day 15, the synthetic activities of all protein species listed showed significant lowering and tubulin and calmodulin also showed considerable reduction in their synthetic activities.

DISCUSSION

In the present experiments, the mixtures of $^{35}$S-labelled translation products and the $^{3}$H-labelled translation products (control) were analyzed by two-dimensional electro-phoresis. The use of $^{3}$H-labelled translation products as an internal standard to correct for the recovery of protein species during the resolving procedure permits precise comparison of the protein patterns on the different gels. The advantage of combination of double labelling with two-dimensional electrophoresis was demonstrated (Choo et al., 1980; Wheeler et al., 1986) and the methodology was applied to analysis of the effect of MeHg on protein synthesis of DRG slices (Kasama et al., 1989).

Inhibitory characteristics of mercury compound in biological systems have been well known and a limited number of studies have been reported for the stimulating action of MeHg on macromolecule synthesis (Brubaker et al., 1973; Omata et al., 1978; Sauve and Nichols, 1981; Frenkel and Randles, 1982; Frenkel et al., 1985). The results of the present study indicate that the effect of MeHg on the protein synthesis in the brain is not uniform for each protein species: (1) the synthetic activities of some protein species were reduced by MeHg, while those of others were elevated or unaffected, (2) the extent of the reduction or elevation of the protein synthetic rate was not uniform for individual protein species, and (3) the patterns of changes in the protein synthetic rates differed during the progress of MeHg intoxication in the animals.

It is worth emphasizing that difference in the synthetic rates of protein species was evident even on Day 4 (Figure 2). In this phase, no quantitative change was found in the total protein synthesis in the brain (Omata et al., 1978; 1982). The number of protein species, the synthetic rates of which were depressed doubled on Day 10, and on Day 15, and those were about 80% of the total protein species (Figure 2 and Table 2). On the other hand, the 25 protein species showed the stimulated synthetic rates on Day 4, but the number was markedly decreased on Days 10 and 15. It is possible that reduction in the synthetic rate of some protein species induce a defect in the amount of proteins necessary to maintain the structure and function of the brain tissues; overproduction of other protein species gives rise to perturbation of normal cellular activities.

Identification of the protein species whose synthetic activities are significantly affected would help clarification of the mechanism underlying the MeHg action on the nervous tissues. Among 120 protein species, only several protein species were identified in the present study (Table 3). The synthesis of heat shock protein 90 decreased only on Day 15. A considerable reduction in the synthesis of the cytoskeletal proteins, actin and tubulin $\alpha$, was observed on Days 10 and 15, while the synthesis of tubulin $\beta$ subunit showed a different pattern of change. It has been mentioned that tubulin heterogeneity appeared most extensively in mammalian brain and there are multifunctional genes for both $\alpha$ and $\beta$ tubulin isotypes (Mohri and Hosoya, 1988). The present results suggest that MeHg acts not only on the step of specific disturbance of microtubule structure (Miura et al., 1984; Miura and Imura, 1974), but also on the step (s) of gene expression. This is in contrast to the suggestion that autoregulation of tubulin synthesis is the result of cytoplasmic events  (Pachter et al., 1987). It is important to note that the synthesis of 14-3-3 proteins in the brain and DRG were considerably reduced on Day 10 and/or Day 15, since the 14-3-3 protein family is known as a regulator of neurotransmitter enzymes (Ichimura et al., 1987). In the present experiments, the protein species were identified with their pI and Mr. In order to identify these and other protein species more precisely, experiments are in progress including, peptide mapping, co-migration with each purified protein and immunoblotting.

In the present study, an equal amount of polyadenylated mRNA from the brains of control and experimental groups added to the reaction mixtures was translated and examined. The fact that the synthesis of individual protein species was not uniformly affected under the experimental conditions employed suggests that (1) the amount of each mRNA species, as well as composition of the mRNA population, was altered by the action of MeHg on gene expression pathway including transcription of DNA, mRNA processing, etc., and (2) MeHg directly interacted *in vivo* with mRNAs to produce the molecules with the modified template activities, although these possibilities are not mutually exclusive. In connection with the first possibility, RNA synthesis in isolated nuclei (Frenkel and Randles, 1982) as well as the template activity of DNA for RNA polymerase (Frenkel et al., 1985) by MeHg treatment was demonstrated. These authors showed later that the increased template activity of MeHg-exposed DNA was not the sole mechanism to explain the enhanced RNA synthesis in the isolated nuclei (Frenkel and Ducote, 1987). MeHg may act on other factors that affect the amount of available mRNA species, such as RNA polymerase, intranuclear metabolism of RNA, processing of mRNA and polyadenylation. Changes in the structure and function of nucleic acids after interaction with mercury (Barton and Lippard, 1980; Frenkel et al., 1985; Gruenwedel and Cruikshank, 1990) support the second possibility.

The effect of MeHg on translation machinery *per se*, if any, seems to affect rather similarly the synthesis of individual protein species by direct modification of activities of enzymes or protein factors necessary for initiation, elongation and termination of peptide bond formation. An exception is the step of aminoacylation of tRNA. Table 4 shows the different effect of MeHg on aminoacyl tRNA synthetase activities of the

Table 3  Changes in synthetic activities of protein species in the brain and DRG of rats after MeHg treatment

| Protein species | Synthetic activities (% of control) | | | | | |
|---|---|---|---|---|---|---|
| | Brain | | | | DRG | |
| | No. | Day 4 | Day 10 | Day 15 | No. | Day 10 |
| HSP 90 | 9 | 113 | 105 | 72.8 | 31 | 67.0 |
| Tubulin α | 28 | 98.3 | 72.3 | 57.5 | 71 | 55.3 |
| β | 29 | 130 | 95.1 | 36.5 | | |
| NSE (14-3-2) | 34 | 119 | 88.8 | 64.7 | 92 | 62.4 |
| Creatine kinase | 37 | 100 | 91.4 | 86.0 | - | - |
| Actin | 43 | 107 | 84.9 | 68.5 | 106 | 79.1 |
| 14-3-3 proteins | 71 | 81.2 | 47.8 | 77.0 | 145 | 69.2 |
| | 77 | 90.2 | 83.4 | 74.2 | 148 | 47.4 |
| | 84 | 126 | 95.9 | 76.3 | 150 | 55.9 |
| | | | | | 151 | 49.6 |
| | | | | | 152 | 60.0 |
| | | | | | 153 | 46.3 |
| | | | | | 157 | 78.5 |
| Calmodulin | 98 | 64.8 | 70.5 | 51.3 | 186 | 59.2 |

brains from the rats on Day 15. Among 12 enzyme activities examined, significant reduction in six enzymes and elevation in an enzyme were observed, while the other five enzyme activities were unchanged. The lowering of phenylalanyl tRNA synthesis confirmed the report of Cheung and Verity (1985). The results indicate that the synthesis of the protein species was modified differently at the step of aminoacyl tRNA formation depending on their amino acid compositions.

The effect of MeHg on gene expression pathway may be shared between two portions; the effect on the translation step may result in the change uniform for individual protein species except aminoacyl tRNA formation, whereas the effect on the step other than translational machinery may result in the change different for the individual protein species. Investigations are in progress to determine the site(s) of

Table 4  Effect of MeHg on aminoacyl-tRNA synthetase activities in the brains of the rats on Day 15

| Amino acids | Control | MeHg (Day 15) | M/C (%) |
|---|---|---|---|
| Asp | 9. 16 ± 1. 14 | 5. 66 ± 0. 97 | 62* |
| Leu | 19. 65 ± 3. 36 | 15. 53 ± 0. 77 | 79* |
| Tyr | 20. 14 ± 0. 89 | 18. 05 ± 0. 53 | 89* |
| Lys | 45. 13 ± 5. 94 | 49. 97 ± 7. 03 | 110 |
| Met | 50. 17 ± 7. 54 | 45. 27 ± 3. 29 | 90 |
| His | 20. 32 ± 2. 35 | 29. 95 ± 3. 89 | 148* |
| Arg | 60. 02 ± 8. 48 | 42. 15 ± 5. 01 | 71* |
| Phe | 11. 84 ± 3. 28 | 8. 01 ± 1. 04 | 68* |
| Val | 32. 67 ± 2. 71 | 26. 94 ± 2. 76 | 82* |
| Ser | 1. 38 ± 0. 19 | 1. 45 ± 0. 33 | 107 |
| Glu | 5. 13 ± 1. 39 | 5. 54 ± 1. 33 | 108 |
| Gln | 4. 70 ± 1. 03 | 4. 52 ± 0. 47 | 96 |

\* Significant changes ( $p < 0.05$). Each value ( fmol aminoacylated/ug protein of pH 5 fraction/10 min) represents the mean ± SD for five to six rats.
The data for upper six aminoacids (Asp-His) were previously reported (Hasegawa et al., 1988) and enzyme activities for lower six amino acids (Arg-Gln) were determined in the present study.

action of MeHg in these steps, as well as examination of the protein species not detected within the present pI range and activities of poly (A)⁻ mRNA species, in order to elucidate the mode of action of MeHg on the nervous system.

ACKNOWLEDGEMENT

This work was supported in part by a grant from the Japanese Environmental Agency. We thank Miss Kazuko Hasegawa for co-operation in preparation of the manuscript.

REFERENCES

Aviv, H and Leder, P., 1972, Purification of biologically active globin messenger RNA by chromatography on oligothymidylic acid-cellulose, Proc. Natl. Acad. Sc. U.S., 69:1408-1412.

Barton, J.K. and Lippard, S.I., 1980, Heavy metal interactions with nucleic acids, in: Nucleic Acid-Metal Ion Interactions, Spiro, T. G., ed., 31-114, J. Wiley & Sons, New York.

Brubaker, P.E., Klein, R., Herman, S.P., Lucier, G.W., Alexander, L.T., and Long, M.D., 1973 DNA, RNA, and protein synthesis in brain, liver, and kidney of asymptomatic methylmercury treated rats, Exptl. Mol. Pathol., 18: 263-280.

Cheung, M. and Verity, M.A., 1985, Experimental methylmercury neurotoxicity: locus of mercurial inhibition of brain protein synthesis in vivo and in vitro, J. Neurochem., 44:1799-1808.

Chirgwin, J. M., Przybyla, A. E., MacDonald, R. J., and Rutter, W.J., 1979, Isolation of biologically active ribonucleic acid from source enriched in ribonuclease, Biochemistry,18: 5294-5299.

Choo, K.K., Cotton, G.H., and Danks, D.M., 1980, Double-labelling and high precision comparison of complex protein patterns on two-dimensional polyacrylamide gels, Anal. Biochem. 103: 33-38.

Cosgrove, J.M. and Brown, I.R., 1983, Heat shock protein in mammalian brain and other organs after a physiologically relevant increase in body temperature induced by D-lysergic acid diethylamide, Proc. Natl. Acad. Sci. U.S., 80: 569-573.

Darmon, M.C. and Paulin, D.J., 1985, Translational activity of mRNA coding for cytoskeletal brain proteins in newborn and adult mice: A comparative study, J.Neurochem., 44:1672-1678.

Frenkel, G.D. and Randles, K., 1982, Specific stimulation of α-amanitin sensitive RNA synthesis in isolated HeLa nuclei by methyl mercury, J. Biol. Chem., 257: 6275-6279.

Frenkel, G. D., Cain, R., and Chao, E.S. -E., 1985, Exposure of DNA to methyl mercury results in an increase in the rate of its transcription by RNA polymerase II, Biochem. Biophys. Res. Commun., 127: 849-856.

Frenkel, G.D. and Ducote, J., 1987, The enhanced rate of transcription of methyl mercury-exposed DNA by RNA polymerase is not sufficient to explain the stimulatory effect of methyl mercury on RNA synthesis in isolated nuclei, J. Inorgan. Biochem. 31: 95-102.

Fujio, N., Hatayama, T., Kinosita, H., and Yukioka, M., 1987, Induction of four Heat-shock proteins and their mRNSs in rat after whole-body hyperthermia, J. Biochem. (Tokyo), 101: 181-187.

Gruenwedel, D.W. and Cruikshank, M.K., 1990, Mecury-induced DNA polymorphism: Probing the conformation of Hg(II)-DNA via staphylococcal nuclease digestion and circular dichronism measurements, Biochemistry, 29: 2110-2116.

Hall, M.E., 1982, Changes in synthesis of specific proteins in axotomized dorsal root ganglia, Exptl, Neurol., 76: 83-93.

Hall, C., Mahadevan, L., Whatley, S., Biswas, G., and Lim., L, 1984, Characterization of translation products of the polyadenylated RNA of a free and membrane-bound polyribosomes of rat forebrain, Biochem. J., 219: 751-761.

Hasegawa, K., Omata, S., and Sugano, H., 1988, In vivo and in vitro effects of methylmercury on the activities of aminoacyl-tRNA synthetases in rat brain, Arch. Toxicol. 62: 470-472.

Ichimura, T., Isobe, T., and Okuyama, T., 1987, Brain 14-3-3 protein is an activator protein that activates tryptophan 5-monooxygenase and tyrosine 3-monooxygenase in the presence of Ca, calmodulin-dependent protein kinase II, FEBS Lett. 219: 79-82.

Kasama, H., Itoh, K., Omata, S., and Sugano, H., 1989, Differential effects of methylmercury on the synthesis of protein species in dorsal root ganglia of the rat, Arch.Toxicol,, 63: 226-230.

Klein, R., Herman, S.P., Brubaker, P.E., Lucier, G.W., and Krigman, M.R., 1972, A model of acute methylmercury intoxication in rats, Arch. Pathol., 93: 408-418.

Miura, K. and Imura, N., 1987, Mechanism of methylmercury cytotoxicity, CRC Critical Rev. Toxicol., 18: 161-188.

Miura, K., Inokawa, M., and Imura, N., 1984, Effects of methylmercury and some metal ions in microtubule networks in mouse glioma cells and in vitro tubulin polymerization, Toxicol. Appl. Pharmacol., 73: 218-231.

Mohri, H. and Hosoya, N., 1988, Two decades since the naming of tubulin. "The multi-facets of tubulin", Zool. Sci., 5: 1165-1185.

Namba, S., Hikawa, A., Kitoo, N., Horigome, T., Omata, S., and Sugano, H., 1984, Interaction of secretory protein precursors with phospholipids in liposomes, J. Biochem.(Tokyo), 96:1133-1142.

O'Farrell, P. H., 1975, High resolution two-dimensional electrophoresis of proteins, J. Biol. Chem., 250: 4007-4021.

Omata, S., Sakimura, K., Tsubaki, H., and Sugano, H., 1978, In vivo effects of methylmercury on protein synthesis in brain and liver of the rat, Toxicol. Appl. Pharmacol., 44: 367-378.

Omata, S., Horigome, T., Momose, Y., Kambayashi, M., Mochizuki, M., and Sugano, H., 1980, Effect of methylmercury chloride on the in vivo protein synthesis in the brain of the rat: Examination with the injection of a large quantity of [$^{14}$C]valine, Toxicol. Appl. Pharmacol., 56: 207-215.

Omata, S., Momose, Y., Ueki, H., and Sugano, H., 1982, In vivo effect of methylmercury on protein synthesis in peripheral nervous tissue of the rat, Arch. Toxicol., 49: 203-214.

Omata, S. and Sugano, H., 1985, Methylmercury: Effects on protein synthesis in nervous tissue, in: Neurotoxicology, Blum, K. and Manzo, L., eds., 369-383, Marcel Dekker, Inc., New York.

Pachter, J.S., Yen, T.J., and Cleveland, D.W., 1987, Autoregulation of tubulin expression is achieved through specific degradation of polysomal tubulin mRNAs, Cell, 51: 283-292.

Pelham, H.R.B. and Jackson, R.J., 1976, An efficient mRNA-dependent translation system from reticulocyte lysates, Europ. J. Biochem., 67: 247-256.

Sauve, G.J. and Nicholls, D.M., 1981, Liver protein synthesis during the acute response to methylmercury administration, Int. J. Biochem., 13: 981-990.

Sevaljevic, L., Boskovic, B., Glibetic, M., and Tomic, M., 1989, Effect of soman intoxication on the organization of rat brain ribosomes and the translational activity of mRNA in a cell free system., Arch. Toxicol., 63: 244-247.

Wheeler. T. T., Loong, P.C., Jordan, T.W., and Ford, H.C., 1986, A double-label two-dimensional gel; electrophoresis procedure specifically designed for serum or plasma protein analysis, Anal. Biochem., 159: 1-7.

# MICROTUBULES: A SUSCEPTIBLE TARGET OF METHYLMERCURY CYTOTOXICITY

Kyoko Miura[1,2] and Nobumasa Imura[2]

[1]Department of Environmental Sciences
Wako University
Machida-shi, Tokyo, Japan

[2] Department of Public Health
School of Pharmaceutical Sciences
Kitasato University
Minato-ku, Tokyo, Japan

## ABSTRACT

Alteration of many kinds of cellular functions by methylmercury (MM) exposure *in vitro* have been reported. However, the causal mechanism for the specific dysfunction of nervous systems has not been clarified yet. We have previously demonstrated that microtubules in cultured mouse glioma cells are specifically disrupted by MM before the morphological disorders of the other organelles are detected. This specific impairment of microtubules inhibited cell growth.

In addition, we have recently found that the increased cellular pool of tubulin subunits by microtubule depolymerization with MM resulted in an inhibition of tubulin biosynthesis. Since the protein bands other than tubulin on gradient urea-PAGE gel appeared to remain unchanged under the experimental condition used, the inhibition of tubulin synthesis by MM was specific. This reduction in tubulin synthesis was well associated with the specific decline of mRNA level of $\beta$-tubulin. On the other hand, the transcription rate of tubulin gene was confirmed to be unchanged in isolated nuclei derived from MM-treated cells.

*Advances in Mercury Toxicology*, Edited by T. Suzuki *et al.*
Plenum Press, New York, 1991

These results indicate that the growth inhibitory concentration of MM depolymerizes microtubules in mouse glioma cells and, in addition, inhibits tubulin synthesis through the autoregulatory control by the increased pool of tubulin subunits probably at the post transcriptional stages as in the case of colchicine treatment.

INTRODUCTION

Methylmercury (MM) is known to cause a wide variety of neurological disturbances in humans and experimental animals. Although a large number of biochemical studies on the toxic action of MM have accumulated (Miura and Imura, 1987), the mechanisms involved in MM toxicity have not yet been elucidated. Cell culture systems are useful for studying the mechanisms of MM cytotoxicity, since cultured cells are free from the influence of toxicokinetic factors, which complicate the interpretation of results obtained by *in vivo* experiments.

We have carried out a series of investigations of the toxic action of MM on cultured mammalian cells derived from brain and nervous system: mouse glioma and neuroblastoma cells, respectively. This paper deals with MM cytotoxicity, focusing on microtubules as a susceptible target of MM.

Effect of MM and inorganic mercury on macromolecule biosyntheses in mouse glioma and neuroblastoma cells

Major scientific interest in the toxicity of MM at the cellular and subcellular levels has focused upon its adverse effects on macromolecule biosyntheses (Gruenwedel and Cruik-shank, 1979; Chao et al., 1984; Sarafian et al., 1984). Then, the effects of MM ($CH_3HgCl$) and mercuric chloride ($HgCl_2$) on DNA, RNA and protein syntheses were studied using mouse glioma and neuroblastoma cells. Mouse glioma, $SR-CDF_1DBT$ and neuroblastoma C-1300 (N-18 clone) were grown as monolayer cultures in Eagle's minimum essential medium (E. MEM), and Dulbecco's modified essential medium (D. MEM), respectively, supplemented with 10% fetal bovine serum (FBS).

Mercuric chloride enhanced DNA synthesis at concentrations up to $2 \times 10^{-5}M$ and depressed at $5 \times 10^{-5}M$ (Figure 1A) (Nakada and Imura, 1980). [$^3$H]-Thymidine transported into the cells also increased up to $2 \times 10^{-5}M$, then decreased at $5 \times 10^{-5}M$ (Figure 1B). However, MM completely inhibited DNA synthesis at $5 \times 10^{-5}M$ and no stimulation of thymidine transport into the cells was observed in the presence of $10^{-6}M$ - $10^{-5}M$ MM (Figures 1A and 1B). These results indicated that the inhibition of DNA synthesis by mercurials is not a secondary effect caused by the inhibition of transport systems for the precursor nucleosides. In neuroblastoma cells, DNA synthesis was completely inhibited by both mercurials at $5 \times 10^{-5}$M. Synthesis of RNA was inhibited at $2 - 5 \times 10^{-5}$M of both mercurials in mouse glioma cells (Nakada and Imura, 1980). Protein synthesis was markedly depressed by MM and $HgCl_2$ at concentrations above $2 \times 10^{-5}$M and completely depressed at $5 \times 10^{-5}$M (Figure 2) (Nakada et al., 1980). Inhibition of [$^3$H] -leucine transport by MM and $HgCl_2$ was 50 and 80% at $2 \times 10^{-5}$M and $5 \times 10^{-5}$M, respectively. Thus, when mercurials were added into the culture medium, both mercurials showed similar inhibitory effects on macromolecule biosynthesis.

Growth inhibition and microtubules disruption by MM in mouse glioma cells.

The growth of mouse glioma and neuroblastoma cells were much more sensitive to MM than to $HgCl_2$ (Figure 3) (Miura et al., 1978; Miura et al., 1979). Complete growth inhibition of mouse glioma cells was produced by $5 \times 10^{-6}$M MM, while about a 10 fold higher concentration of $HgCl_2$ was required for the same extent of growth inhibition. Similar growth inhibitor effects of the two mercurials were also observed in neuroblastoma cells.

The apparent difference in susceptibility of cell proliferation toward the two mercurials suggested an existence of a more vulnerable target to MM other than macromolecule synthesis.

During the microscopic examination of cells treated with mercurials, an accumulation of mitotic cells was noticed. As shown in Figure 4, a remarkable increase in mitotic index was observed in the cells treated with a growth inhibitory concentration ($5 \times 10^{-6}$M) of MM for 3 to 6 h (Miura et al., 1978). In contrast, an increase of mitotic cells was not observed by treatment with mercuric chloride ($10^{-8} - 10^{-4}$M) (Miura et al., 1979).

The ability of MM to increase the mitotic cell population suggested an alteration of mitotic process by MM treatment. Electron microscopic observation confirmed that microtubules, which act as the mitotic spindle fibers, were absent in cells treated with MM at a growth inhibitory concentration ($5 \times 10^{-6}$ M) for 4 h (Miura et al., 1978). No remarkable morphological change in any cell organelles other than microtubules was observed during 4 h of MM treatment.. Furthermore, indirect immunofluorescence studies using rabbit antiporcine tubulin antibody demonstrated that microtubular networks in mouse glioma cells in interphase were also completely disrupted within 1 h of exposure to MM at a growth inhibitory concentration (Figure 5) (Imura et al., 1980;

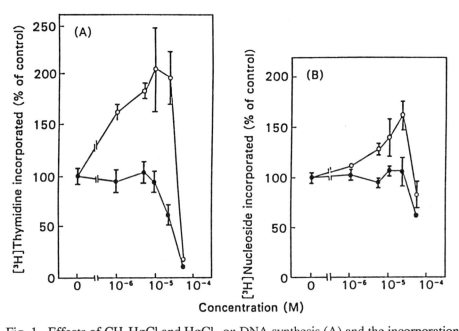

Fig. 1. Effects of $CH_3HgCl$ and $HgCl_2$ on DNA synthesis (A) and the incorporation of deoxyribonucleosides into total cell material (B) in mouse glioma cells. (●) $CH_3HgCl$, (○) $HgCl_2$.

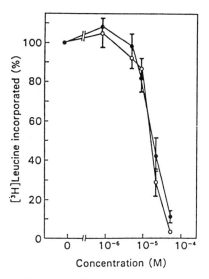

Fig. 2.  Effects of CH$_3$HgCl and HgCl$_2$ on protein synthesis of mouse glioma cells
( ● ) CH$_3$HgCl, ( ○ ) HgCl$_2$

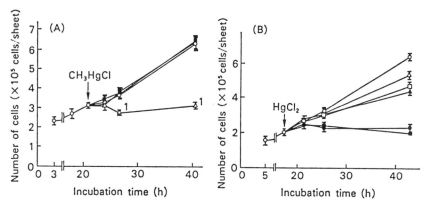

Fig. 3.  Growth inhibition of mouse glioma cells by CH$_3$HgCl (A) and HgCl$_2$ (B)
A: ( ○ ) control, ( □ ) 10$^{-8}$M, ( ■ ) 10$^{-7}$M, ( ▲ ) 10$^{-6}$M, ( △ ) 5 x 10$^{-6}$M

B: ( ○ ) control, ( △ ) 5 x 10$^{-6}$M, ( □ ) 10$^{-5}$M, ( ▲ ) 2 x 10$^{-5}$M,
( ● ) 5 x 10$^{-5}$M, ( ■ ) 10$^{-4}$M.

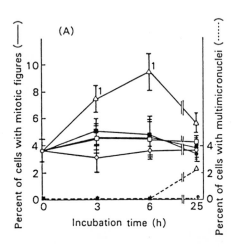

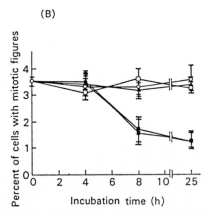

Fig. 4. Effects of $CH_3HgCl$ (A) and $HgCl_2$ (B) on mitotic figure in mouse glioma cells.

A: (○) control, (□) $10^{-8}M$, (■) $10^{-7}M$, (▲) $10^{-6}M$, (△) $5 \times 10^{-6}M$

B: (○) control, (□) $10^{-5}M$, (▲) $2 \times 10^{-5}M$, (●) $5 \times 10^{-5}M$, (■) $10^{-4}M$.

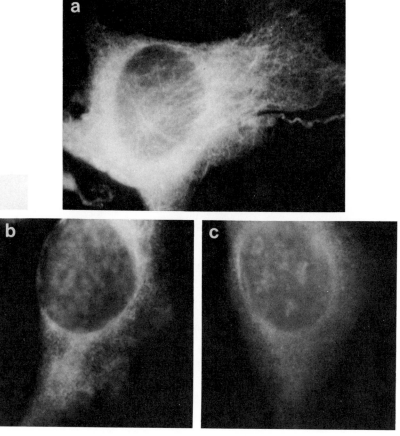

Fig. 5.  Immunofluorescence staining of microtubules in mouse glioma cells exposed to CH₃HgCl

(a) control, (b) 0.1μg/ml colcemid, $10^{-6}$M CH₃HgCl for 60 min.

Miura et al., 1984). These results suggest that MM blocks tubulin polymerization or disrupts existing microtubules at an early stage of growth inhibition of mouse glioma cells.

In contrast, inorganic mercury at a growth inhibitory concentration ($5 \times 10^{-5}$M) seemed to cause widespread lesions in multiple cell organelles including microtubules. (Miura et al., 1979). These results suggest that the specific effect of MM on microtubules described above may account for its higher growth inhibitory activity than that of inorganic mercury.

However, the direct effects of $HgCl_2$ on *in vitro* tubulin polymerization were more potent than those of MM (Figure 6) (Imura et al., 1980; Miura et al., 1984; Imura and Miura, 1986). Binding experiments of both mercurials to tubulin protein demonstrated that 1.3 moles of MM per tubulin dimer inhibited the polymerization of tubulin, while 0.7 mole of mercuric chloride per tubulin dimer blocked the polymerization (Figure 7). Furthermore, inhibition of tubulin polymerization by MM was reversed by dithiothreitol addition, in contrast to the case of mercuric chloride where no recovery of polymerization was observed (unpublished data). In spite of the more potent effect of mercuric chloride than MM on tubulin protein in the cell free system, microtubules in the cells were preferentially damaged by MM when the mercurials were added into the culture medium. Thus, it was concluded that the difference in sensitivity of cell proliferation toward these two mercurials may be ascribed to differences in their membrane permeability. The amount of MM taken up by mouse glioma or neuroblastoma cells was about 10 times larger than that of inorganic mercury at $10^{-6}$ - $10^{-5}$M concentrations (Nakada and Imura, 1987). At concentrations higher than $2 \times 10^{-5}$M, barrier function of the plasma membrane was disrupted by both MM and $HgCl_2$ (Nakada and Imura, 1987). In the presence of $10^{-5}$M MM, substantial amount of MM was taken up by the cells, however, no release of [$^3$H]2-deoxyglucose from the cells was observed. Thus, MM seems to be able to penetrate the plasma membrane without breaking down its barrier function. On the other hand, inorganic mercury is likely to profoundly disorder cellular functions at the comparatively high concentrations which interfered with membrane barrier functions.

$Cd^{2+}$, $Cu^{2+}$ and $Cr^{3+}$ inhibited *in vitro* tubulin polymerization (Miura et al., 1984). However, indirect immunofluorescence studies using rabbit antiporcine tubulin antibody revealed that in the presence of $Cd^{2+}$, $Cu^{2+}$ or $Cr^{3+}$ at their growth inhibitory concentrations no effect on microtubular networks was observed in mouse glioma cells (Miura et al., 1984).

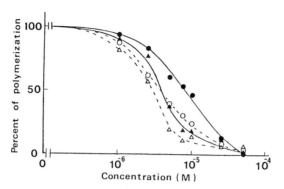

Fig. 6. Effects of $CH_3HgCl$ and $HgCl_2$ on *in vitro* tubulin polymerization
Reaction mixtures were preincubated with various concentrations of
$CH_3HgCl$ ( ○ , ● ) or $HgCl_2$ ( ▲ , △ ) for 10 min at 0°C, and then
polymerization was measured after 5 min ( ○ , △ ) or 30 min ( ● , ▲ )
incubation at 35°C.

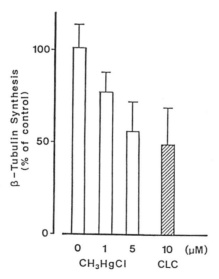

Fig. 7. Amount of mercury bound to tubulin and extent of tubulin polymerization
at various concentrations of mercurials.
( ○ , ● ) $CH_3HgCl$, ( □ , ■ ) $HgCl_2$

The results described above clearly show that the microtubule is the most susceptible cell organelle to MM and its disruption is responsible for the growth inhibition observed. In addition to our investigations, mitosis has been demonstrated as one of the most sensitive stages for MM cytotoxicity by the kinetic analysis of MM-induced cell cycle disruption in CHO cells (Vogel et al., 1986). Sager et al. (1983) and Sager and Syversen (1984) reported the time and dose dependency of microtubule disruption by MM in relation to MM accumulation in the cells. Furthermore, it was confirmed by *in vivo* experiments that cells in the early stage of mitosis accumulated, however, the number of late mitotic figures decreased in several regions of the developing brains in mice exposed to MM *in utero* on day 12 of gestation (Rodier et al., 1984).

## Effect of microtubule disruption by MM on tubulin synthesis

Recently, it was demonstrated using drugs affecting tubulin polymerization that tubulin synthesis was controlled by the concentration of unpolymerized tubulin subunits in cytoplasm (Ben Ze'ev et al., 1979; Cleveland et al., 1983). Colchicine, which depolymerizes microtubules, depresses the rate of tubulin synthesis, whereas taxol, which is known to stabilize microtubules, slightly enhances the rate of tubulin protein synthesis. Therefore, the effect of depolymerization of microtubules by MM on tubulin synthesis was studied (Miura and Imura, 1989). Mouse glioma cells were incubated with various concentrations of MM for 3 h, then pulse-labeled with [$^{35}$S]-methionine for 15 min. Colchicine was used as a positive control for this series of experiments. The pulse-labeled proteins were analyzed by two-dimensional gel electrophoresis, followed by autoradiography and densitometry. The radioactivity of ß-tubulin was markedly reduced in MM treated cells compared with that in control cells. Inhibition of ß-tubulin synthesis in cells treated with $5 \times 10^{-6}$M MM was 50 - 70% (Figure 8). Protein bands other than tubulin on gradient urea-PAGE gel remained unchanged for 3 h in the cells under the same condition as above, although total protein synthesis monitored by measuring TCA-precipitable counts was slightly inhibited. Similar results were obtained in neuroblastoma cells.

The mRNA levels of tubulin and actin were examined using mouse ß5-tubulin cDNA and ß-actin cDNA as probes, respectively. Total RNA was extracted from cell homogenates and Northern blot analysis was carried out. The resulting autoradiogram showed that tubulin mRNA level declined in response to MM, whereas no significant

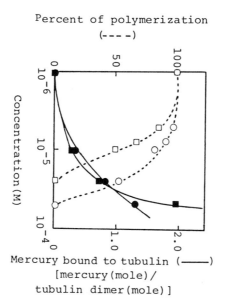

Fig. 8.  Effects of CH₃HgCl on tubulin synthesis in mouse glioma cells. CLC, colchicine

decline was observed for actin mRNA. The extent of decrease in tubulin mRNA level was similar to that of tubulin protein synthesis in $5 \times 10^{-6}$M MM-treated cells. To measure the specific transcription rates of ß-tubulin and ß-actin genes, nuclear run-on experiments were carried out.  Cloned DNAs were linearized and blotted onto nitrocellulose membranes and hybridized *in vitro* to RNAs from isolated nuclei. The relative transcription rate of ß-tubulin in the cells treated with $5 \times 10^{-6}$M MM for 3 h remained unchanged. From these results it was concluded that the reduction in tubulin synthesis by growth inhibitory concentrations of MM was associated with a specific decline of mRNA level but not with the transcription rate of tubulin gene as in the case of colchicine treatment.

It is well known that the axoplasm of the neuronal axon contains a large quantity of microtubules and that neuronal microtubules provide the structure which is essential to the axonal transport.  Considering the functions of microtubules in neuronal cells together with our experimental results, it is suggested that the specific disruption of microtubule structure and the successive inhibition of tubulin synthesis by MM play s important roles in the manifestation of MM neurotoxicity.

REFERENCES

Ben Ze'ev, A., Farmer, S.R., and Penman, S., 1979, Mechanisms of regulating tubulin synthesis in cultured mammalian cells, Cell, 17:319.

Chao, E.S.E., Gierthy, J.F., Frenkel, G.D., 1984, A comparative study of the effects of mercury compounds on cell viability and nucleic acid synthesis in HeLa cells, Biochem. Pharmacol., 33:1941.

Cleveland, D.W., Pittenger, M.F., and Feramisco, J.R., 1983, Elevation of tubulin levels by microinjection suppresses new tubulin synthesis, Nature (London), 305:738.

Gruenwedel, D.W., and Cruikshank, M.K., 1979. Effect of methylmercury (II) on the synthesis of deoxyribonucleic acid, ribonucleic acid and protein in HeLa S3 cells, Biochem. Pharmacol., 28:651.

Imura, N., Miura, K., Inokawa, M., and Nakada, S., 1980, Mechanism of methylmercury cytotoxicity: by biochemical and morphological experiments using cultured cells, Toxicology, 17:241.

Imura, N., and Miura, K., 1986, Mode of toxic action of methylmercury, in: "Recent Advances Minamata Disease Studies," T. Tsubaki and H. Takahashi, eds., pp. 169-188, Kodansha, Tokyo.

Miura, K., and Imura, N., 1987, Mechanism of methylmercury cytotoxicity, CRC Crit. Rev. Toxicol., 18:161.

Miura, K., and Imura, N., 1989, Mechanism of cytotoxicity of methylmercury, with special reference to microtubule disruption, Biol. Trace Element Res., 21:313.

Miura, K., Inokawa, M., and Imura, N., 1984, Effects of methylmercury and some metal ions in microtubule networks in mouse glioma cells and in vitro tubulin polymerization, Toxicol. Appl. Pharmacol., 73, 218.

Miura, K., Nakada, S., Suzuki, K., and Imura, N., 1979, Ultrastructural studies on the cytotoxic effects of mercuric chloride on mouse glioma, Ecotoxicol. Environ. Saf., 3:352.

Miura, K., Suzuki, K., and Imura, N., 1978, Effects of methylmercury on mitotic mouse glioma cells, Environ. Res., 17:453.

Nakada, S., and Imura, N., 1980, Stimulation of DNA synthesis and pyrimidine deoxyribonucleoside transport systems in mouse glioma and mouse neuroblastoma cells by inorganic mercury, Toxicol. Appl. Pharmacol., 53:24.

Nakada, S., and Imura, N., 1987, Uptake of methylmercury and inorganic mercury by mouse glioma and mouse neuroblastoma cells, Neurotoxicology, 3:249.

Nakada, S., Nomoto, A., and Imura, N., 1980, Effect of methylmercury and inorganic mercury of protein synthesis in mammalian cells, Ecotoxicol. Environ., Saf., 4:184.

Rodier, P.M., Aschner, M., and Sager, P.R., 1984, Mitotic arrest in the developing CNS after prenatal exposure to methylmercury, Neurobehav. Toxicol. Teratol., 6:379.

Sager, P.R., Doherty, R.A., and Olmsted, J.B., 1983, Interaction of methylmercury with microtubules in cultured cells and in vitro, Exp. Cell Res., 146:127.

Sager, P.R., and Syversen, T.L.M., 1984, Differential responses to methylmercury exposure and recovery in neuroblastoma and glioma cells and fibroblasts, Exp. Neurol., 85, 371.

Sarafian, T.A., Cheung, M.K., and Verity, M.A., 1984, *In vitro* methylmercury inhibition of protein synthesis in neonatal cerebellar perikarya, Neuropathol. Appl. Neurobiol., 10:85.

Vogel, D.G., Rabinovitch, P.S., and Mottet, N.K., 1986, Methylmercury effects on cell cycle kinetics, Cell Tissue Kinet., 19:227.

# DNA DAMAGE BY MERCURY COMPOUNDS: AN OVERVIEW

Max Costa, Nelwyn T. Christie, Orazio Cantoni*, Judith T. Zelikoff,
Xin Wei Wang and Toby G. Rossman

Institute of Environmental Medicine
New York University Medical Center
550 First Avenue
New York, New York 10016

*Istituto di Farmacologia
Universita di Urbino
61029 Urbino (PS)
Italy

ABSTRACT

HgCl$_2$-induced DNA damage has many similarities to those caused by X-rays; however, the single strand breaks induced by HgCl$_2$ are not readily repaired, in contrast to those induced by X-rays. HgCl$_2$ has also been shown to inhibit the repair of X-ray induced DNA single strand breaks. HgCl$_2$ has been shown to induce frank single strand breaks, not alkali-labile sites. DNA repair was assessed by the disappearance of DNA lesions as evaluated by alkaline elution studies and also by CsCl density gradient analysis. Similar to X-rays, HgCl$_2$ has also been shown to cause the formation of superoxide radicals and the depletion of reduced glutathione in intact cells. The binding of mercury to DNA was shown to be very tight since it resisted extraction with high salt and chelating agents. The binding to DNA depended upon the polynucleotide structure of the DNA, because degradation of DNA to mononucleotides resulted in the release of bound $^{203}$Hg. Methyl-HgCl has been compared with $HgCl_2$ for the induction of DNA strand breaks in cultured rat glial cells, human nerve cells (HTB), and rat or human fibroblasts. Methyl-HgCl was much more effective at inducing DNA single strand breaks in cultured nerve cells compared with the fibroblasts. Methyl-HgCl also reduced the plating efficiency of nerve cells more than fibroblasts. These and other findings suggest that the potent cellular toxicity induced by $HgCl_2$ and methyl-HgCl involve DNA damage. Although methyl-$HgCl_2$ is more potent in producing DNA damage and cytotoxicity in nerve cells than in fibroblasts, it is more toxic to fibroblasts than $HgCl_2$. The lack of repair of the extensive mercury-induced DNA damage may, in part, explain the lack of carcinogenicity of these agents, despite their potential for inducing DNA damage.

## INTRODUCTION

Molecular mechanisms of mercury toxicity are not well-understood. At the level of the whole organism, it is known that the divalent ionic form of inorganic mercury primarily damages the kidney, while both organic and metallic mercury are known to penetrate into the central nervous system (CNS) and produce severe toxicity in this organ (Clarkson et al., 1988). Methyl mercury (methyl-HgCl), for example, greatly impairs sensory input into the CNS, affecting hearing and sight (Clarkson et al., 1988).

At the cellular level, there have been limited studies on the toxicity of mercury compounds. Mercury is an extremely reactive, very soft metal whose ions form coordinate covalent bonds with ligands such as sulfhydryl groups. The reactivity and toxicity of Hg(II) was the highest observed for any divalent metal ion (Williams et al., 1982). Methyl-HgCl is even more toxic than inorganic mercury, particularly to nerve cells (Clarkson et al., 1988). The basis for the nerve cell specific toxicity of methyl-HgCl may reside in its ability to affect microtubule functions and also, due to its high lipid solubility, it readily enters cells (Clarkson et al., 1988).

Previous studies have shown that mercury chloride ($HgCl_2$) is very potent at producing DNA damage in mammalian cells (Cantoni et al., 1982; Cantoni and Costa, 1983; Cantoni et al., 1983; Cantoni et al., 1984; Christie et al., 1986). Despite this potency in damaging the DNA, mercury compounds have not been reported to be carcinogenic or mutagenic in many systems. In the present study, we have reviewed some of our previous work on the mechanisms by which mercury induces DNA damage. We have also presented new data on the effects of methyl-HgCl on DNA in cultured fibroblasts, as well as in cultured nerve cells.

MATERIALS AND METHODS

DNA damage was assessed by the alkaline elution method, as previously described (Cantoni et al., 1982; Cantoni et al., 1984). Superoxide radicals were measured by cytochrome C reduction (Cantoni et al., 1984). DNA was isolated, mercury complexes bound to DNA were quantitated (Cantoni et al., 1984). All other culture procedures have been described previously (Cantoni et al., 1984). Unless otherwise indicated, cultured Chinese hamster ovary cells were utilized in all the experiments described.

RESULTS

DNA damage induced by mercury compounds: Figure 1 shows the striking effects of $HgCl_2$ on the induction of DNA damage in CHO cells, both as a function of concentration and time. Figure 1A illustrates the mercury concentration dependency for the induction of single strand breaks, as measured by alkaline elution, while Figure 1B illustrates the time dependency for strand break generation. Additional work in this area has shown that mercury induces true strand breaks and not alkaline label sites (Cantoni et al., 1984). Alkaline elution studies have also revealed that mercury does not produce substantial amounts of DNA-protein crosslinks, but does yield some DNA-DNA crosslinks (Cantoni et al., 1984). Thus, the most abundant DNA lesions induced by mercury in intact cultured cells are strand breaks. The appearance of DNA strand breaks correlate with the threshold of $HgCl_2$ cytotoxicity (Table 1).

The rapid induction of DNA strand breaks by $HgCl_2$ exhibited considerable similarity to the strand breaks induced by X-rays. A number of our studies have pointed to the similarity in strand breaks induced by $HgCl_2$ to those induced by X-rays (Cantoni et al., 1982; Cantoni and Costa, 1983; Cantoni et al., 1984). Figure 2 compares the induction of strand breaks by $HgCl_2$ and X-rays as a function of dose. There are similarities in the dose dependency for the induction of DNA strand breaks for the two agents. In accordance with the relationship between strand breaks and cytotoxicity for $HgCl_2$, a similar high correlation also exists for the induction of cytotoxicity by X-rays being caused by DNA strand breaks (Cantoni and Costa, 1984). Therefore, one of the mechanisms by which both $HgCl_2$ and X-rays kill cells is probably through the induction of DNA single strand breaks.

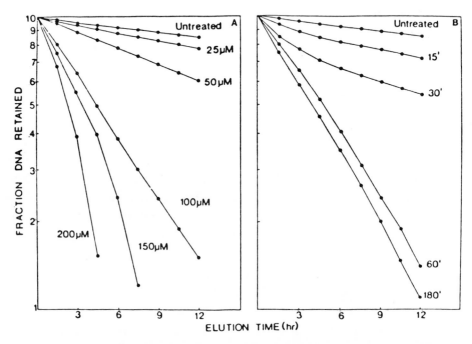

Fig. 1. Concentration and early temporal effects of HgCl$_2$ on DNA strand breaks analyzed by alkaline elution. CHO cells were treated for 1 hr with varying concentrations of HgCl$_2$ (A) or with a 100 μM concentration of HgCl$_2$ for varying time intervals (B). (Reproduced with permission from Chem.-Biol. Interactions. (Cantoni, O., Christie, N.T., Robison, S.H. and Costa, M. Characterization of DNA lesions produced by HgCl$_2$ in cell culture systems. Chem.-Biol. Interact. 49:209-224, 1984).

Table 1  Effect of HgCl$_2$ on plating efficiency and growth of CHO cells

| HgCl$_2$ Concentration ($\mu$M) | Plating Efficiency[b] (% Controls) | Cell Growth[c] (% Controls) |
|:---:|:---:|:---:|
| 1 | N.D.[a] | 110 ± 14 |
| 2.5 | N.D. | 91 ± 7 |
| 5 | N.D. | 34 ± 9 |
| 7.5 | 104 | 3 ± 1 |
| 10 | 93.8 | 0 |
| 25 | 72.3 | N.D. |
| 50 | 18.7 | N.D. |
| 75 | 0 | N.D. |

[a]Not determined.

[b]CHO cells were incubated for 1 h with McCoy's medium containing various concentrations of HgCl$_2$. Following this treatment cells were plated to form colonies for 8 days.

[c]10$^5$ CHO cells were incubated in McCoy's medium containing different concentrations of HgCl$_2$ and, 24 h later, the cell number of each plate was determined with a Coulter particle counter.

(Reproduced with permission from **Chem.-Biol. Interactions.** [Cantoni, O., Christie, N.T., Robison, S.H. and Costa, M. Characterization of DNA lesions produced by HgCl$_2$ in cell culture systems. **Chem.-Biol. Interactions 49**:209-224, 1984]).

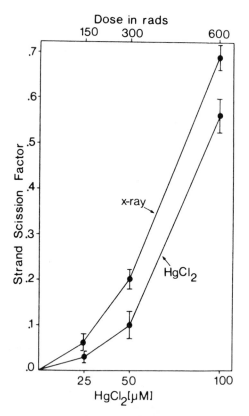

Fig. 2. <u>Similarity in the Induction of DNA strand breaks by HgCl$_2$ and X-rays</u>.
Alkaline elution was employed to determine the incidence of DNA strand
breaks following 1 h of exposure to the indicated concentrations of HgCl$_2$
or X-rays. (Reproduced with permission from Biochem. Biophys. Res.
Commun. (Cantoni, O., Evans, R.M. and Costa, M. Similarity in the acute
cytotoxic response of mammalian cells to mercury(II) and X-rays: DNA
damage and glutathione depletion. Biochem. Biophys. Res. Commun.
208(2):614-619, 1982.

<u>Oxygen radical production by HgCl$_2$</u>: Since HgCl$_2$ mediated-strand breaks
exhibit similar kinetics to those induced by X-rays, we examined whether HgCl$_2$
treatment of cells resulted in the production of oxygen radical species, as are known to
be produced by X-rays. Figure 3 shows the production of superoxide anion radical
in cultured fibroblasts as a function of exposure concentration of HgCl$_2$, while
Figure 4 shows time-dependent, HgCl$_2$-mediated production of superoxide radicals
which disappeared upon the addition of exogenous superoxide dismutase enzyme.

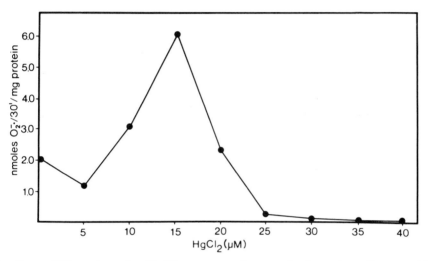

Fig. 3.  Effect of varying HgC1$_2$ concentrations on the formation of superoxide anion radicals. Cells were treated for 30 min with concentrations of HgC1$_2$ shown, and superoxide levels in the medium were measured by the reduction of exogenously added cytochrome C. (Reproduced with permission from Mol. Pharmacol. (Cantoni, O., Christie, N.T., Swann, A., Drath, D.B. and Costa, M. Mechanism of HgC1$_2$ Cytotoxicity in Cultured Mammalian Cells. Mol. Pharmacol. 26:360-368, 1984).

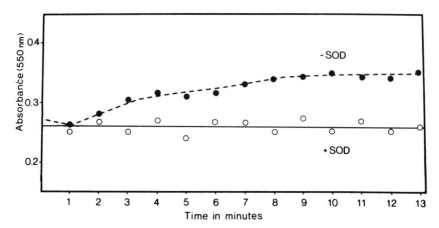

Figure 4. HgCl₂-dependent superoxide formation by CHO cells. Cells (1 X 10⁶)
were incubated in Hanks' balanced salt solution containing 15 μM HgCl₂
in the presence 0-----0 or absence 0-----0 of superoxide dismutase.
Leakage of superoxide anion radicals ($O_2$) in the medium was measured
by the reduction of exogenously added cytochrome at various time
intervals following addition of HgCl₂. (Reproduced with permission
from Mol. Pharmacol. (Cantoni, O., Christie, N.T., Swann, A., Drath,
D.B. and Costa, M. Mechanism of HgCl₂ Cytotoxicity in Cultured
Mammalian Cells. Mol. Pharmacol. 26:360-368, 1984).

The effect of mercury chloride on repair of DNA damage: In contrast to X-ray
induced single strand breaks which are rapidly repaired, those induced by HgCl₂ are
not readily repaired as illustrated in Figure 5. The lack of complete repair of strand
breaks induced by HgCl₂ is likely to explain both its potent toxicity and also its lack of
mutagenicity and carcinogenicity. DNA repair was measured not only by assessing the
repair of strand breaks measured with alkaline elution but also using cesium chloride
density gradient centrifugation. Figure 6 illustrates the low level of repair induced by
HgCl₂ compared to that induced by calcium chromate in cesium chloride gradients. In
fact, addition of HgCl₂ to the media inhibits the normal repair of X-ray induced
damage (Figure 7). This HgCl₂ inhibition of DNA repair of X-ray damage occurs at

very low (10 μM) mercury concentrations. For example, in other studies effects on repair are observed following a 15 minute exposure to 1 to 2.5 μM HgCl$_2$ in cells after a 15 minute exposure (Christie et al., 1986). These represent extremely low, non-cytotoxic levels of mercury (Table 1). In contrast to the HgCl$_2$ inhibition of repair of X-ray induced single strand breaks, HgCl$_2$ does not alter the accumulation of breaks associated with the repair of UV-induced damage (Table 2).

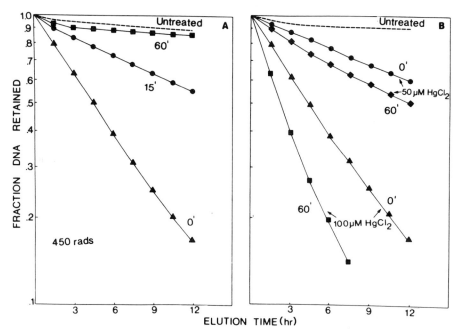

Fig. 5. Induction of DNA single-strand breaks and their repair following treatment with HgCl$_2$ and X-ray. Cells were treated with HgCl$_2$ for 1 h (A) or with X-rays (B), and their DNA was analyzed with alkaline elution immediately following treatment or after incubation of the cultures at 37° for an additional 15-60 min in fresh media. Alkaline elution analysis was performed as previously described (Cantoni et al., 1984). (Reproduced with permission from Mol. Pharmacol. (Cantoni, O. and Costa, M. Correlations of DNA Strand Breaks and Their Repair with Cell Survival Following Acute Exposure to Mercury(II) and X-Rays. Mol. Pharmacol. 24:84-89, 1983).

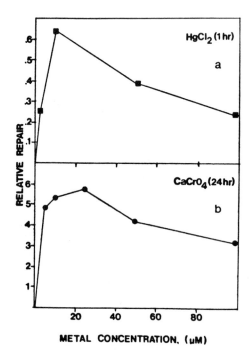

Fig. 6.  DNA-repair replication induced by treatment with mercuric chloride or
calcium chromate. Cells were treated with mercuric chloride in serum free
medium for 1 h and with calcium chromate for 24 h.  Relative repair is the
ratio of $^3$H cpm to $^{32}$P cpm for $10^5$ surviving cells. (Reproduced with
permission from Mutation Res. (Robison, S.H., Cantoni, O. and Costa,
M.  Analysis of metal-induced DNA lesions and DNA-repair replication in
mammalian cells.  Mutation Res. 131:173-184, 1984).

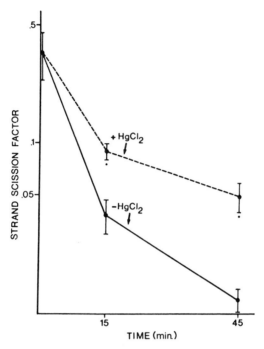

Fig. 7. <u>Effect of noncytotoxic concentrations of HgCl$_2$ on the rejoining of DNA single-strand breaks induced by X-rays.</u> Following exposure to 450 rads, cells were allowed to repair at 37° for 15-45 min in the absence or presence of 10 μM HgCl$_2$. Alkaline elution was performed on these cultures. DNA single-strand breaks were expressed as SSF, calculated from the alkaline elution patterns by the following relationship: SSF - -log A/B, where A - amount of DNA retained in the sixth fraction of the untreated sample and B - DNA retained in the sixth fraction of the treated sample. Asterisks adjacent to HgCl$_2$-treated cell points indicate statistically significant differences from untreated cultures (p < 0.05, Student's t-test). (Reproduced with permission from Mol. Pharmacol. (Cantoni, O. and Costa, M. Correlations of DNA strand Breaks and Their Repair with Cell Survival following Acute Exposure to Mercury(II) and X-Rays. Mol. Pharmacol. 24:84-89, 1983).

Table 2  The effect of HgCl$_2$ on the repair of UV-induced Pyrimidine dimers

| Treatment | SSF[a] | |
|---|---|---|
| | Repair Time After UV[b] | |
| | 1 h | 3 h |
| UV | 0.5 | 0.39 |
| UV + 15 $\mu$M Hg | 0.58 | 0.36 |
| UV + 25 $\mu$M Hg | 0.46 | 0.33 |
| UV + 50 $\mu$M Hg | 0.57 | 0.46 |
| UV + 150 $\mu$M Hg | 0.7 | |

[a]The strand scission factor was calculated as indicated in Materials and Methods.  Each value is an average of two to four determinations and the variation range is from 0.01 to 0.06.

[b]After UV radiation of CHO cell monolayers at a fluence of 5 J/m$^2$, $\alpha$-MEM medium with or without HgCl$_2$ was applied for the times indicated before harvesting cells for alkaline elution.

(Reproduced with permission from **Mol. Pharmacol.** [Christie, N.T., Cantoni, O., Sugiyama, M., Cattabeni, F., and Costa, M., 1986, Differences in the Effects of Hg(II) on DNA Repair Induced in Chinese Hamster Ovary Cells by Ultraviolet or X-Rays, Mol. Pharmacol., 29:173-178]).

Binding of mercury to DNA: In order to further investigate the mechanism of DNA damage induced by HgCl$_2$, we initiated studies of the binding of mercury to DNA (Table 3). These studies indicate that HgCl$_2$ (Hg$^{++}$) exhibits a high rate of exchange reactions in cell lysates making it difficult to study the DNA binding activity of Hg(II). As shown in Table 3, the addition of cytoplasmic complexes from $^{203}$HgCl$_2$-treated cultures to cells that have not been treated with HgCl$_2$ resulted in substantial binding to DNA. Similarly, if cells are treated with unlabeled HgCl$_2$ and, subsequent to this, radioactive HgCl$_2$ cytoplasmic complexes were added to these cells, radioactive mercury was bound to DNA at similar levels, showing either a high rate of exchange or a lack of saturation of sites by the pretreatment with unlabelled HgCl$_2$. However, when intact cells are treated with radioactive mercury, without further addition of the cytoplasmic complexes, there is more binding to the DNA than was present when the cytoplasmic complexes were added to whole cell lysates. These

Table 3  Effect of $^{203}$Hg complexes added to proteinase K lysates on Hg(II) binding to DNA

| Initial Treatment | Addition of Cytoplasmic $^{203}$Hg Complexes from Treated Cells | Amount of Hg Bound |
|---|---|---|
| | | pmol Hg/$\mu$g DNA |
| A. None | $^{203}$HgCl$_2$, 2.5 $\mu$M | 1.0 |
| None | $^{203}$HgCl$_2$, 5 $\mu$M | 5.6 |
| B. HgCl$_2$ 5 $\mu$M | $^{203}$HgCl$_2$, 2.5 $\mu$M | 1.1 |
| HgCl$_2$, 5 $\mu$M | $^{203}$HgCl$_2$, 5.0 $\mu$M | 6.8 |
| C. $^{203}$HgCl$_2$, 2.5 $\mu$M | None Added | 1.95 |
| $^{203}$HgCl$_2$, 5 $\mu$M | None Added | 10.1 |

Initially, cell monolayers were treated for 15 min under each of the following conditions, trypsinized, counted, and mixed with $^{203}$Hg complexes isolated from an equivalent number of cells that had been treated as indicated. Cell monolayers were treated for 15 min with $^{203}$Hg, trypsinized, and counted. Cytoplasmic $^{203}$Hg complexes were isolated after cells were sonicated and centrifuged; the complexes were then added to freshly prepared proteinase K lysates of cells from the series of initial treatments.

(Reproduced with permission from **Mol. Pharmacol.** [Cantoni, O., Christie, N.T., Swann, A., Drath, D.B. and Costa, M. Mechanism of HgCl$_2$ Cytotoxicity in Cultured Mammalian Cells. **Mol. Pharmacol. 26**:360-368, 1984)].

results suggest that, if in fact, Hg binding to DNA can be estimated, but that the actual binding to DNA overestimates the amount of mercury bound by about 50%. These findings were considered in assessing mercury binding.

DNA isolated from cells, treated with mercury, exhibited bound mercury as illustrated in Table 4, where treatment of this DNA with high salt, EDTA and other disassociating agents did not displace the mercury from the DNA. Treatment of the DNA with unlabelled $HgCl_2$ displaced some radioactive mercury, but this effect was not concentration-dependent and 75% of the mercury still remained bound to the DNA. However, when DNA containing bound Hg was degraded in order to study its binding to nucleosides, all of the mercury was released from the DNA. These results suggest that mercury binding to the DNA depended upon its polynucleotide structure.

Table 4   Effect of various agents on the stability of 203Hg: DNA complexes formed intracellularly

| Agents Added in Dialysis | Concentration | % of $^{203}$Hg Remaining Bound to DNA |
|---|---|---|
| None | -- | 87 |
| NaCl | 0.25 M | 90 |
|  | 0.5 M | 91 |
|  | 1.0 M | 89 |
| EDTA | 0.1 M | 82 |
| $HgCl_2$ | $2.5 \times 10^{-6}$ M | 75 |
|  | $5 \times 10^{-6}$ M | 75 |
|  | $1 \times 10^{-6}$ M | 75 |
| Ethanol | 70% | 94 |
| Trichloroacetic Acid | 2% | 100 |
|  | 10% | 72 |

Each agent shown was included in a dialysis solution of 10 mM Tris-HCl and 5 mM EDTA (unless EDTA was added).  The DNA that was added for each dialysis condition at 10 μg/ml had been labeled with $^{203}$Hg in intact cells and subsequently purified.

(Reproduced with permission from **Mol. Pharmacol.** [Cantoni, O., Christie, N.T., Swann, A., Drath, D.B. and Costa, M.  Mechanism of $HgCl_2$ Cytotoxicity in Cultured Mammalian Cells.  **Mol. Pharmacol. 26**:360-368, 1984)].

Comparison of mercury chloride and methylmercury-induced DNA damage in fibroblasts and nerve cells: Figure 8 shows that the induction of strand breaks by methylmercuric chloride in rat glioblastoma cells was considerably greater than in Chinese hamster V79 (fetal lung) cells. Similarly, Figure 9 shows that the induction of strand breaks in human lung cells by methylmercuric chloride is less striking than those in human nerve cells. The difference in sensitivity between the induction of strand breaks is even greater than the difference in sensitivity between the plating efficiencies of the two cell types (Figure 9). Methylmercury chloride is also considerably more potent at inducing strand breaks than $HgCl_2$ (Figure 10).

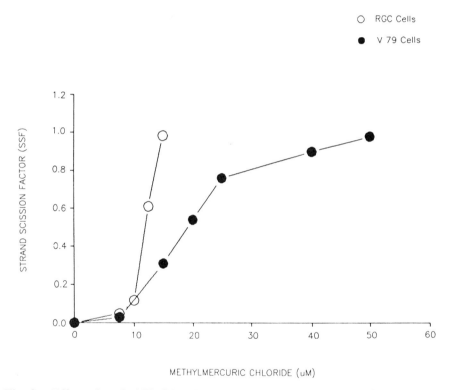

Fig. 8.   Effect of methyl-HgCl on DNA strand break in rat glial or Chinese hamster V-79 cells. Cells were treated with the indicated concentration of methylmercurichloride for 4 h. DNA damage was assessed by the alkaline elution method.

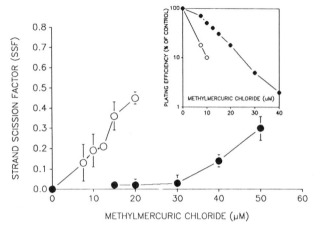

Fig. 9. Effect of methylmercuric chloride on the induction of DNA single strand breaks and cytotoxicity in human brain cells (HTB, open dircle) and human lung cells (HLF, closed circle). Cultured cells were treated with the indicated concentration of methylmercuric chloride for 4 h. DNA damage was assessed by the alkaline elution method.

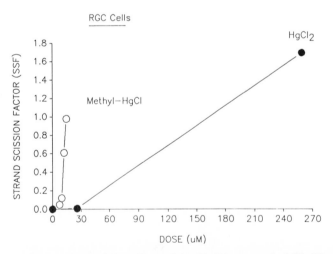

Figure 10. Comparison of the effect of the HgCl$_2$ or methyl-HgCl on DNA strand breaks. Cells were treated with the indicated concentration of the mercury compound for 4 h. DNA damage was assessed by alkaline elution.

DISCUSSION

The mercuric or methyl-HgCl ion, because of its high reactivity, is capable of producing very striking damage to cells at low concentrations. In this study, we have shown that the mercury ion is capable of generating oxygen radicals in cells. We have not uncovered the mechanism by which this occurs, but we have previously shown that it occurs at 4° to a somewhat lesser extent than at 37° (Cantoni et al., 1983; Cantoni and Costa, 1983; Cantoni et al., 1984). Additionally, $HgCl_2$ treatment of intact cells has been shown to induce strand breaks in nucleoids isolated from cells at the same concentration as it induces strand breaks when added directly to the nucleoids *in vitro* (Cantoni et al., 1983). The quantity of oxygen radicals generated by $HgCl_2$ does not appear sufficient to account for the high degree of strand breakage that occurs (Cantoni et al., 1984). We have also shown that mercury binds directly with DNA. This interaction may also contribute to the induction of DNA strand breaks. The rapidity and potency with which mercury produces strand breaks would suggest that it should at some concentrations, be highly mutagenic and carcinogenic. However, mercury is very potent at inhibiting DNA repair and the strand breaks are not repaired. This, we believe, leads to cell death, since irreparable DNA damage is lethal. In fact, we have even shown that mercury is able to inhibit the repair of X-ray-induced damage, but does not alter the accumulation of breaks associated with the repair of UV-induced DNA damage. Therefore, mercury does not affect all DNA repair systems equally.

There are obviously many ways that inorganic mercury and methyl-HgCl could produce cytotoxicity. Other studies show that methyl-HgCl is exquisitely active in disrupting microtubule structures and this may, in fact, be one of the mechanisms for its cytotoxicity. However, both methyl-HgCl and $HgCl_2$ also produce DNA strand breaks in cells. Since these DNA strand breaks are not repaired, this may also be of significant consequence in producing cell death. Some of the data suggest that there may be a dose at which mercury induces repairable strand breaks. These results suggest the possibility that at certain optimal relatively low concentrations, which may be difficult to find, there might be some mutagenic and carcinogenic effects from this metal. A very careful titration of dose and exposure may help find these critical concentrations of mercury which would produce mutagenic and carcinogenic effects. Mercury, very clearly, binds tightly to the DNA and there was no agent found that could dissociate Hg(II) from DNA, however, upon degradation of the DNA, mercury was released from its binding sites. These results suggests that mercury may be binding at the hydrogen binding sites of DNA (Katz, 1963) and its binding depends upon the polynucleotide structure of the DNA. There has not been much work examining mercury binding to DNA and the consequences of this binding. Mercury enters cells rapidly and oxygen radicals form. Mercury is bound to many sites in the cell, exchanges rapidly and, therefore, the consequences of mercury exposure can be quite severe, because it can interact with many ligands.

The more potent effects of methyl-HgCl than $HgCl_2$ on nerve cells, compared to fibroblasts, are interesting because the strand breaks follow the cytotoxic response, suggesting that, in fact, strand breaks are involved in the cytotoxicity of methyl-HgCl, as well as $HgCl_2$. It is unlikely that the DNA of nerve cells is more sensitive to methyl-$HgCl_2$ *per se*, but rather the nerve cell promotes a greater uptake of methyl-HgCl. Strand breaks are induced at exquisitely low concentrations of methyl-HgCl in the nerve cells, sometimes for human nerve cells the concentration is less than 1 μM following a relatively short exposure time. These results argue that DNA strand breaks induced by methyl-HgCl in nerve cells may be important in the ensuing cytotoxic response, but that for other reasons, ( i.e. uptake), the nerve cell is more sensitive than other cells.

## REFERENCES

Cantoni, O., and Costa, M., 1983. Correlations of DNA strand breaks and their repair with cell survival following acute exposure to mercury(II) and X-rays, Mol. Pharmacol., 24:84-89.

Cantoni, O., Christie, N.T., Swann, A., Drath, D.B., and Costa, M., 1984. Mechanism of $HgCl_2$ cytotoxicity in cultured mammalian cells, Mol. Pharmacol., 26:360-368.

Cantoni, O., Evans, R.M., and Costa, M., 1982. Similarity in the acute cytotoxic response of mammalian cells to mercury (II) and X-rays: DNA damage and glutathione depletion, Biochem. Biophys. Res. Commun. 108(2):614-619.

Cantoni, R., Christie, N.T., Robison, S.H., and Costa, M., 1984. Characterization of DNA lesions produced by $HgCl_2$ in cell culture systems, Chem.-Biol. Interact. 49:209-224.

Christie, N.T., Cantoni, O., Sugiyama, M., Cattabeni, F., and Costa, M., 1986, Differences in the effects of Hg(II) on DNA repair induced in Chinese hamster ovary cells by ultraviolet or X-rays. Mol.Pharmacol., 29:173-178.

Clarkson, W.C., Hursh, J.B., Sager, P.R., and Syversen, T.L.M. 1988. Mercury, in: "Biological Monitoring of Toxic Metals," T.W. Clarkson, L. Friberg, G.F. Nordberg, and P.R. Sager, eds., 199-246, Plenum Press, New York.

Katz, S., 1963. The reversible reaction of double-stranded polynucleotides. A step-function theory and its significance. Biochim. Biophys. Acta 68: 240-253.

Robison, S.H., Cantoni, O., and Costa, M. Analysis of metal-induced DNA lesions and DNA-repair replication in mammalian cells. Mutation Res. 131:173-184, 1984.

Williams, M.W., Hoeschele, J.D., Turner, J.E., Jacobson, K.B., Christie, N.T., Paton, C.L., Smith, L.H., Witschi, H.R., and Lee, E.H., 1982. Chemical softness and acute metal toxicity in mice and drosophila. Toxicol. Appl. Pharmacol., 63:461.

# POSSIBLE MECHANISM OF DETOXIFYING EFFECT OF

# SELENIUM ON THE TOXICITY OF MERCURY COMPOUNDS

Nobumasa Imura and Akira Naganuma

Department of Public Health
School of Pharmaceutical Sciences
Kitasato University
Tokyo
Japan

ABSTRACT

Co-administration of selenium compounds with mercuric mercury resulted in remarkable depression of acute toxicity of mercury through 1) reduction of mercury accumulation in the kidneys, a major target of acute toxicity of inorganic mercury, by forming high molecular weight complexes consisting of Hg, Se and proteins in the blood, which were hardly filtered through glomerulus; 2) elongation of retention of mercury and selenium as inert high molecular weight complexes in the blood, especially in the erythrocytes; 3) formation of stable and non-diffusible complexes of Hg and Se in various organs such as the liver and kidneys.

Methylmercury was found to form a complex with selenium, bis(methylmercuric) selenide (BMS) , in various tissues on concurrent administration with selenite. However, the reduction of methylmercury toxicity by selenium as reported in numerous papers seems to be hardly explainable by the formation of BMS which showed a relatively low stability in the blood and other tissues.

Using the assay methods developed in the studies described above to detect interactions between selenium and other metals, we indicated that more than 18 metal ions might be subjected to interactions with selenium in animal body, showing important roles of this essential metalloid as a modifying factor of metal toxicity. Although, according to the experimental data so far obtained, selenium appears not to

*Advances in Mercury Toxicology*, Edited by T. Suzuki *et al.*
Plenum Press, New York, 1991

be clinically applicable as an antidote for mercury intoxication, we could develop a possible application of selenium for reducing dose-limiting toxicity of *cis*-platinum, a platinum complex most widely used in cancer chemotherapy, without compromising its anti-tumor activity.

INTRODUCTION

As reported by Parizek and Ostadalova (1967), and subsequently confirmed by many other researchers, simultaneous administration of selenite dramatically reduces the acute toxicity of inorganic mercury. As a typical example, Figure 1 demonstrates the effect of administration of selenite and/or mercuric chloride on the growth rate of mice. Individual administration of selenite or inorganic mercury markedly depresses the growth rate, however, the simultaneous injection of both compounds resulted in almost the same growth rate as that shown by control animals, suggesting that a mercury-selenium interaction occurred in the animals. This prompted us to investigate the site of primary interaction of mercuric mercury and the mechanism(s) of mutual modification of toxicity of both compounds.

Changes in Tissue Distribution of Mercury and Selenium by their Simultaneous Administration

An *in vitro* experiment showed a remarkable increase in uptake of both mercury and selenium by erythrocytes when mercuric mercury and selenite coexisted in rabbit blood (Naganuma et al., 1978). On the simultaneous i.v. administration of selenite and mercuric chloride to rabbits, the levels of mercury and selenium in blood, liver, spleen and whole body were substantially elevated, while the renal uptake of mercury was markedly decreased, compared with the case of separate administration of these compounds.

Figure 2 shows the time-dependent decrease of mercury levels in the blood of rabbits injected intravenously with 1.5 μmol/kg of $^{203}$Hg-labeled mercuric chloride with or without an equivalent dose of sodium selenite. Compared to the case of injection of only mercuric chloride, mercury which was coadministered with selenite showed a tendency to be retained both in stroma free hemolysate and in plasma for a relatively long period. When selenium was labeled with radioactive selenium ($^{75}$Se) instead of mercury, the selenium showed almost the same pattern of retention in the blood as that of mercury (Imura and Naganuma, 1978).

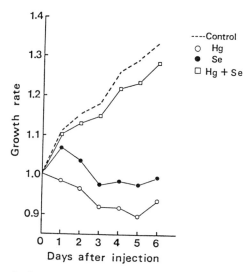

Fig. 1. Growth rate of mice administered with mercuric chloride, sodium selenite or both of the compounds. $HgCl_2$ (25 μmol/kg) and/or $Na_2SeO_3$ (25 μmol/kg) were administered i.v. (Naganuma et al.,1984a).

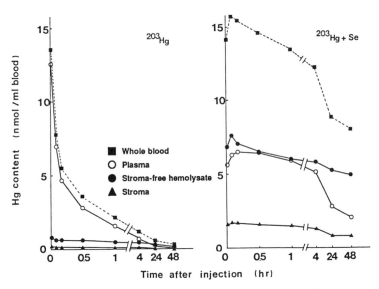

Fig. 2. Content of $^{203}$Hg in blood constituents at various times after i.v. injection of $^{203}$HgCl$_2$ with or without selenite (Imura and Naganuma, 1978).

Gel filtration patterns on Sephadex G-200 of the plasma and stroma-free hemolysate from the rabbits administered mercuric chloride and/or sodium selenite are shown in Figure 3. Intravenous coadministration of inorganic mercury and selenite markedly changed the elution patterns of both the mercury and selenium either in the stroma-free hemolysate or in the plasma. The patterns indicate the formation of high molecular weight complexes containing almost equal amounts of mercury and selenium in the blood of mice taken 1 h after concurrent administration of both the compounds (Naganuma and Imura, 1980b, 1981).

Factors Necessary for Complex Formation in the Blood

Incubation of $^{203}$Hg-HgCl$_2$ and $^{75}$Se-Na$_2$SeO$_3$ with rabbit plasma and subsequent gel filtration of the reaction mixture on Sephadex G-200 column did not result in the formation of any complex of $^{203}$Hg and $^{75}$Se. However, similar high molecular weight complexes of $^{203}$Hg and $^{75}$Se to those observed in the plasma of rabbits administered $^{203}$Hg-HgCl$_2$ and $^{75}$Se-Na$^2$SeO$_3$ were formed in rabbit plasma incubated with both compounds in the presence of glutathione (Naganuma and Imura, 1983). Incubation of the stroma-free hemolysate, which is rich in glutathione, with $^{203}$Hg-

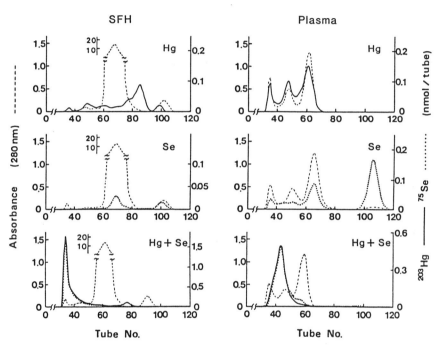

Fig. 3. Sephadex G-200 chromatograms of rabbit plasma and stroma-free hemolysate (SFH) 1 h after administration of $^{203}HgCl_2$ and/or $Na_2{}^{75}SeO_3$ (Naganuma and Imura, 1981)

HgCl$_2$ and $^{75}$Se-Na$_2$SeO$_3$ yielded complexes similar to those in the hemolysate of rabbits injected with both compounds. These results indicate that glutathione isessential for the formation of complexes containing equimolar amounts of Hg and Se with proteins in rabbit blood. Glutathione is probably needed for reducing selenite to the level of selenide, which may easily interact with mercuric ion and proteins to form high molecular weight complexes (Naganuma and Imura, 1983).

Behavior of Mercury and Selenium in the Liver

Gel filtration of the soluble fraction of livers from rabbits injected with mercuric chloride, with or without selenite, revealed that 24 h after the coadministration of the two compounds the formation of high molecular weight complexes was also observed in the liver (Figure 4). However, 1 h after the coadministration this complex could not be detected in the soluble fraction of the liver (Naganuma and Imura, 1980b, 1981).

High molecular weight complexes of Hg and Se with proteins formed in the plasma of mice, as in the case of rabbit plasma, were partially purified by gel filtration and readministered i.v. to other mice. Gel filtration of the plasma of mice administered the complexes clearly indicate that the complexes were readily taken up by the liver (Imura and Naganuma, 1985).

Besides the soluble fraction of the liver (S105 fraction) the insoluble fraction which was obtained by centrifugation of liver homogenates at 105,000 x g for 1 hr was treated with 2% SDS and further separated by centrifugation into SDS-soluble and insoluble fractions. The SDS-soluble fraction (P105 fraction) was then filtered through Sephadex G-200. The gel filtration pattern of the P105 fraction (Figure 4) demonstrated the existence of similar high molecular weight complexes in the liver insoluble fraction to those in the soluble fraction (Naganuma and Imura, 1981; Naganuma et al., 1981).

The high molecular weight complexes were then subjected to treatment with proteinases. The mercury and selenium in the complexes remained as soluble complexes after trypsin digestion probably with the aid of relatively large peptides formed by trypsin treatment. However, these two elements in the high molecular weight complex were turned to an insoluble complex when the proteins in the large complexes were extensively digested with Pronase (Table 1). The properties of the large complexes were further investigated by rechromatography with Sephadex and subsequent ion exchange chromatography, ultracentrifugation and SDS-polyacrylamide gel electrophoresis, demonstrating that the mercury and selenium in the complexes were hardly separable from each other and the ratio of mercury to selenium in the complexes was almost always close to one.

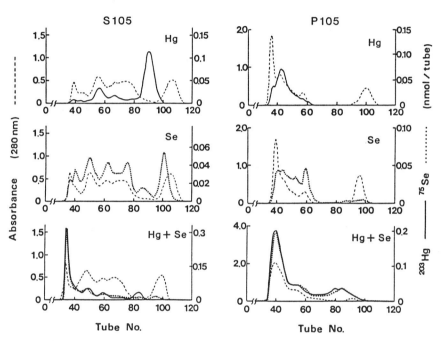

Fig. 4. Sephadex G-200 chromatograms of soluble (S105) and SDS-treated insoluble (P105) fraction of rabbit liver 24 h after administration of $^{203}HgCl_2$ and/or $Na_2{}^{75}SeO_3$ (Naganuma and Imura, 1981).

Table 1 Precipitation (%) of mercury and selenium associated to high-molecular weight Substance after incubation with Trypsin or Pronase (Naganuma and Imura, 1981)

| | Sources of high-molecular weight substance | | | | | | | |
| | Plasma | | SFH | | Liver (S105) | | Liver (P105) | |
| | Hg | Se | Hg | Se | Hg | Se | Hg | Se |
|---|---|---|---|---|---|---|---|---|
| ---- | 1.8 | 3.3 | 8.7 | 7.9 | 1.5 | 2 | 2.2 | 2.5 |
| Trypsin | 3.6 | 4.5 | 9.5 | 9.7 | 5 | 5.6 | 2.4 | 1.7 |
| Pronase | 91.6 | 86.2 | 99.9 | 98.2 | 90.2 | 84.2 | 95.3 | 89.0 |

Although selenium administration prevented acute inorganic mercury toxicity when it was given simultaneously with mercury, the protective effect was markedly altered by other administration sequences shown in Table 2. When selenite was injected 60 min. prior to the mercuric mercury it markedly enhanced its lethality (Naganuma et al., 1984a). These results suggest that selenium compounds may be unsuitable antidotes for acute mercury intoxication.

Interaction of Selenium with Methylmercury

In addition to the antagonism by selenium of the toxicity of inorganic mercury, its detoxifying effect on methylmercury attracted attention of many scientists in heavy metal toxicology. Ganther et al. (1972) and Iwata et al. (1973) reported that coadministration of selenium compounds with methylmercury depressed the toxicity of this metal alkyl. Further, Sumino et al. (1977) found that addition of selenite to the mixture containing protein-bound methylmercury resulted in release of the methylmercury. Since it was hardly believable that the released mercury compound was methylmercury itself as Sumino and his coworkers claimed, we attempted to identify the mercury compound released from proteins by the addition of selenite.

A mercury:selenium (2:1 molar ratio) compound was extracted into benzene (pH7.4) from rabbit blood treated with methylmercury and selenite. Thin layer chromatography, elemental and mass spectral analyses indicated that the compound was bis(methylmercuric) selenide (BMS). The BMS was probably formed in the blood from selenite and methylmercury with the aid of glutathione which is known to reduce selenite to selenide. In fact, BMS has been synthesized in vitro from methylmercury and selenite in the presence of glutathione (Naganuma and Imura, 1980a).

Table 2   Effect of administration sequences of mercury chloride and sodium selenite on their toxicities[a].

| Group of mice | No. of surviving mice | | | | | | | |
| | Days after Hg injection | | | | | | | |
| | 0 | (2hr) | 1 | 2 | 3 | 4 | 5 | 6 |
|---|---|---|---|---|---|---|---|---|
| Control | 10 | 10 | 10 | 10 | 10 | 10 | 10 | 10 |
| $HgCl_2$ | 10 | 10 | 10 | 10 | 9 | 9 | 9 | 9 |
| $Na_2SeO_3$ | 10 | 10 | 10 | 10 | 10 | 10 | 9 | 9 |
| $HgCl_2$ + $Na_2SeO_3$ | 10 | 10 | 10 | 10 | 10 | 10 | 10 | 10 |
| $HgCl_2 \rightarrow Na_2SeO_3$ | 10 | 10 | 10 | 10 | 10 | 10 | 10 | 10 |
| $Na_2SeO_3 \rightarrow HgCl_2$ | 10 | 2 | 2 | 2 | 2 | 2 | 2 | 2 |

Mice were administered $HgCl_2$ (25 μmol/kg), $Na_2SeO_3$ (25 μmol/kg), $HgCl_2$ and $Na_2SeO_3$, simultaneously, $Na_2SeO_3$ 1 hr after $HgCl_2$ injection, or $HgCl_2$ 1 hr after $Na_2SeO_3$ injection.

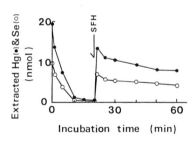

Fig. 5.  Degradation and reproduction of bis(methylmercuric) selenide in rabbit plasma. Bis(methyl-mercuric) selenide was incubated at 37°C in rabbit plasma. Stroma-free hemolysate (SFH) of rabbit was added to the reaction mixture after 20 min. incubation. (Naganuma et al., unpublished date.)

Characterization of BMS revealed that this complex was unstable and degraded quickly in blood *in vitro* (Naganuma et al., 1980). However, BMS was easily generated by the addition of a stroma-free hemolysate to the mixture (Figure 5). Although the possibility of BMS formation in various tissues has been suggested, those properties of BMS thus far elucidated cannot completely account for the ability of selenite to efficiently depress methylmercury toxicity. At present it is uncertain to what extent *in vivo* BMS formation contributes to the modification of methylmercury toxicity following selenium coadministration.

### Interaction of selenium with metals other than mercury

Selenium has been assumed to modify the physiological function of many heavy metals based on its ability to easily form selenides with these metal ions.

The effects of various metal ions on selenium deposition were examined to identify metals capable of interacting with selenium *in vivo*. Several screening systems developed in our laboratory to investigate mercury-selenium interactions were utilized for this purpose (Naganuma et al., 1983b). Based on the results obtained in our laboratory, together with those from other groups, it is likely that selenium interacts with many metal ions (Figure 6).

Although selenium is not an antidote for the acute toxicity of mercury compounds, the apparent interaction of selenium with platinum (Figure. 6) supported an investigation of its utility as an antidote versus the toxicity of the platinum complex: *cis*-diamminedichloroplatinum (cisplatinum), a widely used antineoplastic agent with antitumor activity against a wide range of human neoplasms.

Optimal conditions for reduction of cisplatinum toxicity by sodium selenite administration are shown in Figure 7 (Naganuma et al., 1983a, 1984b; Satoh et al., 1989). The optimized protocol involves simultaneous coadministration of selenite with the cisplatinum dose (at a molar ratio of 3.5:1) to mice followed by daily repetition of the dosing schedule for the next four days. Selenite administration, according to this protocol, dramatically suppressed not only cisplatinum renal toxicity indicated by BUN values, and intestinal toxicity, as indicated by diarrhea frequency, but also improved survival rates of mice (Table 3) (Satoh et al., 1989).

In order to examine the effect of selenite administration on the antitumor activity of cisplatinum, tumor-bearing mice were treated with cisplatinum in combination with selenite. As indicated in Table 4 the limiting cisplatinum dose was approximately 25 μmol/kg. Although the antitumor effect of cisplatinum was more pronounced above 25 μmol/kg, BUN values increased significantly above control values. However, selenite

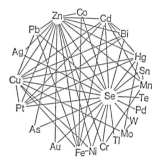

Fig. 6. Interaction of trace elements

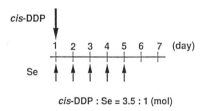

cis-DDP : Se = 3.5 : 1 (mol)

Fig. 7. Administration schedule of cisplatin (*cis*-DDP) and selenite (Satoh et al., 1989).

Table 3  Effect of selenite on toxicity of cisplatin in mice (Satoh et al. 1989).

| cis-DDP (μmol/kg) | Selenite (μmol/kg/day) | BUN (mg/100ml) | Incidence of diarrhea (%) | Survival rate of mice (%) |
|---|---|---|---|---|
| 0 | 0 | 21.8 ± 0.6 | 0 | 100 |
| 35 | 0 | 107.5 ± 16.0 | 100 | 0 |
| 35 | 10 | 28.1 ± 6.0 | 0 | 100 |

Cisplatin (cis-DDP) was administered s.c. on day 0.  Selenite was administered s.c. daily from day 0 to day 4.  BUN level and rate of incidence of diarrhea were determined on day 4.  Survival rate of mice was determined on day 12.

Table 4  Effect of selenite on antitumor activity and renal toxicity of cisplatin in mice inoculated s.c. with Ehrlich ascites tumor cells (Satoh et al., 1989).

| cis-DDP (μmol/kg) | Selenite (μmol/kg/day) | Tumor weight(g) | | BUN on day 8 (mg/100ml) |
|---|---|---|---|---|
| | | on day 8 | on day 15 | |
| 0 | 0 | 0.46 ± 0.19 | 2.30 ± 1.21 | 19.4 ± 2.0 |
| 0 | 10 | 0.15 ± 0.04 | 2.02 ± 0.60 | 23.8 ± 4.1 |
| 25 | 0 | 0.06 ± 0.01 | 0.57 ± 0.19 | 33.2 ± 8.6 |
| 30 | 0 | 0.03 ± 0.01 | 0.49 ± 0.03 | 67.0 ± 39.5 |
| 35 | 0 | 0.02 ± 0.01 | died | 153.0 ± 10.8 |
| 35 | 10 | 0.02 ± 0.005 | 0.17 ± 0.03 | 25.9 ± 1.1 |
| 45 | 12.9 | 0.02 ± 0.004 | 0.05 ± 0.009 | 26.6 ± 5.2 |

Mice were inoculated s.c. with Ehrlich ascites tumor cells on day 0.  Cisplatin (cis-DDP) was administered s.c. on day 3, and selenite was administered s.c. daily from day 3 to day 7.

coadminstration permitted the administration of a larger cisplatinum dose without compromising its anticancer activity (Satoh et al., 1989).

Thus the micronutrient, selenium may be a useful adjunct for cancer chemotherapy with cisplatinum, an anticancer drug widely used against various types of neoplasms. However, the mechanism involved in the protective action of selenium against cisplatinum toxicity, remains to be established by further study.

Thus mercury compounds, in general, are too toxic to be utilized therapeutically. However, as described above, the toxicology of mercury may provide a new approach for development of a useful tool for clinical treatment.

## REFERENCES

Ganther, H.E., Gondie, C., Sunde, M.L., Kopecky, M.J., Wanger, P.A., Oh,.S.H., and Hoekstra,W.G., 1972, Selenium: Relation to decreased toxicity of methylmercury added to diets contain tuna, Science, 175:1122.

Imura N. and Naganuma, A., 1978, Interaction of inorganic mercury and selenite in rabbit blood after intravenous administration, J. Pharmacobio-Dyn., 1:67.

Imura, N. and Naganuma, A., 1985, Mode of modifying action of selenite on toxicity and behavior of mercury and other metals, Nurt. Res. Suppl. I: 499.

Iwata, H., Okamoto, H., and Ohsawa, Y., 1973, Effect of selenium on methylmercury poisoning. Res. Commun. Chem. Pathol. Pharmacol., 5:673.

Naganuma, A. and Imura. N., 1980a, Bis(methylmercuric) selenide as a reaction product from methylmercury and selenite in rabbit blood. Commun. Chem. Pathol. Pharmacol., 27:163.

Naganuma, A. and Imura, N., 1980b, Changes in distribution of mercury and selenium in soluble fractions of rabbit tissues after simultaneous administration. Pharmacol. Biochem. Behav., 13:537.

Naganuma, A. and Imura, N. 1981, Properties of mercury and selenium in a high-molecular weight substance in rabbit tissues formed by simultaneous administration, Pharmacol. Biochem. Behav., 15:449.

Naganuma, A. and Imura, N. 1983, Mode of interaction of mercuric mercury with selenite to form high-molecular weight substance in rabbit blood in vitro, Chem.-Biol. Interact., 43:271.

Naganuma, A., Ishii, Y., and Imura, N., 1984a, Effect of administration sequence of mercuric chloride and sodium selenite on their fates and toxicities in mice, Exotoxicol. Environ. Saf., 8:572.

Naganuma, A., Kojima, Y., and Imura, N., 1980, Interaction of methylmercury and selenium in mouse: Formation and decomposition of bis(methylmercuric) selenide, Res. Commun. Chem. Pathol. Pharmacol., 30:301.

Naganuma, A., Kosugi, K., and Imura, N., 1981, Behavior of inorganic mercury and selenium in insoluble fractions of rabbit tissues after simultaneous administration, Toxicol. Lett., 8:43.

Naganuma, A., Pan, S.K., and Imura, N., 1978, In vitro studies on interaction of mercuric mercury and selenite in rabbit blood, Res. Commun. Chem. Pathol. Pharmacol., 20:139.

Naganuma, A., Satoh, M., and Imura, N. 1984b, Effect of selenite on renal toxicity and antitumor activity of cis-diamminedichloroplatinum in mice inoculated with Ehrlich ascites tumor cell, J. Pharmacobio-Dyn., 7:217.

Naganuma, A., Satoh, M., Yokoyama, M., and Imura, N., 1983a, Selenium efficiently depressed toxic side effects of cis-diamminedichloroplatinum. Res. Commun. Chem. Pathol. Pharmacol., 42:127.

Naganuma, A., Tanaka, T.A., Maeda, K., Matsuda, R., Tabata-Hanyu, J., and Imura, N., 1983b, The interaction of selenium with various metals *in vitro* and *in vivo*, Toxicology, 29:77.

Parizek, J. and Ostadalova, I., 1967, The protective effect of small amounts of selenium in sublimate intoxication, Experientia, 23:142.

Satoh, M., Naganuma, A., and Imura, N., 1989, Optimum schedule of selenium administration to reduce lethal and renal toxicities of *cis*-diamminedichloroplatinum in mice, J. Pharmacobio-Dyn., 12:246.

Sumino, K., Yamamoto, R., and Kitamura, S., 1977, A role of selenium against methylmercury toxicity, Nature, 268:73.

# OVERVIEW ON THE PROTECTION GIVEN BY SELENIUM AGAINST MERCURIALS

Laszlo Magos

MRC Toxicology Unit
Carshalton, Surrey
U.K.

## ABSTRACT

After the publication of "The protective effect of small amounts of selenite in sublimate intoxication" by Parizek and Ostaldalova in 1967, there was a continuous stream of papers on selenium-mercury interactions. This stream has become a flood since the publication of "Selenium: relation to decreased toxicity of methylmercury added to diets containing tuna" by Ganther et al. in 1972. Follow-up research initiated by these two reports and the use of selenium-mercury interaction as a research tool contributed considerably to knowledge on the biological behavior and action of selenium and mercurial compounds. Contrary to the scientific return, the benefit of practical application has remained non-existent. In the light of this experience the practical value of protection by selenium against inorganic mercury and methylmercury toxicities is scrutinized here.

There is hardly any problem to deal with the effect of selenite on sublimate intoxication. The protection is real when optimum experimental conditions are chosen, but in human intoxication the use of selenium as an antidote is impractical and even dangerous. Selenium must be given within a narrow span of time in near equimolar dose with sublimate. When it is given late, it does not prevent renal damage, and when it is given 1-2 hours before mercury, the metabolite of selenite, dimethylselenide, in the presence of mercury becomes a killer. Moreover, while thiol complexing agents increase the urinary excretion of mercury, selenite promotes retention in the the form of HgSe which later can become the secondary source of mercury.

Contrary to acute sublimate intoxication, the timing of selenium administration is not crucial against chronic methylmercury intoxication. This advantage is balanced by the disadvantage that selenite delays, but does not prevent methylmercury intoxication. This makes the use of selenite as a preventive antidote a hardly acceptable proposition. Other contraindications are the following selenium-methylmercury interactions: (1) selenite temporarily increases the brain content of methylmercury; (2) methylmercury intensifies the growth retarding effect of selenite; (3) selenite at low doses increases the foetotoxicity and teratogenicity of methylmercury.

There are two other problems which require consideration. The first is that methylmercury increases the exhalation of selenium (as dimethylselenide) and this effect becomes more pronounced as the dose of selenite or the body burden of selenium is increased. Thus methylmercury progressively decreases the bioavailability of selenide and through this mechanism may contribute to the failure of selenite to prevent intoxication. The second point is concerned with the protective potential of selenium in food. Because a protein rich diet without selenium supplementation also delays onset, the role of natural selenium in the effect of dietary fish high in natural selenium cannot be ascertained. The extrapolation of experiments with selenite most probably overestimates the protective potency of natural selenium. The bioavailability of selenium in different chemical forms is different, and e.g. the bioavailability of natural selenium for HgSe formation is less than 20% of the bioavailability of selenite selenium.

## INTRODUCTION

### Selenium against sublimate intoxication

There is hardly any problem to deal with the effect of selenite on sublimate intoxication. Sublimate intoxication is an acute condition which has been successfully treated with thiol complexing agents Thus dimercaptopropanol saved many lives and animal experiments demonstrated that its protective action includes increased clearance from the whole body and the kidneys (Swensson and Ulfvarson, 1967). On these two accounts selenite is very much inferior. Though it is able to decrease the accumulation of mercury in the target kidney, contrary to dimercaptopropanol it does not accelerate but decelerate clearance from the whole body (Magos and Webb, 1976) or the kidneys (Naganuma et al, 1984). Moreover selenium must be given within a much narrower span of time. When it had been given 1-2 hrs before mercury, lethality increased (Parizek et al, 1980), when it had been given to mice 20 min. after sublimate, it did not alleviate growth retardation and a further 40 min. delay cancelled its preventive potency against renal damage (Naganuma et al, 1984). This is hardly a recommendation for clinical trials.

## Selenium against methylmercury intoxication

The interaction of selenium with methylmercury is different from its interaction with sublimate. Table 1 shows that, contrary to acute sublimate intoxication, the timing of selenium administration is not crucial and the adduct is unstable bis-methylmercury selenide (Magos et al., 1979; Naganuma et al, 1980). However these advantages are balanced by the disadvantage that selenite at least temporarily increases the concentration of methylmercury in the target brain (Magos and Webb, 1977), as shown on Table 2, and selenite only delays and does not prevent methylmercury intoxication (Chang et al, 1977). The effect of selenite on the clearance of methylmercury is ambiguous. Komsta-Szumska et al (1983) found that in guinea pigs given a single dose of methylmercury and selenite both organ mercury concentrations and the excretion of mercury were lowered by selenite.

Table 1  Characteristics of mercury-selenium interactions relevant to the protection given by selenite against mercurial intoxications.

|  | Interactions with | |
|---|---|---|
|  | Inorganic Hg | Methyl Hg |
| protection | life saver in acute intoxication | only delays onset: may increase teratogenicity |
| timing of treatment | critical | not critical |
| conc in target | decreased | increased |
| adduct | stable HgSe | unstable MeHgSeHgMe |
| Se toxicity | decreased | may be increased |

Table 2    The effect of selenite (5 µmol/kg) on the brain concentration of mercury in rats treated with 5 µmol/kg methylmercury (Magos and Webb, 1977).

| Timing of Se administration (days after MeHg) | 24 h change in brain mercury concentration |
|---|---|
| 2 | + 134 % |
| 3 | + 103 % |
| 7 | +  88 % |

In dietary experiments both Ohi et al (1976) and Stillings et al (1974) have found selenium without any notable effect on organ mercury concentrations, but rats on selenite supplemented diet excreted a lower proportion of the ingested mercury (Stillings et al, 1974). There is no ambiguity in the effects of thiol complexing agents. Dimercaptosuccinic acid rapidly depleted mercury from the brain and the whole body of rats and stopped the progress of intoxication (Magos et al, 1978).In patients 2,3-dimercaptopropane-1-sulphonate and N-acetyl-penicillamine decreased clearance half time from 60 to 10 and 24 days, respectively (Clarkson et al, 1981).

The protective role of selenite becomes even more dubious during pregnancy. The experiments of Nishikido et al (1988) and Naganuma et al (1979) demonstrated (see Table 3) that selenium deficiency did not influence the teratogenicity of methylmercury and at a high level of methylmercury exposure selenite decreased resorption only at a dose which increased the incidence of cleft palate. As the developing foetus is a sensitive target for methylmercury, these observations must weigh against the antidotal use of selenite when protection is most needed.

The interaction between methylmercury and selenite is a two-way process. Table 4 shows that at least in female rats methylmercury was able to increase the toxicity of selenite. A single dose of 24 µmol/kg selenite arrested weight gain in female rats. The same dose of methylmercury did not have any effect on body weight, but given in combination with selenite it turned the lack of weight gain to weight loss (Yonemoto et al, 1985). The mechanism of this potentiation is not known. Perhaps it may be linked to enhanced selenium methylation, or methylmercury acts only as a selenium carrier through the formation of bis-methylmercury selenide.

Selenium in food versus selenite

Selenium is both an essential and toxic element, with a wide gap between daily intakes associated with selenium deficiency or selenosis. The utilization of dietary selenium for the prevention of methylmercury intoxication is an attractive idea.

Table 3 The effects of selenite in food or drinking water on the teratogenicity and foetotoxicity of methylmercury in mice (Naganuma et al., 1979; Nishikido et al,. 1988).

| | | % of cleft palate (litter size) | | |
|---|---|---|---|---|
| μmoles Se/kg food | none | 5.0 | | |
| μmoles Se/l water | | | 11.4 | 22.8 |
| Methylmercury | | | | |
| none | 0 (9.2) | 0 (10.0) | 0 (9.3) | 2.4 (8.8) |
| 75 μmol/kg (single oral dose on day 11) | 42.9 (8.2) | 46.7 (10.0) | - | - |
| in food: | | | | |
| 31.9 μmol/kg | 55.4 (5.3) | - | 60.7 (4.0) | 85.5 (8.4) |

Food or water supplementation with the test chemicals started 30 days before gestation and thereafter up to day 18.

Table 4 The effect of treatment with 24 μmol/kg selenite and methylmercury alone or in combination on the body weight of female rats (Yonemoto et al., 1985).

| treatment | % weight change | in the first week |
|---|---|---|
| none | + 7 | |
| | Se | + 0.5 |
| | MeHg | + 7.0 |
| | Se with MeHg | - 10.0 |
| | Se 1 h after MeHg | - 10.0 |
| | Se 2 h after MeHg | - 14.5 |

In support of the protective role of biological selenium several investigators have found that diet supplemented with seafood high in selenium delayed the onset of methylmercury intoxication (Ganther et al, 1972; Ohi et al, 1976; Friedman et al, 1978). However, these experiments did not define the contribution of other dietary constituents present in the selected seafood and, therefore, the role of biological selenium in the protective effect remained unquantified or even uncertain. Ohi et al.

(1976) found that a tuna diet was half as effective in preventing the neurological manifestations of methylmercury as a selenite supplemented casein diet which contained as much selenium as the tuna diet. This finding indicates that the efficacy of selenium in tuna for the protective mechanism is at lest 50% lower than the efficacy of selenite and it may be substantially less. The same team also demonstrated that selenium is only one of the dietary factors which influence the onset of methylmercury intoxication. Thus, increase in the dietary concentration of casein delayed neurological manifestations without supplementation with selenium. Stillings et al (1974) also observed that an increase in the dietary level of casein or fish protein concentrate from 10 to 20% increased growth and survival of methylmercury exposed rats.

The lower efficiency of biological selenium is most likely the consequence of its lower bioavailability for the protective interaction. For both methylmercury and inorganic mercury the first chemical interaction between selenium and the mercurial is an adduct formation: bis-methylmercury selenide and mercuric selenide. These reactions are conditional on the formation of selenide anion. Table 5 shows that, judged from the concentration of HgSe in blood, the bioavailability of selenium for this reaction is high after the administration of selenite, lower after selenomethionine and even lower after the administration of selenium incorporated *in vivo* into liver soluble fraction (Magos et al, 1984). Table 5 also shows that the protection given by selenite against $HgCl_2$ induced renal damage follows the order of HgSe formation (Magos et al, 1987) and, therefore, compared with selenite biological selenium is a very inferior antidote. Though the toxicologic importance of bis-methylmercury selenide is most likely different from the toxicologic importance of HgSe, nevertheless the bioavailability of selenium species for HgSe formation may correlate with its protective efficacy against methylmercury. This view is supported by the observed lower protective effect of selenium in tuna compared with selenite added to casein (Ohi et al, 1976).

Loss of selenium through exhalation

The methylation of selenide and the exhalation of dimethylselenide decrease the availability of selenium for other reactions. It is a point of interest that the antidotal use of selenite against methylmercury is inherently associated with increased selenium methylation. This boomerang effect was first observed by Yonemoto et al (1985) who have shown, that methylmercury increased the exhalation of selenium. In a follow-up to this work Tandon et al (1986) observed, as Table 6 shows, that exhalation depends

Table 5  Effects of different forms of selenium on the presence of HgSe and renal
damage 48 h after oral administration (Magos et al., 1984, 1987).

| Forms of Se | HgSe in nmol/l blood in rats dosed with 0.75 μmol/kg HgCl$_2$ | alkaline phosphatase in 48 h urine samples (log U/kg) in rats dosed with 2.5 μmol/kg HgCl$_2$ |
|---|---|---|
| control | 2.2 | 2.74 |
| biological selenium | 59 | 2.36 |
| selenomethionine | 102 | 2.14 |
| selenite | 229 | 1.80 |

Biological selenium was selenium *in vivo* incorporated into liver soluble fraction, the other
two selenium species were administered with control liver soluble fraction.

Table 6  Exhalation of selenium in the first 24 h after the simultaneous s.c.
administration of selenite and methylmercury to rats (Tandon et al., 1986).

| | exhalation of Se in % of dose | |
|---|---|---|
| Selenite --> | 0.25 μmol/kg | 24.0 μmol/kg |
| Methylmercury | | |
| none | 0.8 | 24 |
| 6.0 μmol/kg | 1.5 | 36 |
| 24.0 μmol/kg | 2.9 | 47 |

on both the dose of selenite and the dose of methylmercury. An approximately ten-
fold increase in the dose of selenite resulted in a 16 to 30-fold increase in exhalation
depending on the dose of methylmercury. When rats were given subcutaneously 24
μmol/kg selenite with equimolar doses of methylmercury, nearly half of the
administered selenium was exhaled in the first 24 h. These values are further
increased by pretreatment with selenite.

The conversion of selenite to exhalable dimethylselenide also depends on the body burden of selenium. Table 7 shows that when treatment was repeated, daily exhalation showed an upward trend and only a smaller part of the exhaled selenium came from deposited selenium, the higher part from the last dose. Thus three days treatment with 12 µmol/kg selenite caused a four-fold increase in exhalation from a subsequent dose (Magos et al, 1987).

The extrapolation of these findings to daily exposure to methylmercury and selenite gives the following scenario. On the first day animals exhale part of the administered selenium. Exhalation is less in animals exposed to low doses and more from animals exposed to high doses of methylmercury. This means that more selenium is lost when more is needed. As the experiment proceeds and the body burden (and brain concentration) of methylmercury becomes nearer to the toxic threshold, the daily exhalatory loss of selenium becomes progressively larger and the availability of selenium for the protective reaction declines. This mechanism limits the efficacy of selenium and may contribute to the failure of selenite to prevent manifest intoxication.

Table 7  Exhalation of selenium after daily doses of 12 µmol/kg selenite (Magos et al., 1987).

| 24 h collection periods | | Se exhalation in µmol/kg | |
| --- | --- | --- | --- |
| | total | from last dose | from previous dose |
| 1st | 1.7 | - | - |
| 2nd | 4.1 | - | - |
| 3rd | 5.9 | - | - |
| 4th | 7.7 | 6.8 | 0.9 |

## CONCLUSION

The interaction between selenium and inorganic mercury or methylmercury has been proved a profitable experimental research tool without enhancing, or even permitting, the clinical use of selenite against mercurial intoxication. The efficacy of selenium treatment in sublimate intoxication is limited to such a narrow time, that no responsible medical team would think about substituting BAL or other dithiols for selenite. Other contraindications are higher body burden and the difficulty to mobilize mercury from HgSe deposits.

The use of selenite against methylmercury intoxication is also contraindicated on several grounds. First, it does not prevent but delays intoxication and therefore does not replace preventive measures which decrease exposure. Second, it does not protect the vulnerable foetus. Third, it increases temporarily methylmercury levels in the target brain. Fourth, as exposure proceeds the bioavailability of selenium for the protective reaction declines. Fifth, methylmercury may increase the toxicity of selenite.

As selenium is a natural constituent of several foodstuffs, the dietary intake of selenium, as a possible antidote against methylmercury, does not pose the same dilemma as medicinal selenium. One problem with natural selenium is that its most abundant source (fish) is also the most abundant source of methylmercury. The second problem is that its efficacy is unquantified and therefore the most elementary condition for its antidotal use is missing. Contrary to selenium, dithiol and monothiol complexing agents have the proven efficacy to remove methylmercury from the body, including, the brain.

## REFERENCES

Chang, L., Dudley Jr., A.W., Dudley, M.A., Ganther, H.E., Sunde, M.L., 1977, Modification of the neurotoxic effects of methylmercury by selenium, in: "Neurotoxicology", L.Roizin, H.Shiraki, N.Grcevic, eds. 275-282, Raven Press, New York.

Clarkson, T.W., Magos, L., Cox, C., Greenwood, M.R., Amin-Zaki, L., Majeed, M.A., Al-Damluji, S.F., 1981, Tests of efficacy of antidotes for removal of methylmercury in human poisoning during the Iraq outbreak. J. Pharmacol. Exp. Therap., 218:74.

Freidman, M.A., Eaton, L.R., Carter, W.H., 1978, Protective effects of freeze dried swordfish on methylmercury chloride toxicity in rats. Bull. Environ. Contam. Toxicol., 19:436.

Ganther, H.E., Goudie, C., Sunde, M.L., Kopecky, M.J., Wagner, P., Oh, S.-H., Hoekstra, W.G., 1972, Selenium: relation to decreased toxicity of methylmercury added to diets containing tuna. Science, 175:1122.

Komsta-Szumska, E., Reuhl, K.R., Miller, D.R., 1983, Effect of selenium on distribution, demethylation, and excretion of methylmercury by the guinea pig. J. Toxicol. Environ. Health, 12:775.

Magos, L., Berg, G.G, 1988, Selenium, in: "Biological Monitoring of Toxic Metals", T.W. Clarkson, L.Friberg, G.G.Nordberg, P.R.Sager, eds. 383-405, Plenum Press, New York.

Magos, L., Clarkson, T.W., Hudson, A.R., 1984, Differences on the effects of selenite andbiological selenium on the chemical form and distribution of mercury after the simultaneous administration of $HgCl_2$ and selenium to rats. J. Pharmacol. Exp. Therap., 228:478.

Magos, L., Clarkson, T.W., Sparrow, S., Hudson, A.R., 1987, Comparison of the protection given by selenite, selenomethionine and biological selenium against the renotoxicity of mercury. Arch. Toxicol., 60:422.

Magos, L., Peristianis, G.C., Snowden, R.T., 1978, Postexposure preventive treatment of methylmercury intoxication in rats with dimercaptosuccinic acid. Toxicol. Appl. Pharmacol., 45:463.

Magos, L., Tandon, S.K., Webb, M., Snowden, R., 1987, The effects of treatment with selenite before and after the administration of [$^{75}$Se]selenite on the exhalation of [$^{75}$Se]dimethyl-selenide. Toxicol. Letters, 36:167.

Magos, L., Webb, M., 1976, Differences in the distribution and excretion of selenium and cadmium or mercury after their administration simultaneously in equimolar doses. Arch Toxicol., 36:63.

Magos, L., Webb, M., 1977, The effect of selenium on the brain uptake of methylmercury, Arch. Toxicol., 38:201.

Magos, L., Webb, M., Hudson, A.R., 1979, Complex formation between selenium and methylmercury. Chem. Biol. Interact., 28:359.

Naganuma, A., Ishii, Y., Nakama, A., Endo, R., Imura, N., 1984, Effect of time intervals of selenium administration after injection of mercuric chloride on toxicity and renal concentration of mercury in mice. Indust. Health, 22:91.

Naganuma, A., Kojima, Y., Imura, A., 1980, Interaction of methylmercury and selenium in mouse: formation and decomposition of bis(methylmercuric) selenide. Res. Commun. Chem. Pathol.Pharmacol., 30:301.

Naganuma, A., Satoh, H., Suzuki, T., 1979, Effects of sodium selenite on methylmercury embryotoxicity and teratogenicity in mice. Toxicol. Appl. Pharmacol., 47:79.

Nishikido, N., Satoh, Y., Naganuma, A., Imura, N., 1988, Effect of maternal selenium deficiency on the teratogenicity of methylmercury, Toxicol. Letters, 40:153.

Ohi, G., Nishigaki, S., Seki, H., Tamura, Y., Maki, T., Konno, H., Ochiai, S., Yamada, H., Shimamura, Y., Mizoguchi, I., Yagyu, H., 1976, Efficacy of selenium in tuna and selenite in modifying methylmercury intoxication. Environ. Res., 12:49.

Parizek, J., Kalouskova, J., Benes, J., Pavlik, L, 1980, Interactions of selenium-mercury and selenium-selenium compounds. Ann. N.Y. Acad. Sci. 355:351.

Parizek, J., Ostadalova, I., 1967, The protective effect of small amounts of selenite in sublimate   intoxication. Experientia, 23:142.

Stillings, B.R., Lagally, H., Bauersfeld, P., Soares, J., 1974, Effect of cystine, selenium, and fish protein on the toxicity and metabolism of methylmercury in rats. Toxicol. Appl. Pharmacol., 30:243.

Swensson, A., Ulfvarson, U., 1967, Experiments with different antidotes in acute poisoning by different mercury compounds: effects on survival and on distribution and excretion of mercury. Int. Arch. Gewerbepathol. Gewerbehyg., 24:12.

Tandon, S.K., Magos, L., Webb, M., 1986. The stimulation and inhibition of the exhalation of volatile selenium. Biochem. Pharmacol. 35:2763.

Yonemoto, Y., Webb, M., Magos, L., 1985, Methylmercury stimulates the exhalation of volatile  selenium and potentiates the toxicity of selenite. Toxicol. Letters, 24:7.

# EFFECT OF MERCURIC CHLORIDE ON ANGIOTENSIN II-INDUCED CA++ TRANSIENT IN THE PROXIMAL TUBULE OF RATS

Hitoshi Endou and Kyu Yong Jung

Department of Pharmacology, Faculty of Medicine
The University of Tokyo
Hongo 7-3-1, Bunkyo-ku
Tokyo 113, Japan

## ABSTRACT

Angiotensin II (AII) is a powerful agent in the regulation of renal functions, and possesses its receptor(s) within the kidney. The second messenger of AII is considered to be $Ca^{++}$. Since mercuric chloride is one of nephrotoxic heavy metals, its effects on AII-mediated signal transduction have been investigated.

Rat kidneys were treated with collagenase, and microdissection was made for isolating defined nephron segments. $[Ca^{++}]i$ was determined using fluorescent indicator fura-2. In the freshly isolated early proximal tubule ($S_1$), AII-induced $[Ca^{++}]_i$ rise was biphasic, demonstrating the two peaks corresponding to the $10^{-11}$ and $10^{-7}$M. $HgCl_2$ ($10^{-10}$ - $10^{-8}$M) potentiated the $[Ca^{++}]_i$ increase induced by AII ($10^{-11}$M) in a dose dependent manner, up to the $10^{-9}$M $HgCl_2$. A similar effect was observed with methylmercury. To determine the mechanism of stimulatory effect of $HgCl_2$ on $10^{-11}$M AII-induced $[Ca^{++}]_i$ increase, nephron segments were pretreated with propranolol ($10^{-4}$M), a PLC inhibitor. This stimulatory effect of $HgCl_2$ was completely inhibited by propranolol. Moreover, $10^{-4}$M propranolol completely blocked the stimulatory effect of $HgCl_2$ on AII-mediated inositol 1,4,5-triphosphate ($IP_3$) production.

This study suggests for the first time that mercurial compounds stimulates the $[Ca^{++}]_i$ increment induced by AII, possibly through an activation of phospholipase C (PLC).

## INTRODUCTION

The kidney maintains body fluids and electrolyte balance by regulating excretion of ions, salts, metabolic waste materials, and water into the urine. Kidney cells, therefore require energy to maintain these functions, and the primary energy donor is adenosine triphosphate (ATP). The overall chemical reactions that lead to ATP synthesis have been well documented (Hinkle and McCarty, 1978; Manillier et al., 1986; Moraes and Meis, 1987; Weinberg and Humes, 1986). For the manufacture of ATP in cells, various substrates are metabolized at a high rate in the body. There are many investigations of substrate metabolism within the kidney (Goldstein, 1987; Klein et al., 1981; Silva, 1987; Weidemann and Krebs, 1969; Wirthensohn and Guder, 1986). Recently, we demonstrated intranephron characteristics of substrate utilization for producing ATP (Uchida and Endou, 1988; Jung et al., 1989a). This unique nephron heterogeneity in consuming and producing ATP may well reflect different physiological functions or toxic reactions by chemicals in each of various nephron segments (Jung and Endou, 1989a)

The mammalian kidney receives about 25% of the cardiac output, which exposes this organ to a variety of chemical agents. One event related to a variety of nephrotoxic responses produced by xenobiotics is alteration in active transport systems of the renal tubules. These systems use ATP as the energy source.

The kidney has been well known as one of the target organs of mercurial compounds. It has been generally considered that $HgCl_2$ directly inhibits water channels being present in the kidney proximal tubule. Although $HgCl_2$ has been shown to induce $[Ca^{++}]_i$ transient from intracellular store, a precise mechanism by mercury is still obscure. In this communication, we report effect of mercurial compounds on angiotensin II (AII)-induced $[Ca^+])_i$ transient to elucidate their mechanisms.

### Relationship between cellular ATP and cytosolic free calcium

To establish an adequate experimental condition for the determination of $[Ca^{++}]_i$ using kidney cells, several substrates were tested by monitoring intracellular

ATP contents at the same time (Jung and Endou, 1989b)  Defined nephron segments were isolated as previously described (Uchida and Endou, 1988; Jung and Endou, 1989a).  In short, male sprague-Dawley rats were decapitated, and left kidneys were perfused with cold Hanks' buffer containing 0.1% collagenase.  The collagenase treatment was carried out at 37° C for 15min with the same buffer.  The nephron segments were identified under a stereomicroscope.  ATP content in each nephron segment was determined by the microchemiluminescence method (Uchida and Endou 1988).  [Ca$^{++}$]$_i$ was quantified using a fluorescent dye, fura-2.  For fluorescence loading, single nephron segments were incubated at 25° C with Hanks' solution containing 10mM CH$_3$COONa, 2mM Pyruvate, 10% fetal calf serum (FCS), 10 μM fura-2/AM, and 0.5mM CaC1$_2$ (pH 7.40).  Loaded cells were washed three times with the same solution, excluded 10% FCS.  For fluorescence measurement, intracellular fura-2 fluorescence changes were measured at 25° C using a 2 wave-length microscopic fluorometer (CAM 220, JASCO, Japan) according to Grynkiewicz et al. (1985).

Figure 1 shows the relationship between cellular ATP contents and [Ca$^{++}$]$_i$ (Jung and Endou, 1989b).  In the condition 1 (C$_1$) excluding any substrates, cellular ATP content was very low in both nephron segments (1st point from the left in Figure 1).  As can be expected, the values of [Ca$^{++}$]$_i$ in both the terminal proximal tubule (S$_3$) and the cortical collecting tubule (CCT) was significantly higher than those of C$_4$ (4th point).  This result shows that low intracellular ATP increases [Ca$^{++}$]$_i$.  Moreover, the correlation between cellular ATP and [Ca$^{++}$]$_i$ is different according to different segments.

The results shown in Figure 1 provide evidence of a suitable procedure to measure [Ca$^{++}$]$_i$ using isolated nephron segments.


Effect of  angiotensin II on [Ca$^{++}$]$_i$ in isolated rat early proximal tubule

AII is a powerful agent in the regulation of extracellular volume, and exerts an important effect on renal sodium excretion.  Micropuncture and microperfusion studies (Harris and Navar, 1985; Schuster et al., 1984) have demonstrated that proximal tubular sodium and water transport are stimulated by psysiological concentrations ($10^{-12}$-$10^{-10}$M) of AII from the peritubular side, whereas AII at higher concentration ($10^{-7}$M) causes its inhibition.  Moreover, a mechanism involving changes in ion transport

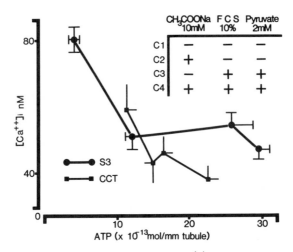

Fig. 1  Relationship between cellular ATP and $[Ca^{++}]_i$ in the terminal proximal tubule
($S_3$) and the cortical collecting tubule (CCT). After the nephron segments were
incubated under various conditions (C1 to C4) containing $10^{-5}$M fura-2/AM,
each papameter was measured. Each point (from the left) in concurrent curve
represents C1 to C4. Values are means ± SE. of 5 experiments.

through cellular membrane implies the existence of some intracellular messenger or
signal to couple with the membrane receptor of AII. The activity of adenylate cyclase in
homogenates of the rat renal cortex is inhibited maximally with $10^{-7}$M AII (Woodcock
and Johnston, 1982). In cultured smooth muscle (Smith et al., 1984) and cultured
mesangial cells (Schlondorff et al., 1987), AII activates PLC, thus increasing $IP_3$ and
diacylglycerol and leading to $[Ca^{++}]_i$ mobilization.

From these backgrounds, we studied effects of AII on $[Ca^{++}]_i$ in isolated rat
early proximal tubule ($S_1$). As shown in Figure 2, $Ca^{++}$ signaling obtained under two
different conditions ($C_1$ and $C_4$ as indicated in Figure 1) was markedly different. When
we fixed our protocol to $C_4$, AII even at $10^{-14}$M showed a transient increase of $[Ca^{++}]_i$.
Figure 3 represents a typical series of fluorescence records from fura-2-loaded S1
stimulated by AII ($10^{-14}$-$10^{-7}$M) in the presence of 0.5mM $CaCl_2$. Figure 4 is the
dose-response curve of AII on $Ca^{++}$ transients in $S_1$. AII at $10^{-11}$ and $10^{-7}$M evoked
peak values in $[Ca^{++}]_i$. To the contrary, $10^{-9}$M AII showed a valley in $[Ca++]_i$,
demonstrating existence of different receptor subtypes having different affinities for
AII.

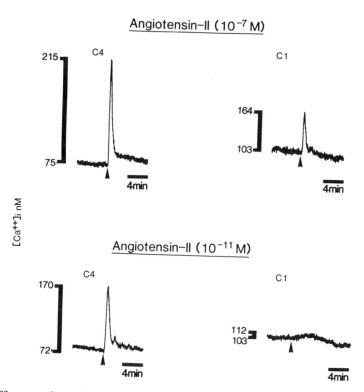

Fig. 2 Differences in angiotensin II-induced Ca$^{++}$ transient according to the conditions of fura-2/AM loading in the isolated rat early proximal tubule. Fura-2 loading was carried out under C1 and C4 as described in Fig. 1. Angiotensin II, at the concentrations indicated, was added to fura-2-loaded cells at the arrow head.

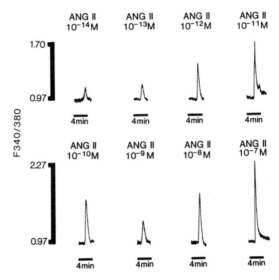

Fig. 3 Angiotensin II-induced Ca$^{++}$ signalings in the isolated rat early proximal tubule. Fluorescence in each record represents a rapid increase and recover to its base line.

These results clearly suggest for the first time that high intracellular ATP is necessary to evaluate a high affinity AII action, and that there are at least two distinct AII receptor subtypes in rat $S_1$ (Jung and Endou, 1989b).

## Effect of mercuric chloride on $[Ca^{++}]_i$ in rat $S_1$

During the past several years, the actions of mercurial compounds have been evaluated at biochemical levels (Brunder et al., 1988; Smith et al., 1987; Trimm et al., 1986). Using cultured rabbit renal tubular cells, HgCl$_2$-induced $[Ca^{++}]_i$ transient originates from intracellular nonmitochondrial calcium storages and most likely from the endoplasmic reticulum, and the IP$_3$ pathway has been suspected as one of its possible mechanisms (Smith et al., 1987). It has been generally considered that HgCl$_2$ directly inhibits water channels being present in the kidney proximal tubule, which are not affected by N-ethylmaleimide (Pratz et al., 1986). Recently, Hoch et al. (1989) have suggested that mercurial compounds act on the vasopressin-sensitive water channel, resulting in inhibition of water flow in the proximal tubule and the toad urinary bladder.

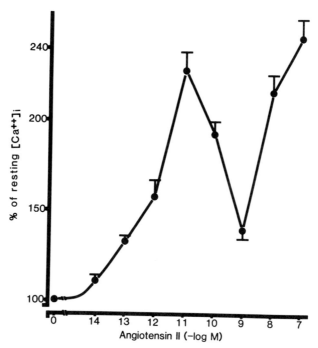

Fig. 4 Dose-response of angiotensin II on $[Ca^{++}]_i$ in the isolated rat early proximal tubule. The ordinate represents the relative change in calculated $[Ca^{++}]_i$ from basal levels at each sample. Values are means $\pm$ SE. of 5 experiments.

When S1 isolated from rat kidneys was incubated with $HgCl_2$ at relatively high concentrations ($10^{-7}$ $10^{-4}$M) of $HgCl_2$, $[Ca^{++}]_i$ increases in a dose-dependent manner (Figure 5). The pattern of changes in $[Ca^{++}]_i$ after $HgCl_2$ differs according to applied doses. In the presence of $10^{-4}$M $HgCl_2$, $[Ca^{++}]_i$ increases transiently, and it is followed by a continuous gradual increase. The minimal concentration of $HgCl_2$ to cause $[Ca^{++}]_i$ change is $10^{-7}$M. A quite similar pattern to $HgCl_2$ could be obtained by using methylmercury, indicating that these single effects of mercurial compounds may be common in inorganic and organic mercurials. Thus, doses of mercurials to manifest their nephrotoxicity are too high to clarify their specific toxicity in the proximal tubule, though the intranephron target being the proximal tubule is in accordance with previous reports (Endou et al., 1986; Houser and Berndt, 1988; Ruegg et al., 1987).

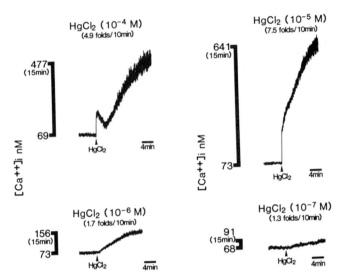

Fig. 5  HgCl$_2$-induced [Ca$^{++}$]$_i$ signaling in the isolated rat early proximal tubule. The increased folds stated in the parentheses mean the ratios between [Ca$^{++}$]$_i$ at 10min after HgCl$_2$ addition and that of the resting level.

## Effect of mercuric chloride on angiotensin II-induced Ca$^{++}$ transient

HgCl$_2$ at lower concentrations than 10$^{-7}$M did not have any effect on the basal [Ca$^{++}$]$_i$ as measured by fura-2. However, pretreatment of S$_1$ with HgCl$_2$ for 5 min stimulated [Ca$^{++}$]$_i$ transient induced by 10$^{-11}$M AII in a dose-dependent manner (Figure 6). Figure 7 depicts the dose-response of HgCl$_2$ on AII-induced [Ca$^{++}$]$_i$. The [Ca$^{++}$]$_i$ increase by 10$^{-11}$M AII alone was observed as 262.8% of resting [Ca$^{++}$]$_i$. This increment was maximally stimulated by 10$^{-9}$M HgCl$_2$ to 328% of the resting [Ca$^{++}$]$_i$ (Jung and Endou, submitted). To evaluate this mechanism, we investigated the involvement of PLC in the action of HgCl$_2$ using propranolol, a PLC inhibitor (Blackwell et al., 1983; Chau and Tai, 1982; Hostetler and Matsuzawa, 1981). As shown in Figures 8 and 9, pretreatment of S$_1$ with propranolol (10$^{-4}$M) for 3 min almost completely inhibited the AII (10$^{-11}$M)-induced [Ca$^{++}$]$_i$ transient both with and without 10$^{-9}$M HgCl$_2$ (Jung and Endou, submitted). Using renal cortical tubule suspension, AII (10$^{-11}$M) significantly stimulated the cellular IP$_3$ production, and the

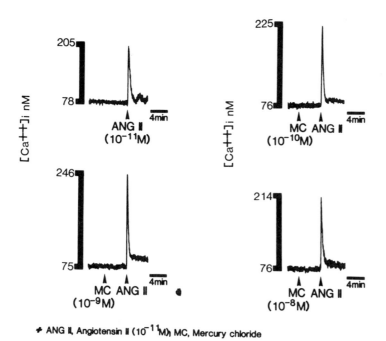

* ANG II, Angiotensin II (10$^{-11}$M), MC, Mercury chloride

Fig. 6 Effect of HgCl$_2$ at various concentrations on angiotensin II (10$^{-11}$M)-induced calcium transient in the isolated rat early proximal tubule.

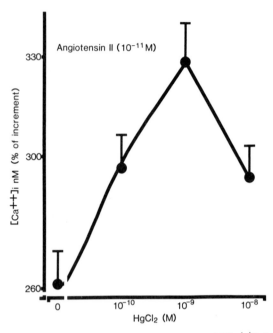

Fig. 7  Dose-response of $HgCl_2$ on $10^{-11}M$ AII-induced $[Ca^{++}]_i$ increase in isolated rat $S_1$. After the pretreatment of $S_1$ with each concentration of $HgCl_2$ for 5 min AII was added. The ordinate represents the relative change in $[Ca^{++}]_i$ calculated from basal levels of each sample. Values are mean $\pm$ SE. of 6 experiments.

mean value of $IP_3$ in the presence of AII and $HgCl_2$ was higher than that with AII alone, though the difference was statistically insignificant (Figure 10). Propranolol at $10^{-4}M$ also completely blocked the stimulatory effect of $HgCl_2$ on AII-mediated $IP_3$ production, whereas $10^{-9}M$ $HgCl_2$ alone did not change the $IP_3$ production.

When AII binds to its receptor in numerous types of cells, at least two different biochemical changes can be observed: an increase in the rate of $Ca^{++}$ influx across the plasma membrane, and the activation of PLC that catalyzes the hydrolysis of phosphatidylinositol 4,5-bisphosphate ($PIP_2$). The hydrolysis of $PIP_2$ leads to the rapid generation of $IP_3$ and diacylglycerol (Benabe et al., 1982; Brock et al., 1985; Griendling et al., 1986; Johnson and Garrison, 1987; Rasmussen, 1986a). On the other hand, $HgCl_2$ at a higher concentration than $10^{-6}M$ increases the $[Ca^{++}]_i$ via $Ca^{++}$ influx through plasma membrane in cultured rabbit renal tubular cells, followed by cell death (Smith et al., 1987). Brunder et al. (1988) have suggested that heavy metals have led to oxidation of sulfhydryl groups. Moreover, PKC differs from the other protein kinases on one hand because of its enriched cysteine, i.e., numerous sulfhydryl groups (Kikkawa et al., 1986; Parker et al., 1986). On the other hand, mercurials have extremely high affinity to sulfhydryl groups of proteins (Inouye et al., 1988; Rabenstein, 1978). Thus, $HgCl_2$ at high concentrations inhibits the PKC activity. Although there has been no information about the effect of $HgCl_2$ at low concentrations on cellular signal transduction, we could demonstrate that $HgCl_2$ at low concentrations ($10^{-10}$ - $10^{-8}M$) significantly potentiated the AII ($10^{-11}M$)-induced $[Ca^{++}]_i$ transient. In this study, we have investigated the involvement of PLC in $HgCl_2$ action. Although a highly specific inhibitor of PLC has not been known, propranolol has been used as one of PLC inhibitors (Blackwell and Flower, 1983; Chau and Tai, 1982; Hostetler and Matsuzawa, 1981). As shown in Figures 8 and 9, $10^{-4}M$ propranolol completely inhibited the stimulatory effect of $HgCl_2$ on AII-induced $[Ca^{++}]_i$ transient, and also blocked the stimulatory action of $HgCl_2$ to AII-Mediated $IP_3$ production (Fig. 10). In addition, it has been known that the initial rapid $[Ca^{++}]_i$ mobilization within a minute by AII application occurs through $IP_3$, and PKC activation mediates the sustained phase of $[Ca^{++}]_i$ in AII response (Kojima et al., 1985; Rasmussen 1986a). As shown in Figure 8, propranolol ($10^{-4}M$) completely diminished the sustained phase of $[Ca^{++}]_i$ caused by both AII and $HgCl_2$. Our results in this study indicate that propranolol inhibits the PLC activity, that is well accorded with the previous suggestions (Blackwell and Flower, 1983; Chau and Tai, 1982; Hostetler and Matsuzawa, 1981; Kojima et al., 1985; Rasmussen 1986a).

As shown in Figure 7, $HgCl_2$ significantly potentiated the AII-induced $[Ca^{++}]_i$ rise. However, we could not obtain a statistically significant increase of $IP_3$ by $HgCl_2$ (Figure 10). This discrepancy may be due to different materials used. For $[Ca^{++}]_i$ measurement, we used a single nephron segment, but for $IP_3$ measurement, the renal cortical tubule suspension had to be used, because of limited sensitivity to $IP_3$ assay. Since AII receptor is enriched in $S_1$, the cortical tubule suspension which is contaminated with the other segments should attenuate to reveal $HgCl_2$ stimulation on PLC through AII receptor. To clarify these results, further studies are needed.

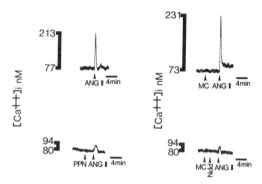

Fig. 8 $[Ca^{++}]_i$ tracings changed by AII $(10-^{11}M)$ in the presence or absence of propranolol $(10^{-4}M)$ and/or $HgCl_2$ $(10^{-9}M)$. After the pretreatment of $S_1$ with $HgCl_2$ for 5 min and/or propranolol for 3min. $10^{-11}M$ AII was applied to each sample.

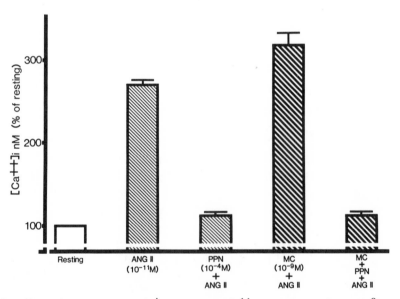

Fig. 9 Effect of propranolol $(10^{-4}M)$ on AII $(10^{-11}M)$ and/or $HgCl_2$ $(10^{-9}M)$ induced- $[Ca^{++}]_i$ increase in isolated rat $S_1$. Each bar represents the relative change in $[Ca^{++}]_i$ calculated from basal levels at each sample. Values are means $\pm$ SE. of 6 experiments.

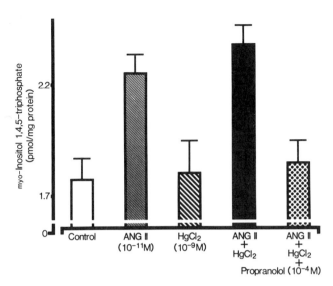

Fig. 10   Effect of propranolol (10$^{-4}$M) on AII (10$^{-11}$M) and/or HgCl$_2$ (10$^{-9}$M) mediated-IP$_3$ production in renal cortical tubule suspension.  Values are means $\pm$ SE. of 5 rats.

Propranolol (10$^{-4}$M) completely diminished the increment of [Ca$^{++}$]$_i$ by AII.  AII: angiotensin II, MC: mercury chloride, PPN: propranolol.

CONCLUSION

  We suggest for the first time that mercuric chloride at low concentrations stimulates angiotensin II-induced cytosolic free calcium increase, probably through an activation of phospholipase C in the isolated rat early proximal tubule.

ACKNOWLEDGMENTS

  The authors are grateful to Profs. F. Sakai and M. Endo for their valuable suggestions and encouragements.  This study was supported in part by grants from the Japanese Ministry of Education, Science and Culture (01480136, 01870111), the Japanese Ministry of Health and Welfare, the Japan Society for the Promotion of Science, the Hoansya, Sankyo Life Science Foundation, and the Shimabara Science Promotion Foundation.

## REFERENCES

Benabe, J.E., Spry, L.A., and Morrison, A.R., 1982, Effect of angiotensin II on phosphatidylinositol and polyphosphoinositide turnover in rat kidney, J. Biol. Chem., 257:7430-7434.

Blackwell, G.J., and Flower, R.J., 1983, Inhibition of phospholipase, Br. Med. Bull., 39:260-264.

Brock, T.A., Rittenhouse, S.E., Powers, C.W., Ekstein, L.S., Gimbrone, M.A., and Alexander, R.W., 1985, Phorbol ester and 1-oleoyl-2-acetylglycerol inhibit angiotensin activation of phospholipase C in cultured vascular smooth muscle cells, J. Biol. Chem., 260:14158-14162.

Brunder, D.G., Dettbarn, C., and Palade, P., 1988, Heavy metal-induced $Ca^{2+}$ release from sarcoplasmic reticulum, J. Biol. Chem., 263:18785-18792.

Chau, L.Y. and Tai, H.H., 1982, Resolution into two different forms and study of the properties of phosphatidylinositol-specific phospholipase C from human platelet cytosol, Biochem, Biophys. Acta, 713:344-351.

Endou, H., Koseki, C., Yamada, H., and Obara, T., 1986, Evaluation of nephrotoxicity using isolated nephron segments, in: "Nephrotoxicity of Antibiotics and Immunosuppressants," T. Tanabe, J.B. Hook, and H. Endou, eds., 207-216, Elsevier, Amsterdam.

Goldstein, L., 1987, Renal substrate utilization in normal and acidotic rats, Am. J. Physiol., 253:F351-F357.

Griendling, K.K., Rittenhouse, S.E., Brock, T.A., Ekstein, L.S., Gimbrone, M.S., and Alexander, R.W., 1986, Sustained diacylglycerol formation from inositol phospholipids in angiotensin II-stimulated vascular smooth muscle cells, J. Biol. Chem., 261:5901-5906.

Grynkiewicz, G., Poenie, M., and Tsien, R.Y., 1985, A new generation of $Ca^{++}$ indicators with greatly improved fluorescene properties, J. Biol. Chem. 260:3440-3450.

Harris, P.J., and Navar, L.G., 1985, Tubular transport responses to angiotensin, Am. J. Physiol., 248:F621-630.

Hinkle, P.C., and McCarty, R., 1978, How cells make ATP, Sci. Amer., 238:104-123.

Hoch, B.S., Gorfien, P.C., Linzer, D., Fusco, M.J., and Levine, S.D., 1989, Mercurial reagents inhibit flow through ADH-induced water channels in toad bladder, Am. J. Physiol., 256(25):F948-F953.

Hostetler, K.Y., and Matsuzawa, Y., 1981, Studies on the mechanism of drug-induced lipidosis: Cationic amphiphilic drug inhibition of lysosomal phosphodipase A and C, Biochem. Pharmacol, 30:1121-1126.

Houser, M.T., and Berndt, W.O., 1988, Unilateral nephrectomy in the rat: Effect on mercury handling and renal cortical subcellular distribution, Toxicol. Appl. Pharmacol., 93:187-194.

Inoue, Y., Saijoh, K., and Sumino, K., 1988, Action of mercurials on activity of partially purified soluble protein kinasec from mice brain, Pharmacol. Toxicol., 62:278-281.

Johnson, R.M., and Garrison, J.C., 1987, Epidermal growth factor and angiotensin II stimulate formation of inositol 1,4,5-and inositol 1,3,4-triphosphate in hepatocytes, J. Biol. Chem., 262:17285-17293.

Jung, K.Y., Uchida, S., and Endou, H., 1989, Nephrotoxicity Assessment by Measuring Cellular ATP Content. Toxicol. Appl. Pharmacol., 100:369-382.

Jung, K.Y., and Endou, H., 1989a, Nephrotoxicity Assessment by Measuring Cellular ATP Content, Toxicol. Appl. Pharmacol., 100:383-390.

Jung, K.Y., and Endou, H., 1989b, Biphasic increasing effect of angiotensin-II on intracellular free calcium in isolated rat early proximal tubule, Biochem. Biophys. Research Communications, 165(3):1221-1228.

Kikkawa, U., Go, M., Koumoto, J., and Nishizuka, Y., 1986, Rapid purification of protein kinase C by high performance liquid chromatography, Biochem. Biophys. Res. Commun., 135:636-643.

Klein, K.L., Wang, M.S., Torikai, S., Davidson, W.D., and Kurokawa, A., 1981, Substrate oxidation by isolated single nephron segments of the rat, Kidney Int., 20:29-35.

Kojima, I., Kojima, K., and Rasmussen, H., 1985, Role of calcium fluxes in the sustained phase of angiotensin II-mediated aldosterone secretion from acrenal glomerulosa, J. Biol. Chem. 260:9177-9184.

Manillier, C., Vinay, P., Lalonde, L., and Gougoux, A., 1986, ATP turnover and renal response of dog tubules to pH changes in vitro, Amer. J. Physiol., 251:F919-F932.

Moraes, V.L.G., and Meis, L., 1987, ATP synthesis by the (Na$^+$+K$^+$) -ATPase in the absence of an ionic gradient, FEBS Lett., 222:163-166.

Parker, P.J., Coussens, L., Totty, N., Rhee, L., Chen, Y.E., Stabel, S., Waterfield, M.D., and Ullrich, A., 1986, The complete primary structure of protein kinase C-the major phorbol ester receptor, Science, 233:853-859.

Pratz, J., Ripoche, P., and Corman, B., 1986, Evideme for proteic water pathways in the luminal membrane of kidney proximal tubule, Biochim. Biophys. Acta., 856:259-266.

Rabenstein, D.L., 1978, The chemistry of methylmercury toxicology, J. Chem.Educ., 55:292-296..

Rasmussen, H., 1986a, Mechanisms of disease: The calcium Messenger System, New Eng. J. Med., 314:1164-1170.

Ruegg, C.E., Gandolfi, A.J., Nagle, R.B., and Bredel, Kl, 1987, Differential patterns of injury to the proximal tubule of renal cortical slices following in vitro exposure to mercuric chloride, potassium, dichromate, or hypoxic condition, Toxicol. Appl. Pharmacol. 90:261-273.

Schlondorff, D., DeCandido, S., and Satriano, J.A., 1987, Angiotensin II stimulates phospholipaye C and A2 in cultured rat mesangial cells, Am. J. Phisiol., 253:C113-C120.

Schuster, V.L., Kokko, J.P. and Jacobson, R.H., 1984, Angiotensin II directly stimulates sodium transport in rabbit proximal convoluted tubules, J. Clin. Invest, 73:507-515.

Silva, P., 1987, Renal fuel utilization, energy requirements and function, Kidney Int., 32:S9-S14.

Smith, L.B., Smith, L., Brown, E.R., Barnes, D., Sabir, M.A., Davis, J.S., and Farese, R.V., 1984, Angiotensin II rapidly increases phosphatidate-phosphoinositide synthesis and phosphoinositide hydrolysis and mobilizes intracellular calcium in cultured arterial muscle cells, Proc. Natl. Acad. Sci. USA, 81:7812-7816.

Smith, M.W., Ambudkar, I.S., Phelps, P.C., Regec, A.L., and Trump, B.F., 1987, HgCl$_2$-induced changes in cytosolic Ca$^{2+}$ cultured rabbit renal tubular cells, Biochim. Biophys. Acta., 931:130-142.

Trimm, J.L., Salama, G., and Abramson, J.J., 1986, Salfhydryl oxidation induces rapid calcium release from sarcoplasmic reticulum vesides, J. Biol. Chem., 261:16092-16098.

Uchida, S., and Endou, H., 1988, Substrate specificity to maintain cellular ATP along the mouse nephron, Amer. J. Physiol., 255:F977-F983.

Weindemann, M.J., and Krebs, H.A., 1969, The fuel of respiration of rat kidney cortex, Biochem. J., 112:149-166.

Weinberg, J.M. and Humes, H.D., 1986, Increase of cell ATP produced by exogenous adenosine nucleotides in isolated rabbit kidney tubules, Amer. J. Physiol., 250:F720-F733.

Wirthensohn, G., and Guder, W.G., 1986, Renal substrate metabolism, Physiol.Rev., 66(2):469-497.

Woodcock, E.A., and Johnston, C.I., 1982, Inhibition of adenylate cyclase by angiotensin II in rat renale cortex, Endocrinol., 111:1687-1691.

# EFFECTS OF METHYLMERCURY ON THE DEVELOPING BRAIN

Ben H. Choi

Division of Neuropathology
Department of Pathology
University of California
Irvine, CA
USA

## ABSTRACT

In an attempt to elucidate the mechanisms of methylmercury (MeHg) action on the developing brain, MeHg effects on neuronal migration and on proliferative neuroepithelial germinal cells within the ventricular zone of telencephalic vesicles were investigated in C57BL/6J mice. $^3$H-thymidine autoradiographic studies following acute and chronic MeHg intoxication in embryonic mice showed that heavily labeled neurons (generated on E-16) within specified regions of the cerebral cortex of prenatally intoxicated offspring were distributed irregularly throughout layers II and III, whereas those in controls were tightly clustered within the upper part of layer II. These findings support the hypothesis that prenatal MeHg poisoning results in anomalous cytoarchitectonic patterning of the cerebral cortex, and provide a possible morphological basis for some of the neurobehavioral abnormalities that may follow sublethal prenatal MeHg intoxication. Whether the irregular distribution of labeled neurons was due to the effects of MeHg on the dividing neuroepithelial germinal cells, on the process of neuronal migration, or on the final positioning of postmigratory neurons within the cerebral cortex remained unclear. Ultrastructural studies of telencephalic vesicles during acute phases of MeHg intoxication revealed the presence of acute degenerative changes within ventricular cells, characterized by spongy change and vacuolization of the cytoplasmic matrix and by loss of microtubules. $^3$H-thymidine autoradiography demonstrated features suggestive of disturbed interkinetic nuclear migration. Also noted were reduction of the mitotic indices of neuroepithelial germinal cells at the ventricular surface at 4 to 12 hours and early-phase mitotic arrest. These findings suggest that MeHg exerts significant effects on proliferating neuroepithelial germinal cells during the acute phases of MeHg poisoning, and may eventually affect the architectonic makeup of the cortical plate as the brain matures. Additional studies in our laboratory demonstrated 1) a failure of histotypic re-organization of dissociated embryonic cerebral cortical cells in rotation-mediated re-aggregation culture, 2) disturbances in neural cell adhesion molecule expression in PC12 cells, 3) modifications in the density of excitatory neurotransmitter L-glutamate receptors, 4) marked disturbances in the glutamate uptake mechanism of fetal astrocytes *in vitro,* and 5) reductions in cholinoceptive muscarinic receptor (M1 and M2) binding in selected regions of the cerebrum. Thus, MeHg may affect the developing brain through diverse pathogenetic pathways. Possible effects of mercurial compounds on the interaction between certain serine proteases and protease inhibitors during brain development are also discussed.

INTRODUCTION

The vulnerability of the developing CNS to the toxic effects of MeHg is now well established (Amin-Zaki et al., 1974, 1976; Choi, 1989; Clarkson, 1987; Harada, 1968, 1976, 1978; Marsh et al., 1980; Reuhl and Chang, 1979). It is apparent that mercurial compounds cause both degenerative changes and developmental deviations in the developing brain, depending upon the degree and duration of exposure as well as the gestational stages at which exposure occurs (Takeuchi, 1977). Thus, MeHg effects may appear immediately and result in cell death or embryonic death, or they may appear much later and be ascertainable only in the postnatal or adult period. It is also probable that MeHg exerts its various effects on the developing brain through diverse pathogenetic mechanisms.

During its protracted ontogenesis, the developing brain undergoes a programmed series of cellular events that includes proliferation of neuroepithelial germinal cells to produce neurons and glial cells; adhesion, dissociation and migration of neurons; aggregation of postmigratory neurons to form organized nuclear groups or to form cortical plates; elongation of axons and dendrites with establishment of synapses; differentiation and maturation of neurons and also programmed cell death. The final outcome of these intricate developmental events is the formation of a mammalian neocortex that is endowed with a remarkable regularity in the geometric and spatial arrangement of different neuronal cell types, despite its seemingly chaotic and complex organization (Choi, 1988a). At the same time, it should be recognized, however, that the brain is structurally and functionally nonhomogeneous, and that the program of development may differ even within the same brain region (Choi, 1989; Rodier, 1983). It is a challenging task indeed to attempt to elucidate specific mechanisms of action or to identify specific target sites for MeHg action among such diverse and continually evolving ontogenetic events in the developing brain. Obviously, precise information regarding the location and timing of insults in relation to ontogenetic events during brain development is absolutely necessary when studying the effects of noxious agents such as MeHg.

The purpose of this paper is to highlight some of our recent findings regarding the neurotoxic effects of MeHg on the developing brain, and to discuss possible mechanisms of MeHg action at the cellular and subcellular levels.

## Effects of methylmercury on neuronal migration

Previously we reported the results of neuropathological studies of the brains of human newborn infants whose mothers had ingested MeHg-contaminated bread during pregnancy (Choi et al., 1978). These brains showed developmental deviations characterized primarily by disturbances in neuronal migration and laminar cortical organization within the cerebrum and cerebellum associated with diffuse astrocytosis of the white matter. Based on these findings, we proposed that damage to the developing human brain following sublethal MeHg intoxication primarily affects neuronal migration and the patterning of cortical cytoarchitectonic organization.

With the use of time-lapse cinematographic and ultrastructural studies, we showed abrupt cessation of neuronal migration in organotypic cultures of human fetal cerebrum following MeHg exposure that appeared to be primarily due to the cytotoxic effects of MeHg. In particular, the damaging effect of MeHg on microtubules and membranes of neurons and astrocytes were quite apparent (Choi et al., 1981). These studies were followed by $^3$H-thymidine radioautographic studies in C57BL/6J embryonic mice subjected to acute and chronic MeHg intoxication. As shown in Figure 1, statistically significant irregularities in the distribution of heavily labeled neurons (generated on E-16) were demonstrated within the cerebral cortical plate of prenatally intoxicated offspring (Peckham and Choi, 1988) at postnatal day 10. These findings support the hypothesis that prenatal MeHg poisoning results in anomalous cytoarchitectonic patterning within the cerebral cortex. Inouye and Kajiwara (1988), using guinea-pigs, detected "developmental disturbances of the fetal brain (e.g., small-sized brain, thinner brain mantle, dilated lateral ventricles, and disarrangement of the cerebral cortical lamination as a result of abnormal migration of neurons" following prenatal MeHg intoxication. On the other hand, Mottet and his colleagues (1987) failed to detect abnormal neuronal migration in the brains of prenatally intoxicated offspring in nonhuman primates at a chronic, low-dose level.

Neuronal circuits and their component neurons form the structural basis of all behavior. Since most neurons within a given layer of cerebral cortex exhibit similar patterns of connectivity, it seems reasonable to suggest that alterations of lamination may adversely affect neuronal circuitry. These findings therefore appear to provide a possible morphological basis for some of the neurobehavioral abnormalities that may follow sublethal prenatal MeHg poisoning. However, whether the irregular distribution of postmigratory neurons in the cerebral cortical plate was due to the effects of MeHg on proliferating neuroepithelial germinal cells, on the migratory process itself or on the final segregation and positioning of post-migratory neurons remained unclear.

### Effects of methylmercury on neuroepithelial germinal cells in ventricular zone of telencephalon of embryonic mice

Timed-pregnant mice were given intraperitoneal injection of methylmercuric chloride (MMC), 20 mg/kg of body weight, at E-14. Ten minutes after the injection of MMC, both control and MeHg groups were injected with $^3$H-thymidine. Animals were sacrificed at 4, 12, 24, and 48 hours. Coronal sections of the cerebrum were processed for light and electron microscopic studies. Ultrastructural and radioautographic analysis were correlated with mitotic indices of ventricular germinal cells and with the percentage of early and late phase mitotic figures per high power field.

Acute degenerative changes within scattered ventricular cells, characterized by edema and spongy changes associated with dissolution of the cytoplasmic matrix and loss of organelles (including microtubules), were identified by electron microscopy (Figure 2). $^3$H-thymidine autoradiography also revealed features suggestive of disturbances in interkinetic nuclear migration (Figure 3). Reduction of mitotic indices in neuroepithelial germinal cells was present at 4 to 12 hours. However, after 24 hours, the mitotic indices were nearly equal both in control and MeHg groups. At 48 hours, the mitotic indices were even higher in the MeHg group as compared to controls. More importantly, at all intervals, MeHg-intoxicated animals showed reductions in the percentage of late-phase mitotic figures as compared to controls (Figure 4). These findings indicate that MeHg exerts considerable effects on the

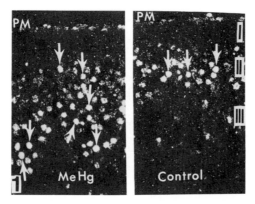

Fig. 1.  Dark-field autoradiographs showing heavily labeled neurons (generated on E-16) in control and MeHg groups. Note irregular distribution of neuronal nuclei (arrows) in the MeHg group as compared to controls.

mitotic activity of germinal neuroepithelial cells. It is suggested that acute degenerative changes, particularly, those affecting microtubules, may have been partly responsible for disturbances in interkinetic nuclear migration and also for mitotic arrest within ventricular cells. From these results, it is quite apparent that MeHg may exert significant effects of proliferating neuroepithelial germinal cells during the acute phase of MeHg poisoning and may eventually affect the architectonic makeup of the cortical plate as the brain matures (Choi, 1991).

Among the various effects of MeHg on cellular and subcellular organelles, those on the microtubules appear to be most consistent and specific (Choi, 1991; Choi et al., 1981; Clarkson, 1987; Ramel, 1969; Sager et al., 1982; Rodier et al., 1984; Vogel et al., 1985). MeHg causes depolymerization of microtubules (Abe et al., 1975) and disrupts the assembly process by reacting with SH groups on tubulin monomers (Margolis and Wilson, 1981). These effects are readily demonstrable within mitotic cells, as microtubules are essential for cell division.

What is more significant is the fact that the effects of MeHg on microtubules, particularly on those within mitotic cells, can be visualized both *in vivo* and *in vitro* at different concentrations (Choi, 1991; Miura et al., 1978; Sager et al., 1982; Rodier et al., 1984). Figure 5 demonstrates a few examples of mitotic figures affected by MeHg. When organotypic cultures of human fetal cerebrum were exposed to MeHg there was an accumulation of large round cells at periphery of the explants at 4 hours (Figure 5a). These proved to be mitotic astrocytes in metaphase-arrest by electron

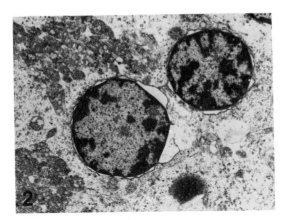

Fig. 2. Electron micrography showing neuroepithelial germinal cells in the ventricular zone of embryonic mouse 12 h after MeHg poisoning. Note nuclear pyknosis and spongy changes and vacuolization of the cytoplasm associated with loss of cytoskeletal elements, including microtubules. Mitochondria are relatively electron-dense and clustered at one poly of the cytoplasm.

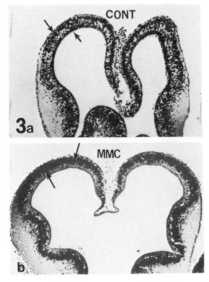

Fig. 3. $^3$H-thymidine radioautography showing labeled nuclei (arrows) in control groups scattered through all layers of the telencephalic vesicle, while those in the MMC group are primarily located within the outer half. Embryonic mice 4 h after MeHg poisoning. CONT: control; MMC:methylmercuric chloride.

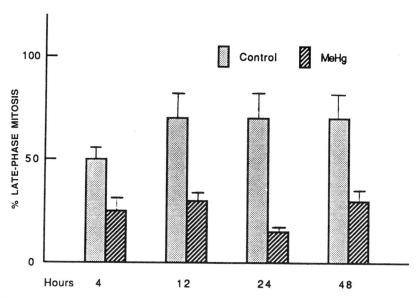

Fig. 4. Percentages of late-phase mitoses at the ventricular surface of the telencephalic vesicles of embryonic mice following MeHg poisoning.

microscopy (Figure 5b). Typically, all of the mitotic figures found in the MeHg group demonstrated a lack of mitotic spindle tubules in addition to degenerative features. Astrocytes in rotation-mediated re-aggregation cultures also showed a lack of mitotic spindle tubules (Figure 5c). Figure 5d represents a mitotic neuroepithelial germinal cell within the telencephalic vesicle of an embryonic mouse. Again, the lack of mitotic spindle tubules was a characteristic finding.

Effects of Methylmercury on cell-to-cell interactions in the developing brain

A.   Studies using rotation-mediated aggregation cultures of embryonic brain cells

In order to test the hypothesis that prenatal MeHg poisoning may cause disturbances in histotypic organization and differentiation in the developing brain, and to gain further insight into the nature and mechanisms of MeHg neurotoxicity, we used a rotation-mediated aggregation culture system to examine the effects of MeHg upon reorganization of mechanically dissociated embryonic rat cerebrum.

It is well established that mechanically dissociated embryonic brain cells can be induced to undergo histotypic reassembly in a rotation-mediated aggregation culture system (Trapp et al., 1979). Furthermore, neurons and glial cells within the aggregates continue to proliferate and to undergo differentiation, mimicking *in vivo* brain development both in morphology and in biochemical differentiation (Honegger

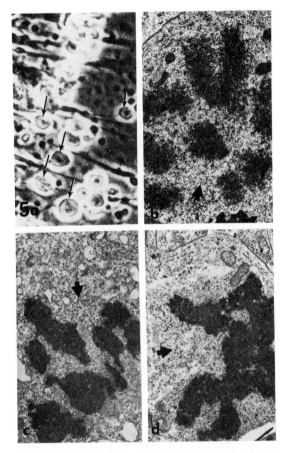

Fig. 5. Mitotic figures showing lack of mitotic spindle tubules in MeHg-exposed cells.

5a:    Metaphase-arrested astrocytes (arrows) accumulated at the periphery of explants after 4 hours of exposure to MeHg of organotypic culture of human fetal cerebrum. Phase-contrast photomicrography taken during time-lapse cinematography.

5b:    Mitotic cell observed in a MeHg-exposed organotypic culture of human fetal astrocytes.

5c:    Mitotic figure observed in embryonic rat cerebral cell grown in rotation-mediated aggregation culture following MeHg exposure.

5d:    Mitotic figure in neuroepithelial germinal cell 12 hours after MeHg poisoning.

and Richelson, 1977; Lu et al., 1980, Orkand et al., 1984). This system thus provides an opportunity to assess some of the cellular events in normal and abnormal brain development under controlled and replicable conditions at both light microscopic and ultrastructural levels.

Aggregating cell cultures were prepared from the cerebra of E-17 Sprague-Dawley rats. Cells were dissociated by filtering through two passages of Nitex cloths (130 and 33 μm mesh, respectively) and known concentrations of cells were placed into Erlenmeyer flasks and kept under constant rotation in synthetic medium. The aggregates were harvested at 1,2,3,7,14 and 21 days-*in vitro*. The final concentrations of MMC in the culture medium were 0.1,0.2 and 0.5 μM, respectively.

Although both dissociated cells in control and MeHg groups formed relatively uniform cellular aggregates initially, those exposed to MeHg underwent almost total disaggregation after 48 h (Choi and Ahn, 1989). Electron microscopy revealed the presence of marked cytotoxic damage in all cells, characterized by spongy change and cytoplasmic vacuolization that could already be observed at 24 h. At this time, mitotic figures in the MeHg group showed an almost total lack of mitotic spindle tubules (Figure 5b). However, when cell aggregates were allowed to form in control medium for 7 days and then exposed to 0.5 μM MMC, the cells remained in clumps, although considerable degeneration occurred. Immunocytochemical staining in the control group demonstrated the presence of glial fibrillary acidic protein (GFAP)-positive astrocytes and their processes more prominently at the periphery of the cell aggregates. These cells extended their processes in a criss-cross pattern among cells within the aggregates. MeHg-exposed cell aggregates still contained astroglial cells and processes at the periphery, as evidenced by immunoreactivity for GFAP. This suggests that the integrity of astroglia play an important role in maintaining cell aggregates in these cultures.

A considerable amount of degenerative change was also seen among cells in the control group, and differentiation of junctional complexes and synapse formation was evident as the cultures matured whereas none of these phenomena took place in MeHg-exposed cultures. The cytotoxic effects of MeHg were extreme after 48 h of exposure, and practically all cells and processes showed almost complete degeneration. Kleinschuster et al (1983) also observed complete inhibition of aggregate formation 24 h after MMC (4μM) exposure in reaggregation cultures of embryonic neural retinal cells.

B.   Effects of MeHg on neural cell adhesion molecule

One of the fundamental characteristics of brain is the formation of cell-to-cell adhesions. Such interactions are believed to induce differentiation and produce specific tissue architectural patterns. Neural cell adhesion molecule (NCAM) is primarily associated with adhesion among neuronal cells and has been purified and characterized (Edelman, 1984). It consists of three glycoproteins with apparent molecular weights of 180, 140 and 120 kDa (Chuong and Edelman, 1984). NCAM expression is believed to be of importance for morphogenetic cell interactions such as neuronal migration and aggregation to form nuclear groups during brain development.

In order to determine whether or not failure of reaggregation of embryonic brain cells following MeHg exposure might be related to alterations in NCAM expression, the effects of MeHg were examined in monolayer cultures of PC12 cells, which normally express NCAM avidly *in vitro*. Phase and electron microscopy were correlated with immunocytochemical studies using indirect immunofluorescence and immuno-gold methods following exposure to varying concentrations of MeHg. As expected, PC12 cells showed dose-dependent cytotoxic effects that eventually led to granular disruption of the membrane and cytoplasm. Alterations in the expression of NCAM appeared to parallel the cytotoxic damage. A noteworthy finding was the preservation of NCAM immunoreactivity in some of the cell clumps even in the presence of severe cytotoxic damage, suggesting that disruption in immunofluorescent images was secondary to membrane damage. Western blots also showed no significant difference between control and MeHg groups (Shen et al., 1989). Molecular interactions between MeHg and NCAM and other adhesion molecules during neurogenesis are deserving of further investigation.

## C. MeHg effects on developing astrocytes

Most studies of MeHg neurotoxicity have focused on the effects on neurons. However, the effects of MeHg on glial cells cannot be overlooked. It is well established that radial glia provide guidance to migrating neurons and also give rise to astrocytes (Choi, 1986; Rakic, 1972). Perivascular ensheathment of astroglial endfeet is accomplished early in CNS development (Choi, 1981; Choi and Lapham, 1978) and probably plays an important role in the transport of materials across blood vessel walls.

### a). MeHg effects on the membranes of astrocytes in culture

#### 1) Alterations in surface charge

The most severe cytotoxic effects of MeHg were apparent along the surface membranes of neurons and astrocytes when we exposed organotypic cultures of human fetal cerebrum (Choi et al., 1981). In order to test the hypothesis that the permeability of the astroglial membrane to HeHg may be related to alterations in surface charge, we examined the distribution of cationized ferritin along the cell membranes of cultured mouse fetal astrocytes following brief exposure to MeHg (Peckham and Choi, 1986). The results showed that, in MeHg-exposed cells, ferritin patches were disrupted in irregular fashion, leaving a larger portion of the cell membrane uncovered, whereas controls showed a fairly even distribution of discontinuous ferritin patches along the cell membrane. Our data show the presence of a negative charge on the surfaces of cultured fetal astrocytes and a marked shift in the distribution of anionic groups as a result of MeHg exposure. The alteration in the surface charge may in turn trigger a cascade of toxic actions on these cells.

### 2) Effects of meso-2,3-dimercaptosuccinic acid (DMSA)

DMSA is a well known chelating agent that is effective in decreasing the body burden and brain content of Hg in experimental animals (Magos, 1976; Kostyniak, 1982). Using time-lapse cinematography, we observed recovery of MeHg-damaged membranes of cultured astrocytes following the addition of DMSA. The exposure of cultures to DMSA and MMC in combination also prevented membrane damage. Thus membrane damage caused by MeHg can be reversed when a proper chelating agent that is not itself toxic is used early enough (Choi and Lapham, 1981).

### 3) Effects of MeHg on membrane fluidity

Electron spin resonance (ESR) studies using 5-doxyl stearic acid as spin label were carried out in cultured mouse fetal astrocytes to examine the molecular interaction between MeHg and the astrocytic membrane. Following exposure of astrocytes to 10 $\mu$M MMC for 3 h, the spin-labeled cell suspensions were placed into an ESR quartz tissue cell and studied with an ESR spectrometer. As shown in Figure 6, MMC produced an increase in the splitting between the outermost peaks from 56 to 57 gauss (Choi, 1988b). The order parameter (S) was determined by observed hyperfine splitting of ESR spectra and calculated according to the formula for 5-doxyl stearic acid. The control group showed an order parameter (S) of 0.68 whereas the MeHg group showed an order parameter (S) of 0.71. These results indicated that there was a significant reduction in the fluidity of the membrane following exposure to MeHg. Further studies using different labels and different exposure times need to be carried out.

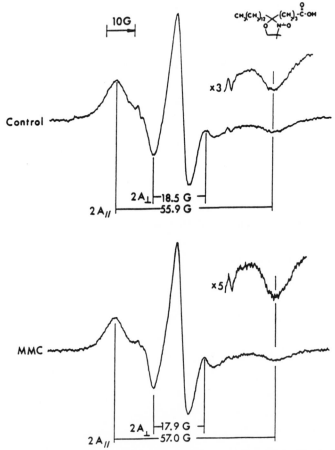

Fig. 6. ESR spectra of spin-labeled mouse fetal astrocytes in control and MeHg groups showing increase in the splitting between the outermost peaks from 56 to 57 gauss in the MeHg group.

b).    MeHg effects on nucleosomes of mouse fetal astrocytes

We examined the interaction of MMC with nuclear proteins with the aid of a fluorescent probe, N-(3-pyrene) maleimide. This probe labels the cysteine groups found primarily in histone H3 of nucleosomes. DNA interactions occur most strongly with the dimer H3-H4 of nuclear histones, which may be of importance in gene regulation. Nucleosomes prepared from MMC-exposed monolayer cultures of astrocytes were subjected to SDS-PAGE as well as acetic acid/urea/Triton X-100 electrophoresis. DNA was extracted from nucleosomes and analyzed on agarose gels. A significant decrease of fluorescence was observed following 6 hours of 10 μM MMC exposure when nucleosomes were reacted with the fluorescent probe at concentrations varying from 0 to 200 nM. This thus appears to be another means by which MeHg may exert its toxic actions on developing fetal astrocytes (Choi and Simpkins, 1986).

c).    MeHg effects on L-glutamate uptake in cultured astrocytes

L-glutamate (GLU) is a ubiquitous excitatory amino acid neurotransmitter that causes depolarization and excitation of mammalian CNS neurons. Due to disturbances in its availability or transport, GLU can also be excitotoxic to cortical neurons and is linked to the pathogenesis of some degenerative diseases of the CNS. Astroglial cells are thought to be involved in the sequestration and control of potentially harmful levels of exogenous GLU. A significant modification in GLU receptor density following MeHg intoxication in young adult mice suggested that excitatory amino acid transmitter pathways may be involved in MeHg neurotoxicity (Choi et al., 1988b). In order to examine this further, monolayer cultures of astrocytes obtained from neonatal mice cerebra were exposed to varying concentrations (0.2 - 10.0 μM) of MMC and mercuric chloride (MC), and the uptake of radiolabeled GLU measured. As shown in Figure 7, exposure to MC (0.2-5.0 μM) for 15 minutes caused selective and dose-dependent inhibiton of GLU uptake to 50% of control levels. Beyond 5.0 μM a sharp decrease (more than 75%) occurred. MMC also inhibited GLU uptake, however, 50% reduction was reached only at 10.0 μM. These results indicate that both inorganic and organic Hg markedly affect the uptake mechanism of GLU by astrocytes and further support the hypothesis that excitotoxicty may play a role in the pathogenesis of Hg neurotoxicity in the CNS (Kim and Choi, 1990). Brookes (1988) also noted 50% inhibition of initial GLU transport in astrocyte cultures exposed to 0.5 μM mercuric chloride, without affecting 2-deoxyglucose uptake. Other heavy metals tested did not inhibit the uptake mechanism.

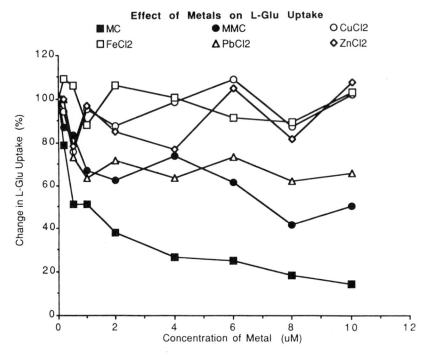

Fig. 7. L-glutamate (GLU) transport in cultured mouse astrocytes following exposure
to various heavy metals. Note the significant reduction in GLU uptake
associated with inorganic mercury intoxication.

It is apparent that developing fetal astrocytes are highly sensitive to the toxic effects of MeHg. Evaluation of the toxic effects of MeHg in the developing brain must, therefore, take into account the role played by damaged astrocytes. Although the results of in-vitro experiments cannot be interpreted literally as phenomena in living organisms, this system allows in-depth analysis of some of the molecular interactions between living brain cells and MMC under controlled conditions.

## EFFECTS OF METHYLMERCURY ON EXCITATORY AMINO ACID RECEPTORS

As noted above, acidic amino acids such as GLU are thought to be excitatory neurotransmitters in the mammalian brain. At least three receptor subtypes of GLU have been characterized on the basis of their selective sensitivity to N-methyl-Daspartate (NMDA), kainate (KA) and alpha-amino-3-hydroxy-5-methyl-4-isoxyazolepropionic acid (AMPA).

Using *in vitro* autoradiography and microcomputer-assisted imaging system, we determined the receptor density in twelve cortical and subcortical regions of the cerebra of mice. A significant increase in density of both AMPA ($2987 \pm 71$ fmol/gm of protein, as compared to $2730 \pm 91$ fmol/mg for controls) and KA ($644 \pm 54$ fmol/mg as compared to $498 \pm 36$ fmol/mg for controls) receptors was demonstrated in the parietal cortex. The density values in the insular cortex ($586 \pm 52$ fmol/mg protein) also exceeded those of controls ($446 \pm 36$ fmol/mg protein). The NMDA receptor density in the frontal cortex of MeHg-treated groups ($1196 \pm 87$ fmol/mg protein), however, was significantly decreased in comparison to that of controls ($1453 \pm 57$ fmol/mg protein) ($p < 0.03$). These findings suggest that excitotoxic damage and/or alterations in plasticity in selected neuronal groups may have occurred following MeHg poisoning (Choi et al., 1988).

## HYDROCEPHALUS FOLLOWING PRENATAL METHYLMERCURY POISONING

The most common malformation produced by prenatal MeHg poisoning has been cleft palate (Harris et al., 1972; Olson and Massaro, 1977; Spyker and Smithberg, 1972; Su and Okita, 1976). Hydrocephalus has also been described by a number of investigators (Fuyuta et al., 1978; Harris et al., 1972; Murakami, 1972; Spyker and Smithberg, 1972). However, its pathogenesis has never been fully explained. We have also observed the development of hydrocephalus in approximately 15 to 25% of offspring of prenatally intoxicated mothers. The characteristic features were extreme dilation of the lateral ventricles; marked narrowing of the aqueduct of Sylvius without complete occlusion; and edema, spongy degeneration and cystic change of the white matter. A remarkable finding was the presence of innumerable Hg grains within the cytoplasm of ependymal and choroid plexus cells, as demonstrated both at the light and EM levels by a modified silver-sulfide method (Choi, 1984).

Although the pathogenesis of hydrocephalus is still not clear, it is possible that the effects of MeHg upon choroid plexus and ependymal cells may have led to alterations in cerebrospinal fluid homeostasis, production and absorption. Choroid plexus is known to be a "sink" for heavy metals. Demonstration of numerous Hg grains in the cytoplasm of both choroid plexus and ependymal cells support such a contention. It is also possible that degeneration of the white matter due to the toxic effects of MeHg may have been aggravated by alterations in CSF homeostasis. Inouye and Kajiwara (1990) suggested that susceptibility to hydrocephalus varies according to the strains of mice used. In their study, C57BL/10 mice treated with 10 mg/kg of MeHg on day 15 of gestation resulted in a 54% incidence of hydrocephalus at postnatal day 30 as compared to 0.8% in controls. The incidence of spontaneous hydrocephalus in C57BL/6J mice is said to be approximately 1%, although in one study an incidence as high as 3% was noted. None of the control animals in our study developed hydrocephalus, however.

## METHYLMERCURY EFFECTS ON ENDOTHELIAL CELLS AND PERIVASCULAR ASTROGLIA OF NEONATAL CEREBELLUM

Neonatal mice were injected with 5 mg/kg body weight of [203] Hg-labeled MMC on postnatal days 3, 4 and 5, totaling 15 mg/kg body weight. Ultrastructural studies on postnatal day 15 demonstrated severe attenuation of endothelial cells with increased electron-density and frequent vacuolization of the cytoplasm. Also noted was marked swelling of pericapillary astroglial endfeet throughout the cerebella. Similar alterations in the endothelium were observed by Chang et al (1977) in the cerebellum of neonatal rats exposed to MeHg prenatally. Vascular injury, of course, lends support to the suggestion that alteration of the blood-brain barrier is one of the mechanisms whereby MeHg damages the nervous system (Chang and Hartmann, 1972; Chang et al, 1977). Steinwall and Olsson (1969) also observed disturbed permeability of cerebral blood vessels in rats following organic and inorganic mercury injection. The effects of mercurial compounds and other neurotoxic heavy metals on the integrity and function of the blood-brain barrier must be further investigated.

## EFFECTS OF METHYLMERCURY ON THE INTERPLAY BETWEEN PROTEASES AND PROTEASE INHIBITORS

Protease nexin-1 (PN-1) is a 43 kDa serine protease inhibitor that was identified in the conditioned medium of cultured human fibroblasts (Baker et al., 1980) and subsequently shown to be secreted by a variety of cultured extravascular cells (Eaton and Baker, 1983), including astrocytes (Rosenblatt et al., 1987). PN-1 rapidly inhibits thrombin, urokinase and plasmin by forming a complex with their catalytic site serine residues. Recent studies showed that the deduced amino acid sequence of PN-1 is identical to that of the glia-derived neurite promoting factor or glia-derived nexin (Gloor et al., 1986; McGrogan et al., 1988) that stimulates neurite outgrowth in cultured neuroblastoma cells. Hirudin, a potent thrombin inhibitor derived from leeches, stimulates neurite outgrowth to the same extent and with the same kinetics. This result, along with the finding that thrombin brings about retraction of neurites in cultured neuroblastoma cells, indicates that the neurite-promoting activity of PN-1/glia-derived nexin depends on the inhibition of thrombin. Recent studies also demonstrated that thrombin reciprocally regulates the stellation of cultured rat astrocytes (Cavanaugh et al., 1990).

Thrombin, in addition to playing a central role in the blood coagulation system, is a serine protease with diverse bioregulatory activity. Minute amounts of extravasated thrombin have the potential for causing CNS injury by altering neuronal and glial processes, leading to disruption of their function. All of these findings underscore the importance of protective mechanisms for inactivating thrombin that may enter the brain when blood-brain barrier function is compromised. PN-1 appears particularly suited for controlling thrombin in the brain, since blood-brain barrier would be expected to exclude protease inhibitors that are abundant in plasma.

Studies of the brains of patients with Alzheimer's disease showed that PN-1 activity and free, uncomplexed PN-1 was reduced approximately 7-fold as compared to age-matched controls (Wagner et al., 1989). This was not due to decreased PN-1 mRNA in the Alzheimer's disease samples. Instead, increased levels of thrombin or thrombin-like protease activity seemed at least partly responsible. These results led to the hypothesis that blood-brain barrier is compromised in Alzheimer's disease, and that reduced levels of PN-1 and increased levels of thrombin or a thrombin-like protease in the brains of Alzheimer patients might lead to neuritic and astroglial alterations that could contribute to the pathogenesis of the disease.

These findings prompted us to examine the cell types with which PN-1 is associated. Using highly specific monoclonal and polyclonal antibodies to PN-1, we carried out immunocytochemical studies and demonstrated that PN-1 was associated with the smooth muscle cells of arteries and arterioles and with astroglial processes within the parenchyma and in the perivascular endfeet of human cerebral cortex (Figure 8). *In situ* hybridization with an [35]S-labeled RNA antisense probe for PN-1 resulted in significant labeling of both astrocytes and blood vessels (Choi et al., 1991). These results suggest that PN-1 around blood vessels may play a major protective role against the extravasation of thrombin, and possible other serine proteases, into the human brain.

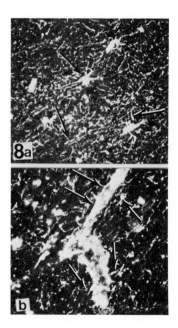

Fig. 8.  Indirect immunofluorescence showing strong PN-1 immunoreactivity in astroglial processes within the cerebral cortex (a).

Immunofluorescence is also noted in the walls of the blood vessels and in perivascular astroglial endfeet (b).

PN-1 in astrocytes and their processes may also play important roles during neurogenesis, such as during neuronal migration and the final positioning of postmigratory neurons.  It is known that migrating cells have more cell surface proteolytic activity than stationary, differentiated cells.  Thus, the interplay between proteases and protease inhibitors may not only influence neuronal-glial interactions during neurogenesis but also may play a crucial role in the constant restructuring of neurites during repair of injury to the nervous system.

As stated earlier, mercurial compounds are known to interfere with blood-brain barrier function.  It is likely, therefore, that extravasation of prothrombin from vessels and its subsequent conversion to thrombin within the brain may occur at certain stages of intoxication.  Preliminary studies in our laboratory indicate that, following injections of MMC (7.5 mg/kg body weight for 6 days) in adult mice, thrombin can escape from blood vessels into the brain.  Furthermore, both inorganic and organic Hg appear to interfere with PN-1-thrombin complex formation (unpublished observation).  If this proves to be the case, then the protective effects of PN-1 against thrombin extravasation would be compromised, leading to further brain injury.  Evaluation of the interplay between proteases and protease inhibitors following exposure to neurotoxic agents should prove to be an exciting area for future research in the field of developmental neurotoxicology.

CONCLUSION

In spite of relatively quiescent research activities dealing with MeHg during the 1980's, the potential for environmental human intoxication still has not diminished. The vulnerability and high sensitivity of the developing brain to the toxic effects of MeHg is now well established. Of most immediate concern is the consumption of MeHg-contaminated food by pregnant women who may sustain relatively minor nervous system damage from MeHg intoxication during pregnancy but whose fetuses may suffer severe consequences. Furthermore, although affected infants may not have grossly apparent brain abnormalities, the internal neuronal circuitry may be deranged, leading to abnormalities in behavior that become more apparent with increasing maturity.

It must be also stressed that there are species differences in MeHg toxicity. For example, in man, cat and rodents, the cerebellum is one of the principal target sites for MeHg toxicity. However, the primate cerebellum suffers no apparent histological damage following MeHg exposure. Therefore, selection of the experimental animal and of the target organ or tissue is critical in evaluating the outcome of MeHg intoxication. Although notable progress has been made during the past decade toward an understanding of the pathogenesis of MeHg neurotoxicity in the developing brain, there is promise of more to come when multidisciplinary collaborative efforts among toxicologists, biochemists, molecular biologists, neurobehavioral scientists and neuropathologists are undertaken. Advances in research techniques and tools will also usher in an exciting new era in the study of MeHg toxicity in the developing brain.

ACKNOWLEDGEMENTS

The author wishes to thank Dr. R. C. Kim for critical reading of the manuscript and Teresa Espinosa, Jann Geddes, Paul Kim, Simon Yee, Paul Amodei and Loan Duong for technical assistance.

REFERENCES

Abe, T., Haga, T. and Kurokawa, M., 1975. Blockage of axoplasmic transport and depolymerication of reassembled microtubules by methylmercury. Brain Res. 86: 504-508.

Amin-Zaki, L., Elhassani, S., Majeed, M.A., Clarkson, T.W., Doherty, R.A. and Greenwood, M.S., 1974. Intra-uterine methylmercury poisoning in Iraq. Pediat. 54:587-595.

Amin-Zaki, L., Elhassani, S., Majeed, M.A., Clarkson, T.W., Greenwood, M.S. and Giovanoli-Jakubezak, T., 1976. Perinatal methylmercury poisoning in Iraq. Amer. J. Dis. Child 130:1070-1076.

Baker, J.B., Low, D.A., Simmer, R.L. and cunningham, D.D., 1980. Protease-nexin: a cellular component that links thrombin and plasminogen activator and mediates their binding to cells. Cell 21:37-45.

Brookes, N., 1988. Specificity and reversibility of the inhibition by $HgCl_2$ of glutamate transport in astrocyte cultures. J. Neurochem. 50:1117-1122.

Cavanaugh, K., Kurwitz, D., Cunningham, D. and Bradshaw, R., 1990. Reciprocal modulation of astrocyte stellation by thrombin and protease nexin-1. J. Neurochem. 54:1735-1743.

Chang, L.W. and Hartmann, H.A., 1972. Blood-brain barrier dysfunction in experimental mercury intoxication. Acta Neuropathol. (Berl.) 21:179-184.

Chang, L.W., Reuhl, K.R., and Lee, G.W., 1977. Degenerative changes in the developing nervous system as a result of in utero exposure to methylmercury. Environ. Res. 14:414-423.

Choi, B.H., 1981. Radial glia of developing human fetal spinal cord: Golgi, immunohistochemical and electron microscopic study. Dev. Brain Res. 1:249-267.

Choi, B.H., 1984. Cellular and subcellular demonstration of mercury in situ by modified sulfide-silver technique and photoemulsion histochemistry. Exp. Mol. pathol. 40:109-121.

Choi, B.H., 1986. Glial fibrillary acidic protein in radial glia of early human fetal cerebrum: A light and electron microscopic immunoperoxidase study. J. Neuropathol. Exp. Neurol. 45:408-418.

Choi, B.H., 1988a. Developmental events during the early stages of cerebral cortical neurogenesis in man: A correlative light, electron microscopic, immunohistochemical and Golgi study. Acta Neuropath. (Berl.) 75:441-447.

Choi, B.H., 1988b. Effects of methylmercury on developing fetal astrocytes. In: The biochemical pathology of astrocytes. M.D. Norenberg, L. Hertz, and A. Schousboe, eds. Alan R. Liss, Inc., New York, pp. 219-230.

Choi, B.H., 1989. The effects of methylmercury on the developing brain. Prog. Neurobiol. 32:447-470.

Choi, B.H., 1991. Effects of methylmercury on neuroepithelial germinal cells in the developing telencephalic vesicles of mice.Acta Neuropath. 81:359-365.

Choi, B.H. and Ahn, B.T., 1989. Effects of methylmercury upon dissociated embryonic rat brain cells in rotation-mediated re-aggregation culture. Fed. Proc. 3:A291.

Choi, B.H. and Lapham, L.W., 1978. Radial glia in the human fetal cerebrum: A combined Golgi, electron microscopic and immunofluorescent study. Brain Res. 148:295-311.

Choi, B.H. and Lapham, L.W., 1981. Effects of meso-2-3-dimercaptosuccinic acid on methylmercury-injured human fetal astrocytes in vitro. Exp. Mol. Pathol. 34:25-53.

Choi, B.H. and Simpkins, H., 1986. Changes in the molecular structure of mouse fetal astrocyte produced in vitro by methylmercury chloride. Environ. Res. 39:321-330.

Choi, B.H., Cho, K.H. and Lapham, L.W., 1981. Effects of methylmercury on human fetal neurons and astrocytes in vitro: A time-lapse cinematographic, phase and electron microscopic study. Environ. Res. 24:61-74.

Choi, B.H., Lapham, L.W., Amin-Zaki, L. and Saleem, T., 1978. Abnormal neuronal migration, deranged cerebral cortical organization, and diffuse white matter astrocytosis of human fetal brain: A major effect of methylmercury poisoning *in utero*. J. Neuropathol Exper. Neurol. 37:719-733.

Choi, B.H., Lincoln, J. and Amodei, P., 1988. Effects of methylmercury on excitatory amino acid receptors in the mouse cerebrum. Soc. for Neurosci 14:886.

Choi, B.H., Suzuki, M., Kim, T., Wagner, S.L. and Cunningham, D.D., 1990. Protease nexin-1: Localization in human brain suggests a protective role against extravasated serine proteases. Amer. J. Pathol. 137:741-747.

Chuong, C. and Edelman, G.M., 1984. Alterations in neural cell adhesion molecules during development of different regions of the nervous system. J. Neurosci. 4:2354-2368.

Clarkson, T.W., 1987. Metal toxicity in the central nervous system. Environ. Heatlh Perspect. 75:59-64.

Eaton, D.L. and Baker, J.B., 1983. Evidence that a varieyt of cultured cells secrete protease nexin and produce a distinct cytoplasmic serine protease-binding factor. J. Cell Physiol. 117:175-182.

Edelman, G., 1984. Cell adhesion and morphogenesis: The regulator hypothesis. Proc. Natl. Acad. Sci. U.S.A. 81:1460-1464.

Fuyuta, M., Fujimoto, T. and Hirata, S., 1978. Embryotoxic effects of methylmercury chloride administered to mice and rats during organogenesis. Teratol. 18:353-366.

Gloor, S., Odink, K., Guenther, J., Nick, H. and Monard, D., 1986. A glia-derived neurite promoting factor with protease inhibitory activity belongs to the proteasee nexins. Cell 47:687-693.

Harada, Y., 1968. Clinical investigations of Minamata disease: C. congenital (or fetal) Minamata disease. In: Minamata Disease. M. Kutsuna, ed. Kumamoto University Press, Japan, pp. 93-117.

Harada, M., 1976. Minamata Disease, Chronology and Medial Report. Bulletin of the Constitutional Medicine. Kumamoto University Press, Japan, pp. 1-78.

Harada, M., 1978. Congenital Minamata disease: Intrauterine methylmercury poisoning. Teratol. 18:285-288.

Harris, S.B., Wilson, J.G. and Prinz, R.H., 1972. Embryotoxicity of methylmercuric chloride in golden hamster. Teratol. 6:139-142.

Honegger, P. and Richelson, E., 1977. Biochemical differentiation of aggregating cell cultures of different fetal rat brain regions. Brain Res. 109:335-354.

Inouye, M. and Kajiwara, Y., 1988. Developmental disturbances of the fetal brain in guinea-pigs caused by methylmercury. Toxicol. 7:227-232.

Inouye, M. and Kajiwara, Y, 1990.. Strain difference of the mouse in manifestation of hydrocephalus following prenatal methylmercury exposure. Teratol. 41:205-210.

Kim, P. and Choi, B.H., 1990. Effects of mercury on glutamate uptake in cultured mouse astrocytes. Fed. Proc. 4:A1014.

Kleinschuster, S.J., Yoneyama, M. and Sharma, R.P., 1983. A cell aggregation model for the protective effect of selenium and vitamin E on methylmercury toxicity. Toxicol. 26:1-9.

Kostyniak, P.J., 1982. Mobilization and removal of methylmercury in the dog during extracorporeal complexing hemodialysis with 2,3-dimercaptosuccinic acid (DMSA). J. Pharmacol. Exp. Ther. 221:63-68.

Lu, E.J., Brown, W.J., Cole, R. and deVellis, J., 1980. Ultrastructural differentiation and synaptogenesis in aggregating rotation cultures of rat cerebral cells. J. Neurosci. Res. 5:447-463.

Magos, L., 1976. The effects of dimercaptosuccinic acid on the excretion and distribution of mercury in rats and mice treated with mercuric chloride and methylmercury chloride. Br. J. Pharm. 56:479-484.

Margolis, R.L. and Wilson, L., 1981. Microtubule treadmills - possible molecular machinery. Nature 293:705-711.

Marsh, D.O., Myers, G.J., Clarkson, T.W., Amin-Zaki, L., Tikriti, S. and Majeed, M.A., 1980. Fetal methylmercury poisoning: clinical and toxicological data on 29 cases. Ann. Neurol. 7:348-353.

McGrogan, M., Kennedy, J., Li, M.P., Hsu, C., Scott, R.W., Simonsen, C.C. and Baker, J.B., 1988. Molecular cloning and expression of two forms of human protease nexin 1. Bio/Technol. 6:172-177.

Miura, K., Suzuki, K. and Imura, N., 1978. Effects of methylmercury on mitotic mouse glioma cells. Environ. Res. 17:453-471.

Miura, K., Inokawa, M. and Imura, N., 1984. Effects of methylmercury and some metal ions on microtubule networks in mouse glioma cells and in vitro tubulin polymerization. Toxicol. Appl. Pharmacol. 73:218-231.

Mottet, N.K., Shaw, C. and Burbacher, T.M., 1987. The pathological lesions of methylmercury intoxication in monkeys. In: The Toxicity of Methyl Mercury. C.U. Eccles, and Z. Annau, eds. Johns Hopkins University Press, Baltimore, pp. 73-103.

Murakami, U., 1972. The effects of organic mercury on intrauterine life. Adv. Exp. Med. Biol. 27:310-336, 1972.

Olson, F.C. and Massaro, E.J., 1977. Effects of methyl mercury on murine fetal amino acid uptake, protein synthesis and palate closure. Teratology 16:187-194.

Orkand, P.M., Linder, J. and Schachner, M., 1984. Specificity of histiotypic organization and synaptogenesis in reaggregating cell cultures of mouse cerebellum. Dev. Brain Res. 16:119-134.

Peckham, N.J. and Choi, B.H., 1986. Surface charge alterations in mouse fetal astrocytes due to methylmercury: An ultrastructural study with cationized ferritin. Exp. Mol. Pathol. 44:230-234.

Peckham, N.H. and Choi, B.H., 1988. Abnormal neuronal distribution within the cerebral cortex after prenatal methylmercury intoxication. Acta Neuropath. (Berl.) 76:222-226.

Rakic, P., 1972. Mode of cell migration to the superficial layers of fetal monkey neocortex. J. Comp. Neurol. 145:61-84.

Ramel, C., 1969. Methyl mercury as a mitosis disturbing agent. J. Japan Med. Assn. 61:1072-1077.

Reuhl, K.R. and Chang, L.W., 1979. Effects of methylmercury on the development of the nervous system. Neurotoxicol. 1:21-55.

Rodier, P.M., 1983. Critical processes in CNS development and the pathogenesis of early injuries. In: Reproductive and Developmental Toxity of Metals. T.W. Clarkson, G.F. Nordberg and P.R. Sager, eds. Plenum Press, New York and London, pp. 455-471, 1983.

Rodier, P.M., Aschner, M. and Sager, P.M., 1984. Mitotic arrest in the developing CNS after prenatal exposure to methylmercury. Neurobehav. Toxicol. Teratol. 6:379-385.

Rosenblatt, D.E., Cotman, C.W., Nieto-Sampedro, M., Rowe, J.W. and Knauer, D.J., 1987. Identification of a protease inhibitor produced by astrocytes that is structurally and functionally homologous to human protease nexin-1. Brain Res. 415:40-48.

Sager, P.R., Doherty, R.A., and Rodier, P.M., 1982. Effects of methylmercury on developing mouse cerebellar cortex. Exp. Neurol. 77:179-193.

Shen, V., Amodei, P. and Choi, B.H., 1989. Expression of neural cell adhesion molecule in PC12 cells following methylmercury exposure: immunofluorescent, phase and EM study. Fed. Proc. 3:A29.

Spyker, J.M. and Smithberg, M., 1972. Effects of methylmercury on prenatal development in mice. Teratology 5:181-190.

Steinwall, O. and Olsson, Y., 1969. Impairment of the blood-brain barrier in mercury poisoning. Acta Neurol. Scandinav. 45:351-361.

Su, M. and Okita, G., 1976. Embryocidal and teratogenic effects of methylmercury in mice. Toxicol. Appl. Pharmacol. 38:207-216.

Takeuchi, T., 1977. Pathology of fetal Minamata disease. Pediatrician 6:69-87.

Trapp, B.D., Honegger, P., Richeson, E. and DeF. Webster, H., 1979. Morphological differentiation of mechanically dissociated fetal rat brain in aggregating cell cultures. Brain Res. 160:117-130.

Vogel, D.G., Margolis, R.L. and Mottet, N.K.. 1985. The effects of methyl mercury binding to microtubules. Toxicol. Appl. Pharmacol. 80:473-486.

Wagner, S.L., Gdeeds, J.W., Cotman, C.W., Lau, A.L., Gurwitz,D., Isackson, P.L. and Cunningham, D.D.. 1989. Protease nexin-1, an antithrombin with neurite outgrowth activity is reduced in Alzheimer's disease. Proc. Natl. Acad. Sci. U.S.A. 86:8284-8288.

# EXPERIMENTAL APPROACHES TO DEVELOPMENTAL TOXICITY

# OF METHYLMERCURY

Minoru Inouye

Research Institute of Environmental Medicine
Nagoya University, Nagoya  464-01
Japan

ABSTRACT

Dozens of babies congenitally affected by methylmercury were born in Minamata, Japan and Iraq.  For the most severe cases, where death occurred either in infancy or in childhood, pathological examinations were performed.  In cases from Iraq the changes in the brains were the outcome of disturbances of development, more specifically, abnormal neuronal migration and derangement in the fundamental structuring of gray matter.  The focal nerve cell destruction typically seen in adult cases of methylmercury poisoning was not encountered (Choi et al., 1978).  In cases from Minamata, on the other hand, brain changes were the result of degeneration and decrease in the number of nerve cells.  These findings were similar in type to those found in adult patients, and in addition were accompanied by developmental changes  (Matsumoto et al., 1965).

Our experiment using guinea pigs demonstrated that developmental disturbances of the brain as a result of impaired neurogenesis and abnormal neuronal migration were induced when dams were exposed to methylmercury in early pregnancy.  When dams were exposed in later pregnancy, neurons of the cerebral cortex were involved in widespread focal degeneration.  These findings confirmed and extended the observations of the human cases. Iraqi cases were affected acutely in the earlier stage of pregnancy, so the fetal brain might be involved in developmental disturbances.  The

*Advances in Mercury Toxicology*, Edited by T. Suzuki *et al.*
Plenum Press, New York, 1991

Minamata cases were exposed to methylmercury chronically throughout the pregnancy, and thus both the developmental disturbances and the focal degeneration of neurons might be induced in the same fetal brain. Accelerated placental transfer and fetal accumulation of methylmercury at the late pregnant stage, as demonstrated in animal experiments, might cause the degenerative changes of neurons resembling the adult forms.

In addition, hydrocephalus has been detected with a low incidence in congenital Minamata disease sufferers. Experimental studies using inbred mice revealed strain difference in susceptibility to postnatal development of hydrocephalus following prenatal methylmercury exposure. B10.D2 is highly susceptible, C57BL/10 and C57BL/6 are moderately susceptible, and DBA/2, C3H/He and BALB/c are resistant. This indicates that the susceptibility to methylmercury-induced hydrocephalus is under genetic control.

INTRODUCTION

As described previously, dozens of babies who were congenitally affected by methylmercury were born in the Minamata area of Japan and in Iraq. In the Minamata area a human epidemic of methylmercury poisoning occurred in 1953 - 1960, because of the consumption of fish contaminated by industrial effluent (Tsubaki and Irukayama, 1977). The exposed population included pregnant women, and 64 of their offspring, born in 1954 - 1959, have been considered to be congenitally affected (Doi et al., 1985). The clinical features of the affected children are mental retardation and widespread neurological lesions. Microcephaly is observed frequently, but other malformations are said to be rare (Harada, 1978). However, multiple minor morphological abnormalities are recognized as a result of variable manifestation of the developmental toxicity of methylmercury (Murakami, 1972). Hydrocephalus has been detected with a low incidence from the clinical follow-up study by the National Institute for Minamata Disease.

Pathological examinations of the brains were performed in the most severe four cases, where death occurred either in infancy or in childhood; two were from Minamata (Matsumoto et al., 1965) and two from Iraq (Choi et al., 1978). In cases from Iraq the pathological changes are the outcome of disturbance in the development of the brain, i.e., abnormal neuronal migration and derangement in the fundamental structuring of gray matter. The focal nerve cell destruction typically seen in adult cases of methylmercury poisoning was not encountered (see the article by Choi; Choi et al., 1978). However, there is some contrast with those reported from Minamata. In Minamata cases, both brains showed disorganization of the cerebral cortical architecture as a result of degeneration and decrease in the number of nerve cells. These pathological findings were similar in type and distribution to those found in adult Minamata disease sufferers, with a few differences between the fetal and adult forms; the cerebral nerve cell involvement was more diffuse in the former than in the latter. In addition to the changes resembling the adult forms, the developmental changes, such as abnormal architectural arrangement of cortical neurons, etc., have also been reported (Matsumoto et al., 1965; Takeuchi, 1968).

Attempts to reproduce neuropathological changes following prenatal methylmercury poisoning in experimental animals have been made for decades. The developmental disturbance of the brain and destruction of neural tissue were induced in fetuses of mice and rats treated with methylmercury (Murakami, 1972; Inouye and Murakami, 1975). In the experiments, derangement of the cerebral cortical plate and dilatation of the lateral ventricles as the result of inhibited production of neuronal cells and their migration, and focal tissue defects as well, were produced in the fetal brain treated even in the early organogenetic stage (Fig. 1). Methylmercury readily crosses the placenta (Suzuki et al., 1967; Ohsawa et al., 1981). However, it is also retained in the maternal body, and a certain period is needed to reach the peak concentration in the fetal brain (Yang et al., 1972; Inouye et al., 1986). Thus, even when methylmercury is given to mice and rats in early pregnancy, fetal brain damage in the middle and late pregnant stage may be produced because of their short pregnancy.

However, the immaturity of the brain of full-term fetuses of mice and rats makes for difficulty in direct comparison of neuropathological features between human infants and animal models (Kameyama, 1985). Newborn mice with striking damage following prenatal methylmercury exposure were unable to survive to a stage at which the maturation of the brain is comparable to that of human infants (Inouye et al., 1985). O'Kusky (1985) and O'Kusky et al. (1988) daily injected young rats with methylmercury, beginning 5 days after birth through 17 - 20 days of age, and produced brain lesions. Their work may provide good animal models for congenital Minamata disease, which is associated with nerve cell destruction resembling the adult forms (Matsumoto et al., 1965).

We adopted guinea-pigs as experimental animals to study the pathogenesis of congenital methylmercury poisoning both in Iraqi and Minamata cases, since gestation of guinea-pigs is more than 9 weeks and the full-term fetus is comparable to the human neonate in terms of the histological development of the brain (Inouye and Kajiwara, 1988a). In order to determine the relationship between the effect of methylmercury and the developmental stage when treated, a single maternal administration dose was employed.

Pregnant guinea-pigs, weighing 500 - 800 g, were administered a single oral dose of 7.5 mg Hg/animal of methylmercuric chloride on one of days 21, 28, 35, 42 or 49 (3 - 7 weeks) of pregnancy. About half of the litters were aborted within 6 days after treatment. Thus the dose was maximum for an experiment employing a single administration. The dams were killed on day 63 (9 weeks) of pregnancy. The brain, liver, kidney and blood from both dams and fetuses were sampled as well as amniotic fluid, urine, gastric content and hair from fetuses.

The effects of methylmercury on the fetal brain varied depending on the period of pregnancy at which dams were treated. The fetal brains from dams treated at 3, 4 or 5 weeks of pregnancy were smaller than those from the controls under gross observation. The lateral ventricles of these brains were dilated, the cerebral cortices were thinned, and the nucleus caudatus putamen and hippocampal formation were reduced in size (Figure 2). The laminar structure of the cerebral cortex was essentially preserved (Figure 3), but a slight disarrangement of the cellular alignment was observed. Pyramidal cells in layer V seemed smaller that those of controls, and those with apical dendrites pointing in different directions were noted (Figure 4).

In cases of congenital Minamata disease, cerebral gyri were reduced in size, cerebral cortex was narrow, lateral ventricles were dilated, and basal nuclei were smaller than normal, although degeneration of cortical nerve cells was prominent. In addition, the large pyramidal nerve cells in the cortical layers III and V were smaller than normal. In the layer VI the nerve cells showed a disorderly arrangement, running in oblique, horizontal and reverse directions (Takeuchi, 1968). Thus the experiment using guinea-pigs was able to reproduce many pathological features of congenital Minamata disease.

In the neopallial portion of the cerebral vesicle of the 3-week guinea-pig fetus, the wall of the vesicle is made up of a ventricular zone, an intermediate zone and a marginal zone; the ventricular zone forms the greatest part of the thickness of the wall, and the intermediate and marginal zones are poorly developed. The cortical plate forms by 4 weeks of gestation, and the ventricular zone declines by 6 weeks (Wanner et al., 1976).

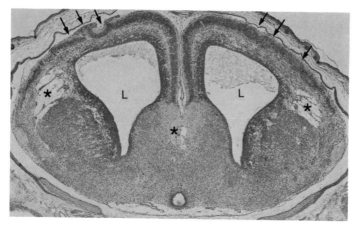

Fig. 1. A frontal section of the brain of a full-term mouse fetus of which the dam was treated with a single oral dose of 24 mg Hg/kg methylmercuric chloride on day 7 of pregnancy. Note dilatation of the lateral ventricles (L) and derangement of the cortical plate (arrows) as a result of inhibited production and migration of neuronal cells. Tissue defects (asterisks) are also found (Inouye and Murakami, 1975).

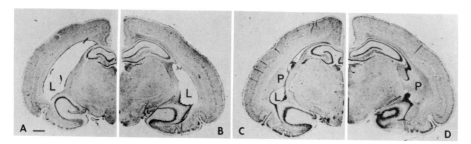

Fig. 2. Paired frontal sections of the brains of full-term fetal guinea-pigs, demonstrating developmental disturbance caused by methylmercury. (A), (B), (C) Brains of fetuses of which dams were treated with a single oral dose of 7.5 mg Hg/animal at 3, 4, and 5 weeks of pregnancy, respectively. (D Control brain. Note gross reduction in size, dilated lateral ventricles (L) and thinned cerebral mantle in the treated brains. The nucleus caudatus putamen (P) and the hippocampal formation are also reduced in size.
Scale bar = 1 mm (Inouye and Kajiwara, 1988a)

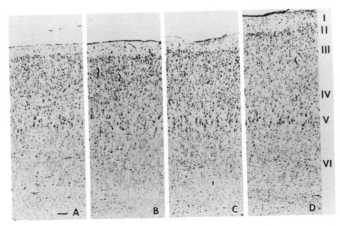

Fig. 3. Photomicrographs of frontal sections through the dorso-lateral cerebral cortices of fetal guinea-pigs. (A), (B), (C) Brains of fetuses of which dams were treated with a single oral dose of 7.5 mg Hg/animal at 3, 4, and 5 weeks of pregnancy, respectively. (D) Control brain. The cerebral cortices of treated fetuses are thinned but the cortical architecture is preserved.
Scale bar = 100 μm (Inouye and Kajiwara, 1988a).

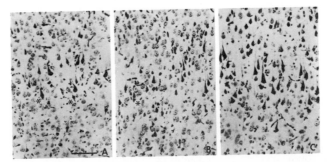

Fig. 4. Sections around layer V of the neocortex, demonstrating neurons with apical dendrites pointing in different directions in fetal guinea-pigs of which dams were treated at 3 (A) and 4 weeks (B) of pregnancy comparing with a control brain (C). Asterisks indicate neurons with apical dendrites oriented toward the white matter. Pyramidal neurons of A and B are smaller than those of C.
Klüver-Barrera's stain.
Scale bar = 100 μm (Inouye and Kajiwara, 1988a).

Therefore, the developmental disturbance of the brain resulting from impaired proliferation and migration of the neuroepithelial cells was induced when dams were given methylmercury at 3, 4 and 5 weeks of pregnancy.

When guinea-pigs were treated at the late pregnant stage (6 or 7 weeks), microscopic spongy degeneration of the cerebral cortex was induced (Figure 5). Extensive cellular necrosis appeared in layers II through VI of the occipital neocortex, mainly in layers III and IV. Degeneration occurred in already defined but not yet fully matured neuronal populations. Thus, the experiment using guinea-pigs demonstrated that developmental disturbance of the fetal brain, including abnormal neuronal migration, was induced when dams were exposed to methylmercury in early pregnancy; and when dams were exposed in later pregnancy, neurons of the cerebral cortex were involved in degeneration.

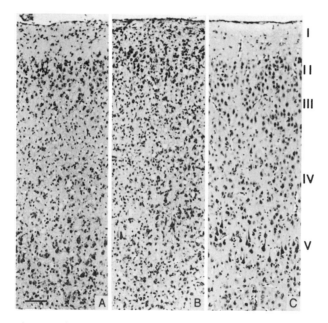

Fig. 5. Photomicrographs demonstrating microscopic spongy degeneration in the striate cortices of fetal guinea-pigs of which dams were treated at 6 (A) and 7 weeks (B) of pregnancy. (C) Control brain. The degenerative changes are prominent in layers III and IV. Scale bar = 100 μm (Inouye and Kajiwara, 1988a).

The regional distribution of mercury in the fetal brain was also quantified in guinea-pigs, and it proved to be similar to that in maternal brain (Figure 6). The mercury concentration was higher in the neopallium and archipallium than in the paleopallium and other ventral and posterior portions, especially following maternal treatment at 6 and 7 weeks of pregnancy, Concomitantly, these areas with higher mercury contents were involved in neuronal degeneration. These areas also corresponded to those reported to be involved in focal degeneration in the human adult Minamata disease patients (Takeuchi, 1968). The nerve cell involvement is similar in type and distribution between congenital and adult cases (Matsumoto et al., 1965), but the result indicates that the nerve cells of fetuses are more vulnerable than those of adults to methylmercury toxicity.

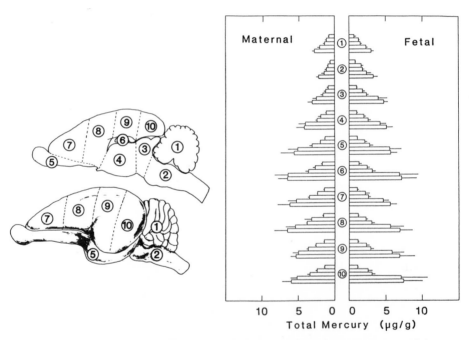

Fig. 6. Schematic drawing of the guinea-pig brain divided into ten parts, and the concentration of mercury in these parts of brains of both maternal and fetal guinea-pigs. (1) Cerebellum, (2) Other rhombencephalon (from pons to decussation of pyramids), (3) Mesencephalon (from superior colliculus to isthmus rhombencephali), (4) Diencephalon (from optic chiasma to corpus mamillare), (5) Paleopallium (ventral to rhinal sulcus), (6) Archipallium, (7-10) Neopallium (7-8 include nucleus caudatus putamen). Columns (top to bottom) indicate specimens treated at 3, 4, 5, 6, and 7 weeks of pregnancy, respectively.

The factors contributing to the discrepancy remain uncertain regarding the pathological effects of congenital methylmercury poisoning between Minamata cases and Iraqi cases, in the latter the focal nerve cell destruction typically seen in adult cases was not oberved (Choi et al., 1978). But the duration of exposure may well be one. In Iraq, mothers of patients consumed methylmercury-contaminated bread for about 10 weeks, when they were either 6 - 8 weeks or 8 - 10 weeks pregnant (Choi et al., 1978). Thus fetuses might be affected more acutely in the earlier pregnant period than that seen in Minamata cases, and the concentration of methylmercury in the fetal brain might be lowered by late pregnancy; the children's brains show defective development without focal nerve cell destruction. In Minamata, however, the mothers consumed mercury-contaminated fish before conception and throughout pregnancy. The fetuses were exposed to methylmercury chronically, so both the developmental disturbance and degeneration of neurons might be induced in the same fetal brain. Placental transfer of methylmercury is greater at the late pregnant stage in rats and mice (Figure 7) (Ohsawa et al., 1981; Inouye and Kajiwara, 1990b). Human pregnancy is physiologically longer than that of the laboratory animals, and methylmercury is accumulated in fetuses with umbilical cord-blood levels in excess of the mother's blood (Skerfving, 1988). In Minamata cases this accelerated accumulation of mercury in fetuses at the late pregnant stage might cause more conspicuous degenerative changes of neurons resembling the adult forms.

When the distribution of mercury in different body parts of the fetus was compared with that of the dam, the blood level of mercury was always higher in the fetus (Table 1). Maternal kidney contained mercury at the highest concentration, reflecting excretion of mercury in the urine (Hirayama and Yasutake, 1986). In the fetus the mercury concentration was low in the kidney, urine, amniotic fluid and gastric content (Table 2). This shows that the fetus is unable to excrete mercury into the urine. The mercury concentration in fetal hair was 100 - 150 times the blood level. Elimination of mercury from the fetus may occur primarily through exchange with the maternal circulation, but accumulation of mercury in the hair may also be an important way to eliminate it from the fetal body. It is practical to analyze the mercury in infant hair, rather than in umbilical-cord blood, to determine whether intrauterine exposure has taken place.

Among congenital Minamata disease sufferers, two cases with hydrocephalus have been recognized from the clinical follow-up study by the National Institute for Minamata Disease. This low incidence makes for difficulty in elucidating whether the hydrocephalus was induced by methylmercury or occurred spontaneously regardless of mercury exposure. We considered that there would be some genetic variability in

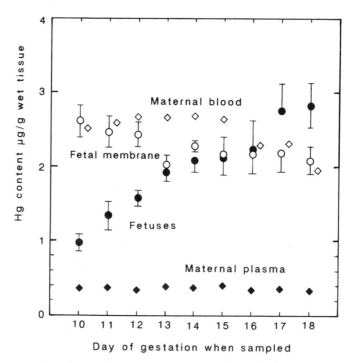

Fig. 7. Concentration of total mercury in the fetus and fetal membrane of the mouse (mean + SD of 9 samples for each point) as well as the maternal blood and plasma (mean of 3 maternal samples) at different days of gestation. Pregnant mice were i.v. injected with 2 μg Hg/g body weight of methylmercuric chloride 24 hours prior to sampling (Inouye and Kajiwara, 1990b).

susceptibility to hydrocephalus; and thus there is a possibility that the hydrocephalus was induced by methylmercury in particularly susceptible victims.

Among laboratory animals many inbred and congenic strains have been established in mice, in particular, and their genetic characteristics have been well documented. A higher incidence of spontaneous hydrocephalus has been observed in congenic B10.D2/n mice. Therefore, we treated pregnant B10.D2/n mice with 8 mg Hg/kg methylmercury by gavage and examined the manifestation of hydrocephalus in their offspring and compared that with mice of genetically relevant C57BL/10 and DBA/2 strains (Inouye and Kajiwara, 1990a).

The results revealed that B10.D2/n strain was highly susceptible to methylmercury-induced hydrocephalus, C57BL/10 was moderately susceptible, and DBA/2 was resistant (Table 3). In addition, following prenatal methylmercury exposure (Choi et al., 1988), hydrocephalus also developed in C57BL/6 mice, which have a common ancestor with C57BL/10 and B10.D2/n mice. However, hydrocephalus never developed in the offspring of C3H/He strain (Inouye and Kajiwara, 1988b). Thus, the susceptibility to methylmercury-induced hydrocephalus must be under genetic control in mice.

Congenic strain B10.D2/n has a genetic background similar to C57BL/10 but differs from that strain in that differential alleles are derived from DBA/2, the histocompatibility-2 (H-2) complex and thymus leukemia antigen (Tla) locus on chromosome 17 (Klein, 1973). B10.D2/n was more susceptible than C57BL/10, but DBA/2 was resistant. Hence, the H-2 and Tla loci are unlikely to have a direct influence on susceptibility. Multifactorial inheritance, rather than single genes, may be involved in regulation of the susceptibility to methylmercury-induced hydrocephalus. The variation of biological half-life of methylmercury is reported in man (Al-Shahristani and Shihab, 1974; also see the article by Doi). Man may have genetic differences also in the teratogenic response to methylmercury.

CONCLUSION

Pathological changes in children's brains congenitally affected by methylmercury have been reported from Minamata and Iraq. However, there is some contrast in the findings between the two reports. Attempts to demonstrate neuropathological changes following prenatal methylmercury poisoning in experimental animals confirmed and extended the observation of the human cases in both areas. Experiments using guinea-pigs, rats and mice have revealed that developmental disturbance of the fetal brain, as reported in Iraqi cases, is induced when dams are exposed to methylmercury in early pregnancy. When pregnant guinea-pigs are exposed in later pregnancy, early postnatal

Table 1 The mercury content (mean ± SD µg Hg/g wet weight) of fetal and maternal tissues of guinea pigs examined at 9 weeks of pregnancy

| Treated at | Samples examined | | Liver | Kidney | Whole blood | Blood plasma |
|---|---|---|---|---|---|---|
| 3 weeks | Fetal | (4) | 2.3±0.3 | 1.1± 0.1 | 1.1±0.2 | 0.035±0.025 |
|  | Maternal | (3) | 2.3±0.4 | 19 ± 3 | 0.7±0.3 | 0.034±0.004 |
| 4 weeks | Fetal | (6) | 3.8±0.3 | 1.9± 0.2 | 2.0±0.2 | 0.063±0.014 |
|  | Maternal | (3) | 3.4±0.5 | 26 ± 9 | 1.2±0.2 | 0.043±0.013 |
| 5 weeks | Fetal | (9) | 4.0±0.4 | 2.4± 0.5 | 2.4±0.3 | 0.084±0.033 |
|  | Maternal | (3) | 4.3±0.8 | 29 ±12 | 1.7±0.4 | 0.051±0.006 |
| 6 weeks | Fetal | (10) | 5.2±1.0 | 3.2± 0.6 | 3.5±0.8 | 0.113±0.043 |
|  | Maternal | (3) | 6.1±2.2 | 19 ± 5 | 2.4±0.6 | 0.061±0.003 |
| 7 weeks | Fetal | (10) | 6.4±1.5 | 4.6± 0.6 | 3.9±0.5 | 0.177±0.090 |
|  | Maternal | (4) | 8.1±1.8 | 73 ±40 | 3.0±0.3 | 0.108±0.038 |
| Untreated | Fetal | (23) | <0.05 | <0.05 | <0.05 | ND |
|  | Maternal | (8) | <0.05 | <0.05 | <0.05 | ND |

Pregnant guinea-pigs weighing 500 – 800 g were treated with a single dose of 7.5 mg Hg/animal of methylmercuric chloride by gavage. They were killed and examined at 9 weeks of pregnancy (Inouye and Kajiwara, 1988a).

**Table 2** The mercury content (mean ± SD μg Hg/g wet weight) of fetal components

| Treated at | Amniotic fluid | Urine | Gastric content | Hair |
|---|---|---|---|---|
| 3 weeks | 0.008±0.002 | 0.008±0.006 | 0.029±0.014 | 119± 21 |
| 4 weeks | 0.013±0.003 | 0.010±0.006 | 0.020±0.013 | 192± 19 |
| 5 weeks | 0.020±0.003 | 0.010±0.003 | 0.036±0.015 | 315± 29 |
| 6 weeks | 0.025±0.009 | 0.015±0.005 | 0.036±0.021 | 535± 73 |
| 7 weeks | 0.036±0.022 | 0.036±0.026 | 0.032±0.017 | 515±112 |
| Untreated | <0.001 | <0.001 | <0.001 | < 5 |

Pregnant guinea-pigs weighing 500 - 800 g were treated with a single dose of
7.5 mg Hg/animal of methylmercuric chloride by gavage. They were killed and
their fetuses were examined at 9 weeks of pregnancy (Inouye and Kajiwara, 1988a).

351

Table 3 Results of prenatal methylmercury (MeHg) exposure of three strains of mice

| Strain Agents | Pregnant day of treatment | Dams | Litters weaned (%) | Implantation sites | Pups weaned (%) | Apparent hydro-cephalus (%) | Dilatation of ventricles (%) |
|---|---|---|---|---|---|---|---|
| B10.D2/n | | | | | | | |
| Untreated | | 35 | 30(86) | 303 | 139(46) | 6 (4) | 6 (4) |
| NaCl | 15 | 20 | 18(90) | 168 | 116(69) | 6 (5) | 5 (4) |
| MeHg | 14 | 20 | 7(35) | 186 | 36(19) | 24(67) | 6(17) |
| MeHg | 15 | 30 | 15(50) | 273 | 67(25) | 59(88) | 4 (6) |
| MeHg | 16 | 20 | 10(50) | 188 | 51(27) | 38(75) | 12(24) |
| MeHg | 17 | 20 | 7(35) | 176 | 29(16) | 14(48) | 8(28) |
| C57BL/10 | | | | | | | |
| Untreated | | 20 | 18(90) | 156 | 123(79) | 1 (1) | 1 (1) |
| MeHg | 15 | 20 | 16(80) | 178 | 102(57) | 55(54) | 14(14) |
| DBA/2 | | | | | | | |
| Untreated | | 20 | 19(95) | 127 | 96(76) | 0 | 0 |
| MeHg | 15 | 20 | 14(70) | 99 | 57(58) | 0 | 0 |

Pregnant mice were treated with a single dose of 8 mg Hg/kg methylmercuric chloride by gavage, and allowed to give birth and rear their litters. The dams and offspring were killed 30 days after birth and examined (Inouye and Kajiwara, 1990a).

period in cases of rats, the cerebral cortices of fetuses or pups are involved in widespread neuronal degeneration, as reported in Minamata cases. These findings confirmed that Iraqi cases were acutely affected in early pregnancy. Thus the infant brain showed developmental disturbance. The Minamata cases were exposed to methylmercury chronically throughout pregnancy, so both the developmental disturbance and degeneration of neurons were induced in the same fetal brain. The accelerated accumulation of methylmercury in fetuses at the late pregnant stage, as demonstrated in laboratory animals, might cause more conspicuous degenerative changes of neurons resembling the adult forms. In addition, hydrocephalus has been detected with a low incidence in congenital Minamata disease sufferers. Experimental studies using inbred strains of mice revealed that the susceptibility to methylmercury-induced hydrocephalus is under genetic control and this might be true of humans.

# REFERENCES

Al-Shahristani, H., and Shihab, K.M., 1974, Variation of biological half-life of methylmercury in man. Arch. Environ. Health, 28:342-344.
Choi, B.H., Lapham, L.W., Amin-Zaki, K., and Saleem, T., 1978. Abnormal neuronal migration, deranged cerebral cortical organization, and difuse white matter astrocytosis of human fetal brain: A major effect of methylmercury poisoning *in utero*. J. Neuropath. Exp. Neurol., 37:719-733.
Choi, B.H., Kim, R.C., and Peckham, N.H., 1988, Hydrocephalus following prenatal methylmercury poisoning. Acta Neuropathol. (Berl.), 75:325-330.
Doi, R., Kobayashi, T., Ohno, H., and Harada, M., 1985, Sex ratio of fetal Minamata disease offspring. Jpn. J. Hyg., 40:306 (in Japanese).
Harada, M., 1978, Congenital Minamata disease: Intrauterine methylmercury poisoning. Teratology, 18:285-288.
Hirayama, K., and Yasutake, A., 1986, Sex and age differences in mercury distribution and excretion in methylmercury-administered mice. J. Toxicol. Environ. Health, 18:49-60.
Inouye, M., and Murakami, U., 1975, Teratogenic effect of orally administered methylmercuric chloride in rats and mice. Cong. Anom., 15:1-9.
Inouye, M., Murao, K., and Kajiwara, Y., 1985, Behavioral and neuropathological effects of prenatal methylmercury exposure in mice. Neurobehav. Toxicol. Teratol., 7:227-232.
Inouye, M., Kajiwara, Y., and Hirayama, K., 1986, Dose- and sex-dependent alterations in mercury distribution in fetal mice following methylmercury exposure. J. Toxicol. Environ. Health, 19:425-435.
Inouye, M., and Kajiwara, Y., 1988a, Developmental disturbances of the fetal brain in guinea-pigs caused by methylmercury. Arch. Toxicol., 62:15-21.
Inouye, M., and Kajiwara, Y., 1988b, An attempt to assess the inheritable effect of methylmercury toxicity subsequent to prenatal exposure of mice. Bull. Environ. Contam. Toxicol., 41:508-514.

Inouye, M., and Kajiwara, Y., 1990a, Strain difference of the mouse in manifestation of hydrocephalus following prenatal methylmercury exposure. Teratology, 41:205-210.

Inouye, M. and Kajiwara, Y., 1990b, Placental transfer of methylmercury and mercuric mercury in mice. Environ. Med., 34: in press.

Kameyama, Y., 1985, Comparative developmental pathology of malformations of the central nervous system, in: "Prevention of Physical and Mental Congenital Defects, Part A: The Scope of the Problem," pp. 143-156, Alan R. Liss, New York.

Klein, J., 1973, List of congenic lines of mice. I. Lines with differences at alloantigen loci. Transplantation, 15:137-153.

Matsumoto, H., Koya, G., and Takeuchi, T., 1965, Fetal Minamata disease. A neuropathological study of two cases of intrauterine intoxication by a methyl mercury compound. J. Neuropath. Exp. Neurol., 24:563-574.

Murakami, U., 1972, The effect of organic mercury on intrauterine life, in: "Drugs and Fetal Development," M.A. Klinberg, ed., Adv. Exp. Med. Biol., 27:301-336.

Ohsawa, M., Fukuda, K., and Kawai, K., 1981, Accelerated accumulation of methylmercury in the rat fetus at the late pregnant stage. Ind. Health, 19:219-221.

O'Kusky, J., 1985, Synaptic degeneration in rat visual cortex after neonatal administration of methylmercury. Exp. Neurol., 89:32-47.

O'Kusky, J.R., Radke, J.M., and Vincent, S.R., 1988, Methylmercury-induced movement and postural disorders in developing rat: loss of somatostatin-immunoreactive interneurons in the striatum. Develop. Brain Res., 40:11-23.

Skerfving, S., 1988, Mercury in women exposed to methylmercury through fish consumption, and in their newborn babies and breast milk. Bull. Environ. Contam. Toxicol., 41:475-482.

Suzuki, T., Matsumoto, N., Miyama, T., and Katsunuma, H., 1967, Placental transfer of mercuric chloride, phenyl mercuric acetate and methyl mercury acetate in mice. Ind. Health, 5:149-155.

Takeuchi, T., 1968, Pathology of Minamata disease, in: "Minamata Disease," 141-228, Study Group of Minamata Disease, Kumamoto University, Kumamoto.

Tsubaki, T., and Irukayama, K., ed., 1977, "Minamata Disease. Methylmercury Poisoning in Minamata and Niigata, Japan," Elsevier, New York.

Wanner, R.A., Edwards, M.J., and Wright, R.G. 1976, The effect of hyperthermia on the neuroepithelium of the 21-day guinea-pig fetus: histologic ultrastructural study. J. Pathol., 118:235-244.

Yang, M.G., Krawford, K.S., Garcia, J.D., Wang, J.H.C., and Lei, K.Y., 1972, Deposition of mercury in fetal and maternal brain. Proc. Soc. Exp. Biol. Med., 141:1004-1007.

# NEUROPATHOLOGY OF METHYLMERCURY INTOXICATION

Takeshi Sato and Yasunori Nakamura*

Department of Neurology, Juntendo University
School of Medicine, Tokyo 113, Japan

*Department of Dental Pharmacology
Nippon Dental University
Niigata 951, Japan

## ABSTRACT

The clinical symptoms of methylmercury intoxication, known as Hunter-Russell's disease and Minamata disease, are characterized by sensory and visual disturbance and cerebellar ataxia. Various animal models have been introduced to study the toxico-pathological effects on the nervous system (Berlin et al., 1975). In our previous study (Sato et al., 1976; Sato and Ikuta, 1976), the selective vulnerability of the central nervous system was also demonstrated in monkeys intoxicated with methylmercury. We reported that early changes detected in the nerve cells by electron microscopy were a very sensitive indicator in determining the minimal toxic dose of methylmercury. In this paper we review the minimal toxic dose of methylmercury in monkeys and the selective vulnerability of nerve tissue in experimental animals.

(1)     To determine the minimal toxic dose of MeHg in monkeys we administered daily 0.2 mg/kg to 4 monkeys for 56-133 days and 0.03 mg/kg to 5 monkeys for 87-331 days. It was demonstrated that 0.03 mg/kg daily (total 3.6-9.0 mg/kg) of MeHg is a toxic enough level to produce ultrastructural changes in the central nervous system.

(2)     To observe  early changes in the central nervous system of monkeys and cats

*Advances in Mercury Toxicology*, Edited by T. Suzuki *et al.*
Plenum Press, New York, 1991

intoxicated with low doses of MeHg, an electron microscopic study was carried out and revealed specific ultrastructural changes in the neurons of the lateral geniculate body, calcarine cortex and in the granule cells in the cerebellum.

(3)    In order to ascertain whether the transmitter binding receptor in the central nervous system is affected in cats with or without symptoms, which were administered toxic and subclinical doses of MeHg, we examined the histological alterations associated with MeHg intoxication and then compared these with the influence of MeHg upon the binding of [$^3$H] WB-4101, a potent alpha 1 norepinephrine receptor antagonist, to the occipital cortexes of 7 cats. We found a markedly decreased receptor affinity for alpha 1 norepinephrine in the membrane fraction of the occipital cortex intoxicated with MeHg.

## INTRODUCTION

In Japan, major outbreaks of methylmercury intoxication occurred in Minamata in 1953 (Tokuomi, 1960) and in Niigata in 1965 (Tsubaki et al., 1977). Since then, methylmercury has been considered a hazardous environmental pollutant. Methylmercury intoxication is characterized by sensory disturbance, concentric constriction of the visual field and cerebellar ataxia. The clinico-pathology is as follows: visual field disturbances are correlated with neuronal changes in the occipital lobes, predominantly in the calcarine regions; sensory disturbances are correlated with loss of peripheral nerve fibers and neuronal degeneration in the postcentral gyri of the parietal cortex; cerebellar ataxia is correlated with loss of granule cells in the cerebellar cortex and astrocytosis is more or less pronounced (Takeuchi et al., 1972).

Various animal models have been introduced to study the toxico-pathological effects on the nervous system (Berlin et al., 1975). In our previous study (Sato et al., 1976; Sato and Ikuta, 1976), the selective vulnerability of the central nervous system was also demonstrated in monkeys intoxicated with methylmercury. We reported that early changes detected in the nerve cells by electron microscopy were a very sensitive indicator in determining the minimal toxic dose of methylmercury. In this paper we review the minimal toxic dose of methylmercury in monkeys and the selective vulnerability of nerve tissue in experimental animals.

1. Methylmercury intoxication in monkeys with long-term, low-dose administration

The purpose of this experiment was to ascertain the minimal toxic dose of methylmercury in cynomolgus monkeys (*Macaca fascicularis*). We administered a daily dose of 0.9 - 0.23 mg/kg to four monkeys for 56-133 days (group A) and 0.02 ~ 0.04 mg/kg to five monkeys for 87-331 days (group B). the monkeys were fed methylmercury contained in pellets prepared as previously described (Ikeda et al., 1973). The monkeys were fed the designated amount of pellets daily. Dose and duration of methylmercury administration varied in each animal and are summarized in Table 1.

The animals were sacrificed following intraperitoneal sodium pentobarbital injection; the brains were removed and fixed in 10% formalin for light microscopy and in 3% glutaraldehyde for electron microscopy. Tissues were treated with chloroform-benzene and methylmercury concentration was determined by gas chromatography assay with an electron capture detector (Shimazu GC-5A).

Table 1. Methylmercury Intoxication in Monkeys

| No. | Body Weight (Kg) | Hg/day mg/kg | Total Hg mg/kg | Administration period (days) | Post-Administration period | Methylmercury Concentration Calcarine | Liver (µg/g) |
|---|---|---|---|---|---|---|---|
| A1 | 1.0 | 0.21 | 12.8 | 62 | 21 | 21.7 ± 0.4 | 31.7 ± 3.6 |
| A2 | 2.9 | 0.23 | 12.8 | 56 | 8 | 16.9 ± 0.3 | 13.7 ± 1.8 |
| A3 | 2.1 | 0.23 | 14.8 | 63 | 15 | 9.9 ± 0.7 | 13.4 ± 3.5 |
| A4 | 2.4 | 0.09 | 12.1 | 133 | 30 | 10.6 ± 0.2 | 11.9 ± 0.7 |
| B1 | 1.6 | 0.04 | 3.7 | 87 | 10 | 8.0 ± 0.9 | 13.4 ± 2.4 |
| B2 | 1.6 | 0.02 | 3.6 | 166 | 14 | 1.6 ± 0.3 | 2.8 ± 0.1 |
| B3 | 1.0 | 0.03 | 8.7 | 327 | 10 | 1.6 ± 0.0 | 6.8 ± 0.7 |
| B4 | 2.3 | 0.03 | 9.0 | 331 | 100 | 0.4 ± 0.0 | 1.4 ± 0.2 |
| B5 | 3.5 | 0.02 | 6.1 | 261 | 236 | 0.1 + 0.0 | 0.01 ± 0.0 |

Monkeys in group A developed clinical symptoms of methylmercury intoxication after administration of a total dose of 12.8 mg/kg or more, characterized by weakness of grip, an unsteady and ataxic gait, fine tremor and anorexia. Visual and hearing impairments were detected through sluggish response to light and sound. Monkeys in group B did not show any remarkable clinical symptoms. However, abnormal nystagmus was demonstrated in an electronystagmogram.

The concentration of methylmercury in the calcarine cortex of monkeys in group A was 9.9 - 21.7 µg/g and in group B, 0.1 - 8.0 µg/g (Table 1). Within groups, there were no differences in methylmercury concentration between the calcarine cortex and other brain areas. Neuropatho-logical examination revealed marked atrophy of the calcarine cortex in group A monkeys. Light microscopic findings showed severe neuronal degeneration with astrocyte proliferation predominantly in the occipital, parietal and temporal lobes. These

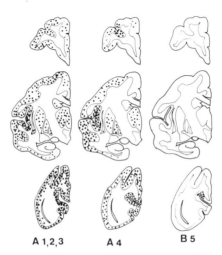

A 1,2,3          A 4          B 5

Fig. 1  Distribution of neuronal changes in coronal brain sections of monkeys intoxicated
with methylmercury.

findings were very similar to those seen in Minamata disease (Figure 1). Severe neuronal
changes were apparent in the calcarine cortices associated with spongy degeneration. The
neuronal loss was characterized by pyknosis or karyolysis with ultimate cytolysis. There
was some involvement of neurons in other areas of the cortex, basal ganglia, thalamus and
lateral geniculate nuclei. Loss of granule cells and mild atrophy of Purkinje cells was seen
in the cerebellum. Moderate degeneration of nerve cells could be fund occasionally in the
dentate nuclei, cranial nerve nuclei in the brain stem, inferior olivary nuclei, and dorsal
nuclei in the spinal cord as well as in some spinal dorsal root ganglion neurons.

Electron microscopic observation revealed remarkable degeneration of nerve cells in
the cerebral cortex, especially in the calcarine cortex. There was marked loss of
mitochondria and endoplasmic reticulum as well as a breakdown of the nuclear membrane.
The necrotic nerve cells were surrounded by astrocytes or macrophages (Figure 2). A
similar degeneration of granule cells was seen in the cerebellum.

In group B monkeys, light microscopy revealed slightly atrophic nerve cells in the
deep layer of the calcarine cortex. Electron microscopy demonstrated many myelin-like
membranous structures in the perikaryon of the neurons. Cisternae of Golgi apparatus
appeared to encircle a portion of cytoplasm (Figures 3,4). The walls of these structures had
increased in electron density and changed into a myelin-like membranous structure. In
other words, ultrastructural changes were apparent in the nerve cells of monkeys
administered methylmercury for one year at a daily dose of 0.03 mg/kg.

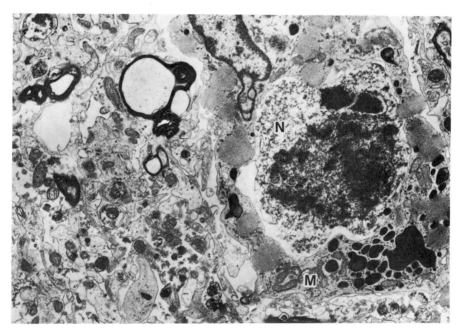

Fig. 2 Electron micrography of calcarine cortex in monkey A1 showing neuronal necrosis (N) and macrophage (M) x 9,000.

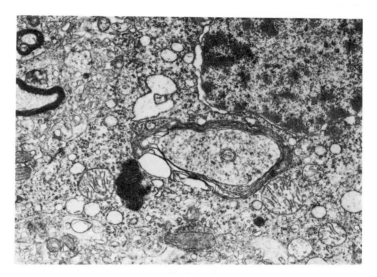

Fig. 3  Portion of neuron in calcarine cortex of monkey B1 showing whirl formation in membrane of endoplasmic reticulum x 13,000.

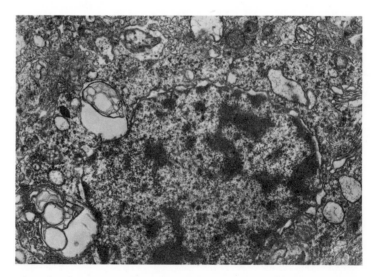

Fig. 4  Neuron in calcarine cortex of monkey B3 showing myelin-like membrane structure in endoplasmic reticulum x 13,000.

2. Selective vulnerability of the nervous system to methylmercury intoxication

To ascertain the selective vulnerability to methylmercury in the calcarine cortex, lateral geniculate bodies and cerebellum, electron microscopy was carried out in monkeys and rats intoxicated with methylmercury. Many degenerated neurons were found in the calcarine cortex of affected monkeys. Presynaptic terminals were seen to be less affected and contained many synaptic vesicles. Nerve cells in the lateral geniculate bodies of monkeys showed a well-preserved fine structure, while the axons of myelinated nerve fibers revealed remarkable degeneration, characterized by an accumulation of mitochondria and dense bodies (Figure 5). Golgi type 2 neurons also underwent marked degeneration. Since the postsynaptic membranes were severely degenerated in the calcarine cortex, it has been suggested that this axonal degeneration might be caused by retrograde degeneration from the axon terminal.

In rats administered a daily dose of 0.5 mg/kg/day methylmercury for 50 days, electron microscopy revealed that granular cells and their parallel fibers were selectively affected in the cerebellum. On the other hand, Purkinje cells were well-preserved showing naked spines in their dendrites (Figure 6).

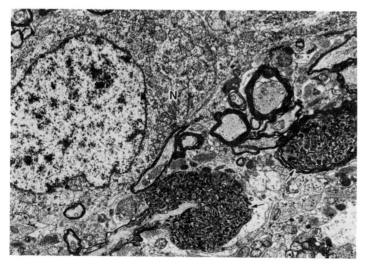

Fig. 5  Lateral geniculate body of monkey A2 showing well-preserved neuron (N) and axonal swelling (arrow) x 8,000.

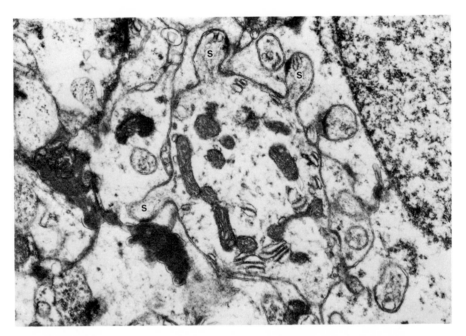

Fig. 6 Cerebellum of a rat intoxicated with methylmercury (0.5 mg/kg/day for 50 days). Dendrite in Purkinje cell showing unattached spine (naked spine) (S) surrounded by astrocyte process x 10,000.

3. Changes of alpha-norepinephrine binding activity to the receptor in cat brains with methylmercury

In order to ascertain whether the transmitter binding receptor in the central nervous system is affected in cats, the animals were administered toxic and subclinical doses of methylmercury. We examined the histological changes in cats with or without symptoms and compared these with the influence of methylmercury on alpha-norepinephrine receptors in the occipital cortices. The early stages of the disturbances caused by methylmercury with regard to the neurotransmitter and its receptors in the cerebral cortex are not thoroughly understood. It has been reported that muscarinic acetylcholine receptors in the rat brain were inhibited by methylmercury in vitro and in vivo (VonBurg, Northington and Shamoo, 1980) Several studies concerned with cerebral cortical norepinephrine fibers have demonstrated the extensive distribution of this axonal network throughout all parts of the mammalian cerebral cortex (Lapierre et al., 1973).

To determine the binding activity of alpha-norepinephrine to the receptor, [3H] WB-4101, a potent alpha-norepinephrine antagonist was used. Cats administrated methylmercury were divided into two groups. Group A received a total dose of 19-21 mg/kg methylmercury for 47 days; cats in group B received 14-15 mg/kg for 39 days. Methylmercury chloride was dissolved in water and administrated by gastric tube to cats two or three times a week for 30 to 40 days. Five to seven days after the final dose, the cats were sacrificed by intraperitoneal barbital injection. The brains were quickly removed and sectioned coronally. Alternate sections were stored at -70°C prior to assay by the [$^3$H] WB-4101 binding method. Specific WB-4101 binding was carried out by the filtration method.

Cats in group A developed symptoms shortly after the administration of methylmercury. These cats showed a decrease in their daily activity and gross tremor of the trunk. Visual impairment was also detected by sluggish response to light. These symptoms progressed, and the cats showed weakness in the hind limbs progressing finally to paralysis. None of the cats in group B showed any evidence of symptoms. Neuropathological examination of the central nervous system of the group A cats revealed marked neuronal loss with astroglial proliferation in the calcarine cortex. the cerebellar cortex also showed a reduction of granular cells. These changes were similar to those in monkeys and patients with Minamata disease. No histological abnormalities were apparent in cats in group B.

A binding assay was performed using [$^3$H] WB-4101 to determine alpha-norepinephrine binding activity to receptors in the occipital cortex. A Scatchard plot of saturation data demonstrated that maximal binding (Bmax) was significantly reduced in group B cats. Moreover, receptor binding affinity (Kd) was markedly reduced in both groups of cats (Figure 7). this study indicated that receptor affinity for alpha-norepinephrine was markedly reduced in the occipital cortex of cats administered subtoxic doses of methylmercury. These cats showed neither clinical symptoms of intoxication, nor histological changes in the cerebral cortex.

As mentioned before, clear histological neuronal changes occurred selectively in the visual cortex, somatosensory cortex and cerebellum of monkeys chronically administrated low doses of methylmercury. Early ultrastructural changes in neurons of the brains of monkeys without clinical symptoms were noted in the endoplasmic reticulum and the Golgi apparatus. Based on the results of both experiments, it is thought that changes in receptor affinity in the occipital cortex of cats might be related to the early ultrastructural alterations of cortical neurons in these animals.

Methylmercury and methylchloride block muscarinic acetylcholine receptors in the rat brain (Lapierre et al., 1973). Furthermore, the activity of choline acetyltransferase was significantly reduced, while the activity of dopamine-β-hydroxylase did not change in the

Table 2. Methylmercury Concentration in Tissues of Cats

| No. | Administrated Total MeHg (mg/kg) | Methylmercury Concentration | | |
| --- | --- | --- | --- | --- |
| | | Cerebrum | Cerebellum | Liver (µg/g) |
| CA1 | 19 | 19.7 | 28.5 | 51.8 |
| CA2 | 21 | 16.6 | 18.0 | 31.7 |
| CA3 | 19 | 13.6 | 17.5 | 73.1 |
| CA4 | 20 | 10.7 | 11.4 | 36.3 |
| CB1 | 15 | 11.6 | 9.9 | 19.5 |
| CB2 | 14 | 10.8 | 11.5 | 27.7 |
| CB3 | 14 | 10.3 | 9.1 | 27.8 |

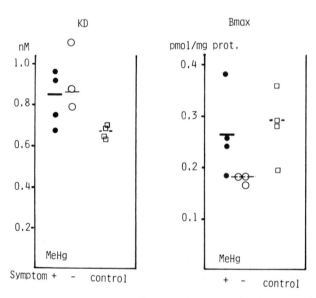

Fig. 7 Administration of methylmercury [$^{3}$H] WB-4101 in binding assay in occipital cortex of cats.

brain of rats after chronic administration of methylmercury (Tsuzuki, 1981). Methylmercury did not inhibit directly the activity of dopamine-β-hydroxylase, which is the synthetic enzyme of norepinephrine. The present *in vitro* study demonstrated that when methylmercury was added directly to the brain homogenate, the binding activity of [$^3$H] WB-4101 was not inhibited.

The mechanisms by which methylmercury reduces the affinity of norepinephrine binding is unclear. It has been suggested that free methyl radicals formed as a result of the breakdown of methylmercury may stimulate the peroxidation of various lipids in cell membranes (Yonaha, Saito and Sagi, 1983)

Numerous reports have suggested that biochemical "membrane lesions" may be the first point of attack for methylmercury. It is thus reasonable to assume that changes in norepinephrine receptor affinity and early ultrastructural changes in membranes might be attributable to selective biochemical changes in cell membranes.


## ACKNOWLEDGEMENTS

This work was supported in part by grants from the Japan Ministry of Environmental Health and the Tokyo Metropolitan Bureau of Hygiene. The authors wish to thank Drs. Ikuta Fusahiro, Sadamitsu Yamamura (Department of Neuropathology, Brain Institute, Niigata University), Gui Lin Zhang and Ruriko Kaise (Department of Neurology, Juntendo University) for their assistance in conducting the animal experiments.


## REFERENCES

Berlin, M., Grant, C.A., Hellberg, J. et al., 1975. Neurotoxicity of methylmercury in squirrel monkeys. Arch. Environ. Health 30:340.

Ikeda, Y., Tobe, M., Kobayashi, K. et al., 1973. Long-term toxicity study of methylmercuric chloride in monkeys. Toxicology 1:361.

Lapierre, Y., Beaudet, A., Demianczuk, N. et al., 1973. Noradrenergic axon terminals in the cerebral cortex of rat. II. Quantitative data related by light and electron microscope radioautography of the frontal cortex. Brain Res. 63:175.

Sato, T., Makifuchi, T., Ikuta, F. et al., 1976. Long-term studies on the neurotoxicity of small amounts of methylmercury in monkeys, Igaku no Ayumi (in Japanese) 97:75.

Sato, T. and Ikuta, F., 1976. Neuropathology of methylmercury intoxication in Niigata and chronic effects in monkeys. In: Neurotoxicology. L. Roisin, H. Shiraki and N. Grcevic, eds. Raven Press, New York, pp. 261-269.

Takeuchi, T., Etoh, K., Kojima, H. et al., 1972. Minamata disease: Pathological findings in 10-year survivors. Kumamoto Med. J. 46:666.

Tokuomi, H., 1960. Minamata disease: Clinical observation and pathologic physiology. Psychiatr. Neurol. Jpn 62:1816.

Tsubaki, T., Sato, T. et al., 1977. Epidemiology of methylmercury poisoning in Niigata. In: Minamata Disease. T. Tsubaki and K. Irukayama, eds. Kodansha and Elsevier, Tokyo, pp. 56-95.

VonBurg, R., Northington, F.K. and Shamoo, A., 1980. Methylmercury inhibition of rat brain muscarinergic receptors. Toxicol. Appl. Pharmacol. 53:282.

Tsuzuki, Y, 1981. Effects of chronic methylmercury exposure on activities of neurotransmitter enzymes in rat cerebellum. Toxicol. Appl. Pharmacol. 60:379.

Yonaha, M., Saito, M. and Sagai, M., 1983. Stimulation of lipid peroxidation by methylmercury in rats. Life Science 32:1507.

# BEHAVIORAL TOXICOLOGY OF MERCURY COMPOUNDS

Hiroshi Satoh

Department of Environmental Health Sciences
Tohoku University School of Medicine
Sendai 980
Japan

## ABSTRACT

The spectrum of mercury toxicity is vastly wide. Much concern has been given to behavioral effects of mercury compounds. Among the mercury compounds, the effects of mercury vapor or methylmercury compounds on behavior have been well recognized in both human intoxication cases and animal experiments. Because of the readiness of mercury vapor and methylmercury crossing the placental barrier, as well as the blood brain barrier, *in utero* exposure of fetuses to mercury vapor or methylmercury is possible. The newborns, thus exposed, may demonstrate behavioral consequences of the exposure during the postpartum period. Epidemiological data on Minamata disease suggest that fetuses are at greater risk from methylmercury poisoning than adults. The first animal model of methylmercury poisoning after *in utero* exposure was demonstrated by Spyker in 1972. The behavioral consequences of *in utero* exposure to a toxic substance is the research interest of <u>behavioral teratology</u>. The term <u>behavioral teratology</u> was coined in 1963 to refer to the postnatal effects of prenatal exposure to drugs. Since the pioneer work by Spyker, numerous papers have been published on the behavioral teratological effects of methylmercury in experimental animals.

The work reported here is mainly concerned with behavioral teratology of methylmercury with special reference to the possibility of modifying effects of selenite on methylmercury toxicity. (1). Pregnant mice were given 30 micromoles/kg of methylmercury with (MeHgxSe) or without (MeHg) equimolar selenite on day 9 of gestation. Pups were observed postpartum days 1, 3 and 8, and evaluated for the development of righting reflex and walking activity. The results indicated decreased developmental scores in the MeHg group on days 1 and 3, and statistical analyses suggested the interaction of selenite. (2). Pregnant mice fed selenium deficient pellets from day 3 of gestation through the gestational period were allowed to litter and the deficient food was given until the weaning of the pups. The pups were examined at the age of 3 months for shock avoidance. Numbers of avoidance in a series of repeated sessions were not different from those of the pups born to the maternal mice given selenite-added drinking water. Contrary to anticipations the effects of selenite on methylmercury behavioral teratological toxicity was not clearly demonstrated. More studies are needed to conclude the possibility of modifying effects of selenite on methylmercury toxicity manifested as behavioral teratological consequences.

## INTRODUCTION

The effects of mercury vapor or methylmercury compounds on the nervous system have been well recognized in both human intoxication cases and animal experiments. Because of the readiness of mercury vapor and methylmercury to cross the blood brain barrier, the central nervous system is believed to be the primary target organ system. Typical cases of mercury vapor poisoning are characterized by tremor with the parallel development of severe behavioral and personality changes such as increased excitability, loss of memory and insomnia. In the case of methylmercury poisoning, the earlier effects are paresthesia, malaise and blurred vision. Subsequently, concentric constriction of visual fields, ataxia, deafness and dysarthria appear.

Mercury vapor and methylmercury easily penetrate the placental barrier, as well as the blood brain barrier. Accordingly, *in utero* exposure of fetuses to mercury vapor or methylmercury is possible. Epidemiological data on Minamata disease suggest that fetuses are at greater risk from methylmercury poisoning than adults. The newborns, thus exposed, may demonstrate various symptoms including behavioral changes. In this report, behavioral consequences after mercury exposure are briefly described with the emphases on prenatal exposure to methylmercury and on the possibility of modifying effects of selenite on behavioral toxicity of methylmercury.

The numbers of animal experiments which investigated behavioral effects of mercury compounds are indicated in Table 1. A total of some sixty papers were found by computer information retrieval. These were original articles of experimental studies using animals. Neither case reports of human poisoning nor reviews were included. It is not likely that these numbers include all the experimental investigations done, however, it is clear that most investigations were done with methylmercury compounds. Considering neurotoxicity of methylmercury compounds, which has been discussed elsewhere in this symposium, it is quite reasonable that many investigations were done with methylmercury. The number of mercury vapor studies is, however, amazingly small. It seems curious that investigations with elemental mercury were much fewer than that with methylmercury because mercury vapor is also neurotoxic and various effects including behavioral and personality changes have been reported in human poisoning cases.

Table 1. Investigations on behavioral consequences of exposure to mercury compounds

| Mercury Postnatal Compounds Exposure | Total No. | Prenatal Exposure | Neonatal Exposure |
|---|---|---|---|
| Elemental Hg | 4 | - | - | 4 |
| Mercuric Hg | 1 | - | - | 1 |
| Methyl Hg | 49 | 18 | 2 | 29 |
| Phenyl Hg | 1 | - | - | 1 |
| Other | (1) | - | - | (1) |

Retrieved from MEDLINE, DIALOG FILE 155, 1966-1990
No case reported on human poisoning included
-: not found

One reason for the small number of mercury vapor studies is that reports on human poisoning cases were not included in this table. Another possible reason is that exposing animals to mercury vapor is more difficult than giving contaminated food or injections of the chemicals.

In this table investigations are classified according to not only mercury compounds applied but also the time when animals were exposed. A number of studies with methylmercury were done employing prenatal exposure. But with other mercury compounds, no investigation employed prenatal exposure. Considering human experiences especially in Minamata, Japan, the fetal Minamata disease, it is reasonable that prenatal exposure studies evoked much interest.

The first animal model of behavioral effects after *in utero* methylmercury exposure indicated "subtle" consequences (Spyker et al., 1972): offspring from maternal mice given methylmercury on day 7 or 9 of gestation were apparently normal, as well as their mothers, and seemed to be unaffected during the postnatal development. However, their behaviors observed in an open field and during forced swimming were different from the pups born to non-treated mothers. Since then, numerous papers have been published on the behavioral consequences after *in utero* exposure to methylmercury.

This type of study, where effects of prenatal exposure are examined later in neonatal developmental stages, is called "behavioral teratology". The term behavioral teratology was coined by Weboff in 1963 to refer to the postnatal effects of prenatal exposure to psychotropic drugs. Methylmercury is the environmental pollutant that was demonstrated for the first time to be a potent behavioral teratogen as shown above. Furthermore, it has been considered that behavioral examinations are sensitive to detect biological effects by exposure to environmental pollutants, especially neurotoxic ones. Thus, the idea of behavioral teratology implies that behavioral effects are able to be observed, under a certain stressful condition, in neonatal or postnatal organisms without overt toxic symptoms.

In the studies of behavioral teratology various behaviors are observed, and the examined behaviors, which were found by reviewing original articles, were summarized in Table 2 (Shimai and Satoh, 1985). A variety of effects are observed. Most categories are, however, found also in studies of behavioral toxicology. Only observation of development of reflexive behaviors is specific to behavioral teratology. Two other categories, swimming ability, especially the development of the swimming pattern with age, and ultrasonic vocalization, which is considered to be a highly functional behavior in younger age, are of more importance in behavioral teratology.

High sensitivity of the method of behavioral teratology to effects of methylmercury exposure was demonstrated by Bornhausen et al. (1980). The examined animals were four month old rats. They were exposed to methylmercury *in utero*. Actually their mothers were given by stomach tube, four doses of 0.005, 0.01 or 0.05 mg/kg of methylmercury chloride on the sixth, seventh, eighth and ninth day of gestation. These doses were extremely small. Other studies often employed a hundred times or even a thousand times larger doses (see Shimai and Satoh, 1985).

After preliminary training in a lever box, they were confronted with a test program called "differential reinforcement of high rates" (DRH). They were given a food pellet as a reinforcement when they pressed a lever in predetermined number of

Table 2. Classification of examined behaviors found in experimental studies of behavioral teratology of methylmercury*

---

1) Development of reflexive behaviors

2) Swimming ability

3) Spontaneous ability

4) Open field behavior

5) Maze learning including water T-maze learning

6) Avoidance learning including taste aversion learning

7) Operant learning

8) Susceptibility to convulsion and seizure

9) Ultrasonic vocalization

10) Visual function

---

*Shimai and Satoh, 1985

times within a predetermined interval. Thus, DRH2/1 required two lever presses within one second. While the behavioral tests were repeated, the required task became much harder.

In Figure 1, their performance was evaluated as percentages of the number of reinforcements obtained to the number of lever presses. The number of lever presses required in each DRH session is taken into account and adjusted. Dose-dependent differences are not evident in DRH2/1 session. However, those are more distinct with increased learning demand, DRH4/2 and 8/4. Rats with treatment of intermediate and high doses performed poorly compared with the control.

It is amazing that four doses of 0.01 mg/kg of methylmercury chloride produced such behavioral effects. This is the smallest amount among the doses found in the literature that produced behavioral effects. Thus, it is considered that this experiment indicated sensitivity of the method of behavioral teratology. However, none of the later studies successfully produced behavioral effects with a similar dose.

Behavioral examination clearly demonstrated effects of methylmercury among a wild mouse population (Burton et al. 1977). Wild mice were obtained using live traps at habitats in Utah, U.S.A. Their swimming ability was scored and behaviors in an open field test were observed. Hair mercury concentration was determined as an indicator of mercury exposure.

As shown in Figure 2, the higher the mercury concentration in the hair, the poorer the swimming score. It was also shown that mice with elevated hair mercury concentrations were less active in the open field. Interpretation of these behavioral deviations is difficult and mechanisms involved are unclear, however, it is interesting from an ecotoxicological point of view that wild mice which were overtly normal showed behavioral deviations under stressful situations; forced swimming and the open field test. Moreover, it is suggested for future research to evaluate the environment.

As shown in this symposium, there are a number of factors to modify the toxicity of mercury. Selenium is one of the major factors and perhaps most intensively studied. However, when behavioral effects are considered, few studies describe effects of selenium on mercury toxicity.

In the experiment described here (Satoh et al., 1985), pregnant mice were given 30 micromol/kg of methylmercury alone or with the equimolar of sodium selenite on day 9 of gestation. They were allowed to litter. The neonatal mice were examined for their behavior and sacrificed afterward to determine tissue mercury concentration.

Table 3 shows developmental scores of righting reflex and waking activity, on postpartum day 1, 3 or 8. The scores were evaluated by an observer who was ignorant of the treatment. Both scores were lowest for the MeHg group on days 1 and 3.

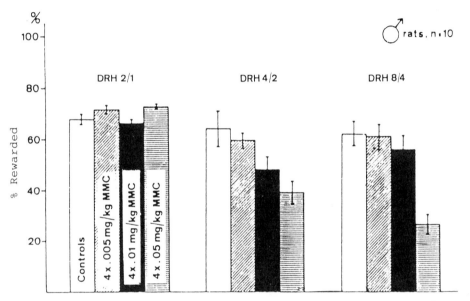

Fig. 1.     Effects of prenatal exposure to methylmercury on the performance of
DRHs (differential reinforcement of high rates) with different difficulties
(Bornhausen et al., 1980)

The "% Rewarded" was calculated as a/b x c x 100 (%),  where
a = Number of reinforcements
b = Lever presses
c = Lever presses required for one reward

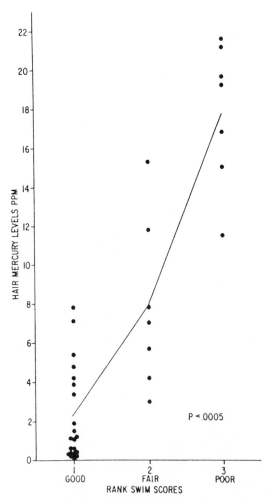

Fig. 2.  Hair mercury concentration and swimming score in a wild mouse
population (Burton et al., 1977)

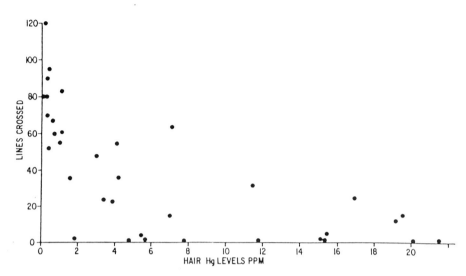

Fig. 3.  Hair mercury concentration and activity in an open field (number of lines crossed) in wild mouse population (Burton et al., 1977).

Table 3

MEANS AND STANDARD DEVIATIONS OF RIGHTING REFLEX AND WALKING ACTIVITY SCORES ON DAYS 1,3,AND 8

|  | Day 1 | Day 3 | Day 8 |
|---|---|---|---|
| Righting reflex |  |  |  |
| MeHg x Se | 3.37±0.65 | 2.62±0.53 | 5.87±0.12 |
| MeHg | 1.50±0.42 | 2.37±0.37 | 5.62±0.27 |
| Se | 2.00±0.33 | 3.75±0.67 | 4.12±0.64 |
| NaCl | 2.25±0.41 | 3.37±0.65 | 4.87±0.35 |
| Walking activities |  |  |  |
| MeHg x Se | 1.37±0.18 | 3.25±0.90 | 7.25±0.80 |
| MeHg | 0.87±0.12 | 1.37±0.53 | 9.12±0.35 |
| Se | 1.62±0.53 | 3.50±0.78 | 9.00±0.30 |
| NaCl | 1.25±0.56 | 2.87±1.04 | 9.25±0.31 |
| N = 8 in all groups |  |  |  |

However, the MeHgxSe group developed, as well as the other two groups, Se and NaCl. Analyses of variance on day 1 for righting reflex and walking activity were significant.

Mercury concentrations in tissues are shown in Table 4. There are some differences between mercury concentrations in tissues from MeHgxSe and MeHg. Statistically significant differences are found in several comparisons. It is considered that coadministered selenite may play a role to alter mercury metabolism and subsequently development of behavior. However, it is not clear that this small difference in the mercury concentrations could produce different behavioral development. Although the mechanisms involved are not known, more precise studies such as determination of mercury concentration at the site of action, molecular mechanism of interaction between mercury and selenium and biochemical changes remain for future studies.

The possible effects of selenium on methylmercury toxicity on behavior were investigated by shock avoidance (Satoh, unpublished data). Pregnant mice were given a selenium deficient diet from day 3 of their gestation through gestational and lactating periods until weaning. They were divided into four groups and three groups were given selenium added water, at the concentrations of 0.2, 1.0 or 4.0 microgram/ml. The rest was given tap water. Intake of selenium under usual feeding conditions is comparable to that of the 1.0 microgram/ml group received. Three weeks after the parturition, the weaning mice were given commercial pellet food and kept until three months old.

They were individually placed in a lever box with a grid floor. The flooring consists of metal rods to deliver electric shock when necessary. The whole system was controlled by a personal computer. Mice examined were requested to press a lever to avoid or escape electric shock, which was delivered after an alarm tone of a buzzer installed in each box. When an animal pressed the lever during the alarm period of ten seconds, no electric shock was delivered and it is referred to as a void. When the animal failed to press the lever during that period, then an electric shock was delivered for fifteen seconds or until the animal pressed the lever. When the animal pressed the lever during the electric shock this is referred to as escape. After the lever press or the electric shock period, a thirty second inter-trial period was inserted. Then the alarm during the electric shock period started again.

The thirty trial sessions were repeated everyday (SA-1 to SA-7). In the beginning, they were unable to avoid shocks, but they readily learned to escape from the electric shock (Figure 4). With repetition of sessions, the operant behavior was

Table 4

MEANS AND STANDARD ERRORS OF TISSUE Hg CONCENTRATION (μg/g) FOR THE MeHg
AND MeHgxSe GROUPS ON DAYS 1, 3, AND 8

|  | Day 1 | Day 3 | Day 8 |
|---|---|---|---|
| Brain |  |  |  |
| MeHg x Se | 5.28±0.28(8) | 2.77±0.14(8)** | 0.83±0.06(8)** |
| MeHg | 6.65±0.86(8) | 3.66±0.23(8) | 1.14±0.07(8) |
| Liver |  |  |  |
| MeHg x Se | 5.95±0.95(8) | 2.89±0.13(7) | 1.07±0.09(8) |
| MeHg | 5.09±0.74(8) | 3.62±0.21(8) | 1.02±0.11(8) |
| Kidneys |  |  |  |
| MeHg x Se | 7.03±0.91(8)* | 2.34±0.18(6)** | 1.32±0.08(8) |
| MeHg | 4.24±0.88(8) | 3.58±0.20(8) | 1.44±0.06(6) |

*P<0.05 and **P<0.01, statistically significant difference compared with the MeHg group. Numbers in parentheses show number of observations.

established to avoid shocks and the number of escapes decreased. After the seven sessions, they were not placed in the boxes but were kept in home cages for a week, then retention was examined. This session is indicated as RT-1 in the figure. After that, the session continued without electric shock, which is called the extinction phase. It is supposed that this experimental schedule examines learning ability, retention and extinction of memory.

In this preliminary result, simply showing group-mean of numbers of avoidances and escapes, there seems to be no distinct difference among the groups with different treatment of selenium. It is considered that this experiment was not enough to reach a conclusion of the mercury-selenium interaction in terms of behavior. In contrast, there is more to be investigated. In addition, more consideration is necessary on the behaviors to be observed, doses to be given, the time of exposure, and species to be examined.

As shown in this symposium, there are many approaches to the goals of mercury toxicology. In different approaches, materials and subjects to be studied are different and techniques used are different. Among these different approaches, behavioral toxicology and teratology are characterized by the need to conduct behavioral studies on each individual subject. Because of this character, experiments of these research areas, especially behavioral teratology, are time-consuming and laborious. Behavioral

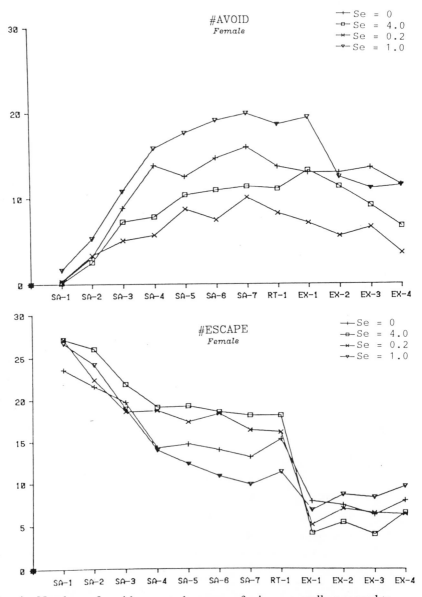

Fig. 4. Numbers of avoidances and escapes of mice prenatally exposed to methylmercury and given a selenium deficient diet with water added with selenite at different concentrations during *in utero* and neonatal periods (Satoh, unpublished data).

changes due to individual differences are often large. Moreover, since even behavioral measurements are affected by treatment, it is not easy to interpret the changes. However, one of the goals of mercury toxicology is to describe the full spectrum of adverse effects found in an individual and to elucidate mechanisms involved. Thus the behavioral approach should be applied.

## ACKNOWLEDGEMENT

Part of the shock avoidance experiment was supported by Grant-in-Aid from the Ministry of Education, Science and Culture of Japan #63570232.

## REFERENCES

Bornhausen, M., Musch and Greim, H., 1980. Operant behavior performance changes in rats after prenatal methylmercury exposure. Toxicol. Appl. Pharmacol. 56:305-310.

Burton, G.V., Alley, J., Rasmussen, G.L., Orton, P., Cox, V., Jones, P. and Graff, D., 1977. Mercury and behavior in wild mouse populations. Environ. Res. 24:30-34.

Satoh, H., Yasuda, N. and Shimai, S., 1985. Development of reflexes in neonatal mice prenatally exposed to methylmercury and selenite. Toxicol. Let. 25:199-203.

Shimai, S. and Satoh, H., 1985. Behavioral teratology of methylmercury. J. Toxicol. Sci. 10:199-216.

Spyker, J.M., Spaber, S.B. and Goldberg, A.M., 1972. Subtle consequences of methylmercury exposure: Behavioral deviation in offspring of treated mothers. Science 621-623.

Weboff, J. and Gottlieb, J.S., 1963. Drugs in pregnancy: Behavioral Teratology. Obstet Gynecol Surv. 18:420-423.

# EFFECT OF METHYLMERCURY ON SLEEP PATTERNS IN THE RAT

Heihachiro Arito and Masaya Takahashi

National Institute of Industrial Health
6-21-1, Nagao, Tama-ku, Kawasaki 214
Japan

## ABSTRACT

In an attempt to find out early and sensitive clinical signs of experimental methylmercury poisoning, physiological indices of wakefulness, slow-wave sleep, paradoxical sleep, body temperature and heart rate were examined during the development of methylmercury toxicity with rats which were chronically implanted with EEG and EMG electrodes or thermistor probes and ECG electrodes. These physiological indices were found to provide early and sensitive signs in experimental methylmercury poisoning with adult rats as compared with the behaviorally-observed neurological signs of motor incoordination. The reduction of paradoxical sleep, its altered circadian rhythm and the lowered body temperature were long-lasting signs as compared with quick recovery from the reduced wakefulness and the lowered heart rate. Analysis of brain monoamines and their metabolites showed that the concentrations of noradrenaline, 3-methoxy-4-hydroxyphenylethyleneglycol and 5-hydroxytryptamine of the frontal cortex of the methylmercury-administered rats were lowered during the dark period, indicating that methylmercury lowered the turnover of central monoamine metabolisms. It was suggested that the reduction of paradoxical sleep and its altered circadian rhythm were mediated through an inhibitory action of methylmercury on central monoaminergic mechanisms, in addition to the well-documented inhibitory effect of methylmercury on central cholinergic mechanisms.

*Advances in Mercury Toxicology*, Edited by T. Suzuki *et al.*
Plenum Press, New York, 1991

## INTRODUCTION

Alterations in reflex, motor coordination and various kinds of behaviors have been recognized as the neurological signs of experimental methylmercury intoxication (Suzuki and Miyama, 1971; Fehling et al., 1975; Ohi et al., 1978; Zimmer and Cater, 1979; Hargreaves et al., 1985). These neurological signs appear with a latent period after administration of methylmercury to animals. However, inhibition of brain protein synthesis (Yoshino et al., 1966; Verity et al., 1977; Omata et al., 1980), electrophysiological changes of the peripheral nerves (Somjen et al., 1973; Misumi, 1979; Miyama et al., 1983) and morphological abnormalities in the central (Klein et al., 1972; Hargreaves et al., 1985; Moller-Madsen, 1990) and peripheral (Herman et al., 1973) nervous system have been reported to occur during the latent period of the neurological signs in the experimental methylmercury intoxication. Alterations in physiological indices such as sleep-wakefulness, heart rate and body temperature may also provide early and sensitive clinical signs of subtle consequences which occur during the development of methylmercury poisoning. Polygraphic measurements of electroencephalographic (EEG) and electromyographic (EMG) activities of the conscious rat implanted with the indwelling electrodes, allow quantitative description of wakefulness (W), slow-wave sleep (SWS) and paradoxical sleep (PS) and their naturally occurring cycles including a circadian rhythm. Body temperature and heart rate (HR) of the rat can also be measured by chronic implantation of a thermistor probe and an electrocardiographic (ECG) electrode, respectively.

In the experiments reported here, we quantified time-course changes in amounts of time spent in W, SWS and PS during the development of the methylmercury intoxication with the adult rats implanted with EEG and EMG electrodes. Secondly, effects of methylmercury on HR, body temperature and their circadian rhythms were examined along the time-course of the toxicity development with the rats implanted with the thermistor probes and the ECG electrodes. Thirdly, monoaminergic mechanisms underlying the methylmercury-induced changes in the physiological indices were explored by HPLC analysis of noradrenaline (NA), its metabolite, 3-methoxy-4-hydroxy-phenylethyleneglycol (MHPG), 5-hydroxytryptamine (5-HT) and its metabolite, 5-hydroxyindoleacetic acid (5-HIAA) of the frontal cortex of the methylmercury- and vehicle-administered rats. Brain concentrations of total mercury were also determined along different days after oral administrations of methylmercury. Some parts of the present results were published previously (Arito et al, 1982, 1983, 1985).

METHODS

<u>Animals</u>: Eight-week-old male rats of the Sprague-Dawley strain were used in the 3 experiments. A colony and a chamber of polygraphic recordings were maintained at a temperature of $24 \pm 1°C$, a relative humidity of $55 \pm 5\%$ with a 12-hr light (8:00-20:00) and 12-hr dark (20:00-8:00) cycle. The rats had free access to food and water in the individual and group cages.

<u>MMC administration</u>: Following the baseline recordings of EEG and EMG for 4 consecutive days, 3 groups of the electrode-implanted rats were administered orally with methylmercury chloride (MMC) at total doses of 30 mg/kg (15 mg/kg/day for 2 consecutive days), 10 mg/kg (5 mg/kg/day x 2 days) and 3.3 mg/kg (1.65 mg/kg/day x 2 days), respectively. Following the baseline recordings of HR and body temperatures for 4 consecutive days, the rats were administered *po* at a total dose of 30 mg MMC/kg (15 mg/kg/day x 2 days). For monoamine analysis, the rats were administered *po* at a total dose of 30 mg MMC/kg (15 mg/kg/day x 2 days) and another group of the rats administered *po* only with a vehicle solution of $Na_2CO_3$ served as controls. Day 1 was defined as the first day when MMC or the vehicle solution was administered.

Experiment 1

<u>Electrode implantation</u>: Under pentobarbital anaesthesia (40 mg/kg), the rats were implanted for EEG recordings with two stainless steel screw electrodes on the frontal and parietal cortex and a reference electrode on the cerebellum and for bipolar EMG recording of the dorsal neck muscle with stainless steel wires. The method of implantation was descrived in the previous paper (Arito et al., 1983). Each electrode-implanted rat was housed in an individual cage for polygraphic recording. Those electrodes were connected with a flexible wire cable to a polygraphic instrument through a slipring connector which guaranteed free movement to the rat in the cage. The rats were allowed a week for recovery from the surgery and habituation to the recording cages.

<u>Polygraphic recording and classification of vigilance states</u>

The two EEG and EMG activities of each rat were continuously recorded at a paper speed of 100 mm/min for 4 consecutive days before and for 3 or 4 consecutive days at an interval of 10 days after the oral administrations of MMC. The polygraphic records were visually scored into W, SWS and PS according to the following criteria

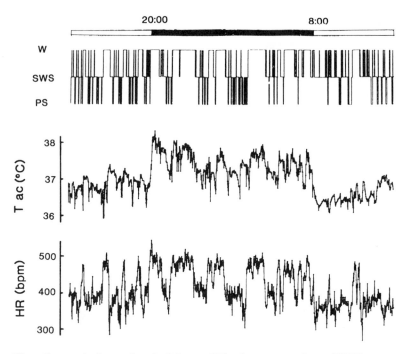

Fig. 1.  Circadian variations of wakefulness (W), slow-wave sleep (SWS)
and paradoxical sleep (PS), abdominal temperature ($T_{ac}$) and heart rate (HR) of
a normal rat.

(Figure 1 of Arito et al., 1988).  The W episode was characterized by low amplitude
fast wave EEG and enhanced EMG tone.  SWS was characterized by large amplitude
slow wave EEG and progressive reduction of the EMG tone.  The PS episode was
identified by W-like EEG in the frontal  cortex, regular theta wave EEG in the parietal
cortex and lack of the EMG tone.  As to reproducibility of the visual scoring, it was
confirmed previously (Arito et al., 1988) that the amounts of W, SWS and PS scored
by the first inspection deviated by less than $\pm$ 7% from those scored by the second
inspection.  The 1-hr, 12-hr and 24-hr amounts of time spent in W, SWS and PS of the
MMC-administered rats were statistically compared with the respective baseline values
obtained during the pre-administration period by T-test.

Experiment 2

Implantation of thermistors and ECG electrode: The rats were chronically implanted for
ECG recording with a teflon-insulated stainless steel wire on the epicostal tissue above
the heart and a reference ECG electrode on the cerebellum.  Two thermistor probes
(Techno-Seven, Japan) were surgically implanted:  one stereotaxically into the striatum

of the brain and another into the abdominal cavity. The rats were housed in the individual cages and those electrodes were led through the slipring connector to a HR counter and a temperature data analyzer. The rats were allowed a week for recovery from the surgery and habituation in the recording cages.

Measurement of heart rate and body temperature: Number of ECG R waves were registered for every 10-sec epoch with a HR counter and stored into a micro-computer during every 4-day period. The stored HR data of each rat were averaged over every 1-, 12- or 24-hr period and expressed as beats/min. The temperatures of the brain ($T_{br}$) and the abdominal cavity ($T_{ac}$) were registered for every 1-min epoch and stored into a micro-computer during every 4-day period. The stored temperature data of each were also time-averaged for every 1-, 12- and 24-hr epoch. The time-averaged HR and body temperatures during the post-administration period were statistically compared with the respective baseline values during the pre-administration period by T-test.

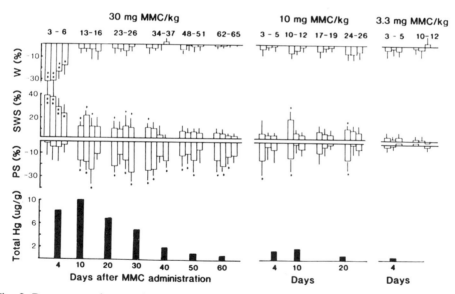

Fig. 2 Percentage changes (mean ± SD) in daily amounts of W, SWS and PS of the MMC-administered rats relative to the respecitve baseline values. Number of the rats used were 5 for the high dose, 6 for the medium dose and 5 for the low dose. The time-course changes in total mercury level of the frontal cortex were given in the lower part. Each filled bar indicates a mean of 5-6 rats. Single and double asterisks indicate statistical significance at the levels of p<0.05 and p<0.01, respectively.

Experiment 3

Analysis of brain monoamines and their metabolites

At 10:00, 14:00, 18:00, 22:00, 2:00 and 6:00 of Day 23, both MMC- and vehicle-administered rats were sacrificed by decapitation. The brains were quickly removed and sectioned into 7 regions according to the method of Glowinski and Iversen (1966). The frontal cortex which contained the preoptic basal forebrain area according to their dissecting method was submitted to HPLC analysis of 5-HT, 5-HIAA, NA and MHPG. The sample preparation and the method for HPLC analysis were described in a previous paper (Arito et al., 1985).

Determination of total mercury in the brain

A group of 5 - 6 rats were sacrificed on different days after oral administrations of 1.7 mg, 5 mg and 15 mg MMC/kg/day for 2 consecutive days. The frontal cortex of each rat was sectioned according to the method of Glowinski and Iversen (1966) and digested with a mixed solution of $HNO_3$, $HC1O_3$ and $H_2SO_4$. The mercury content was determined by atomic absorption spectrometry.

RESULTS

Sleep-wakefulness patterns

The sleep patterns of normal rats were characterized by frequent changes in W, SWS and PS (a stage diagram in Figure 1) and predominance of SWS during the light period and W during the dark period (Figure 4A). W, SWS and PS exhibited clear circadian rhythmicities (upper 3 graphs in Figure 4A). Lighting-off and -on induced an abrupt increase and decrease in hourly amount of W, respectively, which were mirrored by the time-course changes in hourly amount of SWS. On the other hand, time-course changes in hourly amount of PS exhibited a circadian pattern of a cosine curve.

Magnitude and duration of percentage changes in daily (24-hr) amounts of W, SWS and PS of MMC-administered rats relative to the respective baseline values were increased dose-dependently (Figure 2). A total of 30 mg MMC/kg significantly reduced the daily amount of PS from Days 13 through 65, whereas the daily amount of SWS recovered to the baseline level by Day 34. A transient but marked decrease in daily amount of W at the expense of the increase in SWS was observed only during Days 3 - 6 after the high MMC dose. A total dose of 10 mg MMC/kg did not decrease the daily amount of W on Days 3 - 5 but induced a long-lasting reduction in the daily amount of PS at the expense of the increase in SWS thereafter.

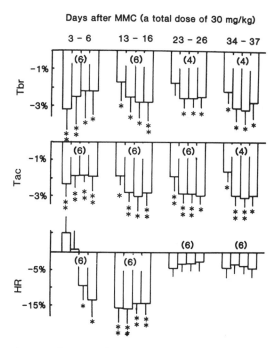

Days after MMC (a total dose of 30 mg/kg)

Fig. 3  Percentage changes (mean ± SD) in 24-hr averaged values of $T_{br}$, $T_{ac}$ and HR of the MMC-administered rats relative to the respective baseline values during the pre-administration period.  Number of rats used wre indicated in the parenthesis.  Single and double asterisks indicate statistical significance at the levels of p<0.05 and p<0.01, respectively.

387

The effects of MMC on circadian W, SWS and PS rhythmicities were represented by time-course changes in 1-hr amounts of W, SWS and PS for Days 3-6 (Figure 4-B) and Days 23-26 (Figure 4-C) after oral administration of 30 mg MMC/kg. The early period (Days 3-6) was characterized by total disappearance of circadian rhythmicities of W and SWS, resulting primarily from a marked decrease in W at the expense of the increased SWS during the dark period. The circadian PS rhythm was maintained without any reduction in daily amount of PS (Figure 2) during the early period but an acrophase of the PS rhythm was delayed by about 5 hrs. On Days 23 - 26, the delayed acrophase of the circadian PS rhythm still persisted. The 12-hr amounts of W were significantly decreased only during the dark period at the expense of the significant increases in SWS and PS. The delayed acrophase by 5 hr of the circadian PS rhythm was still observed for Days 62 - 65, whereas the 12-hr amounts of W and SWS during the dark period returned to the respective baseline values (Data not shown).

Heart rate and body temperature

A control rat exhibited an abrupt elevation of HR and $T_{ac}$ in close association with sustained waking episodes during the dark period (Figure 1). HR, $T_{br}$ and $T_{ac}$ exhibited clear circadian rhythmicities (Figure 4A). Transitions of lighting-off and -on induced an abrupt increase and decrease in HR, $T_{br}$ and $T_{ac}$, respectively, in close association with the circadian W and SWS rhythms.

As shown in Figure 3, the percentage changes in 24-hr averaged values of $T_{br}$ and $T_{ac}$ of the MMC-administered rats relative to the respective baseline values were affected to the same extent through Day 37. This indicates that the body temperatures did not recover to the respective baseline levels. On the other hand, the 24-hr averaged values of HR recovered earlier to the baseline value by Day 23. The circadian rhythms of $T_{br}$ and $T_{ac}$ of the MMC-administered rats were totally masked during the early period (Days 3 - 6), resulting primarily from suppression of the dark-phase rise in the body temperatures (Figure 4B). Bradycardia was followed by transient tachycardia on Days 3 and 4. This early stage of the intoxication was characterized by complete disruption of harmonic circadian rhythmicities of sleep-wakefulness and those autonomic responses. HR was lowered significantly only during the dark period on Days 23 - 26 but the abdominal temperature was significantly decreased in not only the dark but also during the light period.

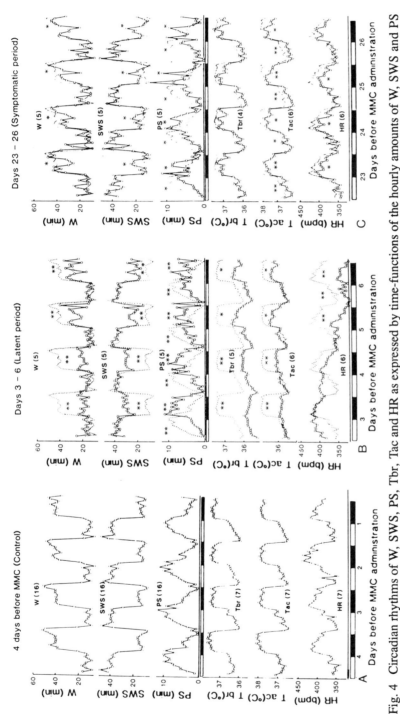

Fig. 4  Circadian rhythms of W, SWS, PS, Tbr, Tac and HR as expressed by time-functions of the hourly amounts of W, SWS and PS and the 1-hr averaged values of Tbr, Tac and HR.  Number of rats used were indicated in the parenthesis.  (A) 4 consecutive days before, (B) Days 3 - 6 after and (C) Days 23 - 26 after oral administration of 30 mg MMC/kg.  The dotted lines in B and C represent the rhythms of the baseline values obtained during the pre-administration period.  Single and double asterisks indicate statistical significance of the 12-hr values at the level of $p < 0.05$ and $p < 0.01$, respectively, as compared to the baseline values by T-test.

389

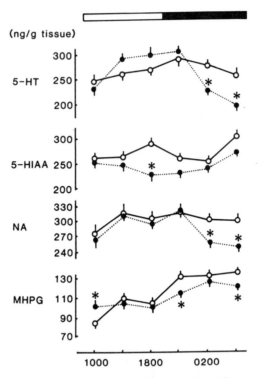

Fig. 5 Circadian variations of 5-HT, 5-HIAA, NA and MHPG concentrations of the frontal cortex containing the basal forebrain area of the MMC-administered (open circle) and the vehicle-administered rats (closed circle) sacrificed at 6 different times of Day 23. Each plot with a vertical bar indicates a mean and S.E.M. of 7 rats. An asterisk shows a significant difference at a level of p<0.05 by T-test.

### Brain monoamines and their metabolites

As shown in Figure 5, the concentrations of 5-HT, NA and MHPG of the frontal cortex of the MMC-administered rats were significantly lowered during the dark period. The MHPG level was significantly elevated at the early light period and the 5-HIAA level was lowered at the late light period.

### Mercury concentrations in the frontal cortex

The brain mercury reached its maximum levels of 10 μg Hg and 2.8 μg Hg/g tissue on Day 10 after a total dose of 30 mg and 10 mg MMC/kg, respectively, and declined thereafter. Less than 1 μg Hg/g was accumulated in the frontal cortex by a total dose of 3.3 mg/kg.

## DISCUSSION

In the present study of experimental methylmercury intoxication with adult rats, it was found that the marked changes in the psysiological indices of W, SWS, PS, HR, $T_{br}$ and $T_{ac}$ were induced by the maximum brain mercury level of 10 μg/g, that the dark-phase levels of W, SWS and HR were more markedly affected than light-phase levels and also that a threshold for the significant changes in the vigilance states existed in the range of 1 to 3 μg Hg/g brain. It was also found that the MMC-induced changes in W, SWS and HR recovered earlier than did those of PS and body temperature. The present findings can be taken to indicate that these physiological indices provides the earliest and the most sensitive clinical signs of subtle consequences of exposure of adult rodents to methylmercury, in consideration with the reports that the early and sensitive neurological signs of motor incoordination such as tail rotation (Ohi et., 1978), head-maintenance inability (Suzuki and Miyama, 1971) and ambulatory activity (Fehling et al., 1975; Zimmer and Carter, 1979) appear at a brain mercury level of 8-10 μg Hg/g after a latent period. The reduction in the daily amount of PS and the altered circadian PS rhythm found in the present study are presumed to be mediated through an inhibitory action of methylmercury on the central cholinergic and monoaminergic mechanisms which are responsible for generating PS and modulating its rhythm, respectively. First, in agreement with the report by Taylor & DiStefano (1976), the present neurochemical finding that the levels of NA, MHPG and 5-HT of the MMC-administered rats were lowered in the frontal cortex area can be taken to indicate the reduced turnover rate of central monoamine metabolisms. The central monoaminergic system was reported to play a modulatory role in the cholinergic PS mechanism (Steriad and Hobson, 1976) and the internal coupling of some circadian oscillators (Honma, 1985). Secondly, the inhibitory effects of methylmercury on central cholinergic receptors and acetylcholine turnover (Von Burg et al., 1980; Kobayashi et al., 1980) might as well be involved in the MMC-induced reduction of PS, because both "sleep center" of the basal forebrain area (Sterman and Shouse, 1985) and PS generator neurons of the brainstem (Sakai, 1985) are cholinergic. Another line of evidence that methylmercury inhibits the brain protein synthesis during the latent and symptomatic period (Yoshino, 1966; Verity et al., 1977; Omata et al., 1980) seems to be in favor of the possibility that the MMC-induced reduction of PS at the expense of the increased SWS is CNS origin, because those changes are in close resemblance with a pharmacological action of protein synthesis inhibitors, cycloheximide and puromycin which were reported to elicit suppression of PS and a dark-phase augumentation of SWS (Uezu et al., 1981). However, we can not totally rule out the possibility of non-neurological origin as the MMC-induced disruption of the physiological indices during the early stage of intoxication, because greater content of mercury is accumulated in the kidney, blood and liver than in the brain during the early period (Hargreaves et al., 1985) and because methylmercury induces extensive lesions in the kidney before the neurological manifestation (Magos and Butler, 1972).

The MMC-induced hypothermia which would reflect an imbalanced control of heat production and dissipation is considered to result from an alteration in the CNS thermoregulatory setpoint, an impairment of the peripheral sensory and motor systems and/or an alteration in thermoregulatory vasomotor control. The dark-phase increase in SWS and PS found in the present study is also presumed to contribute to hypothermia, because SWS and PS are reported to suppress heat production and augment heat dissipation (Roussel and Bittel, 1979). A possibility of the CNS origin of the MMC-induced hypothermia can be suggested by the lowered turnover of 5-HT and NA metabolism in the frontal cortex containing the basal forebrain area where the thermoregulatory center is located, because NA and 5-HT of the preoptic forebrain area were reported to play a permissive role in central thermoregulation (Myers, 1980). Besides, no recovery from the hypothermia and the reduced daily amount of PS in comparison with W, SWS and HR seems to favor this possibility, because circadian rhythms of PS and body temperature are reported to be regulated by a strong endogenous oscillator resistant to the external environment, as compared with the rhythms of sleep-wakefulness and rest-activity (Czeisler and Guilleminault, 1980). Further study will be needed to look into underlying mechanisms of the MMC-induced hypothermia.

## REFERENCES

Arito, H., Sudo, A., Hara, N., Nakagaki, K. and Torii, S., 1982, Changes in circadian sleep-waking rhythms of rats following administration of methylmercury chloride. Industr. Health, 20: 55.

Arito, H., Hara, N. and Torii, S., 1983, Effect of methylmercury chloride on sleep-waking rhythms in rats. Toxicol., 28: 335.

Arito, H., Oguri, M. and Tanaka, S., 1985, Diurnal variations of brain noradrenaline metabolism of the methylmercury chloride-administered rat with reference to an altered circadian rhythm of paradoxical sleep. Industr. Health, 23: 245.

Arito, H.,Tsuruta, H. and Oguri, M., 1988, Changes in sleep and wakefulness following single and repeated exposures to toluene vapor in rats. Arch. Toxicol., 62: 76.

Czeisler, C.A. and Guilleminault, C. (eds.), 1980, A workshop on REM sleep; Its temporal distribution, Sleep, 2: 285.

Fehling, C., Abdulla, M., Brun, A., Dictor, M., Schutz, A. and Skerfving, S., 1975, Methylmercury poisoning in the rat: A combined neurological, chemical, and histophathological study, Toxicol. Appl. Pharmacol., 33: 27.

Glowinski, J. and Iversen, L.L., 1966, Regional studies of catecholamine in the rat brain. I. The disposition of $^3$H-norepinephrine, $^3$H-dopamine and $^3$H-dopa in various regions of brain, J. Neurochem., 13: 655.

Hargreaves, R.J., Foster, J.R., Pelling, D., Moorhouse, S.R., Gangolli, S.D. and Rowland, I.R., 1985, Changes in the distribution of histochemically localized mercury in the CNS and in tissue levels of organic and inorganic mercury during the development of toxication in methylmercury treated rats, Neuropath. Neurobiol., 11: 383-401.

Herman, S.P., Klein, R., Talley, F.A. and Krigman, M.R., 1973, An ultrastructural study of methylmercury-induced primary sensory meuropathy in the rat, Lab. Invest., 28: 104.

Honma, K., 1985, Circadian rest-activity cycles in rats: Its modulation by humoral factors affecting the central transmitter systems, in: "Endogenous sleep substances and sleep regulation," S. Inoue and AA. Borbely, Ed., 89-100, Jpn. Sci. Soc. Press, Tokyo.

Klein, R., Herman, S.P., Brubaker, P.E. and Lucier, G.W., 1972, A model of acute methylmercury intoxication in rats. Arch. Path. 93: 408.

Kobayashi, H., Yuyama, A., Matsusaka, N., Takeno, K. and Yanagiya, I., 1980, Effect of methylmercury on brain acetylcholine concentration and turnover in mice. Toxicol. Appl. Pharmacol. 54: 1.

Magos, L and Butler, W.H., 1972, Cumulative effects of methylmercury dicyandiamide given orally to rats. Food Cosmet. Toxicol., 10: 513.

Misumi, J., 1979, Electrophysiological studies in vivo on peripheral nerve function and their application to peripheral neuropathy produced by organic mercury in rats. III Effects of methylmercury chloride on compound action potentials in the sciatic and tail nerve in rats. Kumamoto Med. J. 32: 15.

Miyama, T., Minowa, K., Seki, H., Tamura, Y., Mizoguchi, I., Ohi, G. and Suzuki, T., 1983, Chronological relationship between neurological signs and electrophysiological changes in rats with methylmercury poisoning, special reference to selenium protection. Arch. Toxicol., 52: 173.

Moller-Madsen, B., 1990, Localization of mercury in CNS of the rat. II. Intraperitoneal injection of methylmercuric chloride and mercuric chloride. Toxicol. Appl. Pharmacol., 103: 303.

Myers, R.D., 1980, Hypothalamic control of thermoregulation, in: "Handbook of the hypothalamus, behavioral studies of the hypothalamus", P.J. Morgane and J. Panksepp, eds., 83-210, Marcel Dekker Inc., New York.

Ohi, G., Nishigaki, S., Seki, H., Tamura, Y., Mizoguchi, I., Yagyu, H. and Nagashima, K., 1978, Tail rotation, An early neurological sign of methylmercury poisoning in the rat. Environ. Res., 16: 353.

Omata, S., Horgome, T., Momose, Y., Kambayashi, M., Mochizuki, M. and Sugano, H., 1980, Effect of methylmercury chloride on the in vivo rate of protein synthesis in the brain of the rat: Examination with the injection of a large quantity of ($^{14}$C)valine. Toxicol. Appl. Pharmacol., 56: 207.

Roussel, B.B. and Bittel, J., 1979, Thermogenesis and thermolysis during sleeping and waking in the rat. Pflugers Arch., 382: 225.

Sakai, K., 1985, Anatomical and physiological basis of paradoxical sleep. In: "Brain mechanism of sleep," D.J. McGinty, ed., 111-135, Raven Press, New York.

Somjen, G.G., Herman, S.P. and Klein, R., 1973, Electrophysiology of methylmercury poisoning. J. Pharmacol. Exp. Ther., 186: 579.

Steriad, M. and Hobson, J.A., 1976, Neuronal activity during the sleep-waking cycle. Prog. Neurobiol., 6:155.

Sterman, M.B. and Shouse, M.N., 1985, Sleep "centers" in the brain: the preoptic basal forebrain area revisited. in : "Brain mechanism of sleep." D.J. McGinty, ed., 277-299, Raven Press, New York.

Suzuki, T. and Miyama, T., 1971, Neurological symptoms and mercury concentration in the brain of mice fed with methylmercury salt. Industr. Health, 9: 51-58.

Taylor, L.L. and DiStefano,V., 1976, Effects of methylmercury on brain biogenic amines in the developing rat pup. Toxicol. Appl. Pharmacol., 38: 489.

Uezu, E., Sano, A. and Matsumoto, J, 1981, Effects of inhibitors of protein synthesis on sleep in rats. Tokushima J. Exp. Med., 28: 9.

Verity, M.A., Brown, W.J., Cheung, M. and Czer, G., 1977, Methylmercury inhibition of synaptosome and brain slice protein synthesis: *In vivo* and *in vitro* studies. J. Neurochem., 29: 673.

Von Burg, R., Northington, F.K. and Shamoo, A., 1980, Methylmercury inhibition of rat brain muscarinic receptors. Toxicol. Appl. Pharmacol., 53: 285.

Yoshino, Y., Mozai, T. and Nakao, K., 1966, Distribution of mercury in the brain and its subcellular units in experimental organic mercury poisoning. J. Neurochem., 397.

Zimmer, L. and Cater, D.E., 1979, Effects of complexing treatment administered with the onset of methylmercury neurotoxic signs. Toxicol. Appl. Pharmacol., 51: 29-38.

# EFFECT OF INORGANIC MERCURY ON THE IMMUNE SYSTEM

Philippe Druet

INSERM U 28
Hôpital Broussais
96, rue Didot
75014 Paris
France

## ABSTRACT

Mercuric chloride (HgCl$_2$) injected subcutaneously three times weekly at a dose of 100 μg/100g body weight to rats has different effects on the immune system depending upon the strain used. Brown-Norway (BN) rats develop an autoimmune disease characterized by a lymphoproliferation of B and CD4+ T cells and by an hyperimmunoglobulinemia affecting mainly IgE. Antibodies of various specificities are produced including autoantibodies (anti-laminin, anti-collagen IV, anti-nuclear) and antibodies to non-self antigens such as TNP or sheep red blood cells. As a consequence they develop an autoimmune glomerulonephritis, a Sjogren's syndrome and a mucositis. This disease is due to a T-dependent polyclonal activation of B cells probably mediated by the appearance of autoreactive anti-self class II T cells. The disease can be transferred to syngeneic rats by T cells from HgCl$_2$-injected rats. Indirect evidence for a role of interleukin 4 (IL-4) was recently obtained since an increase of Ia expression was observed on B cells. Interestingly this increase was observed as soon as 3 days following the first injection of HgCl$_2$ and represents therefore a very early marker of the effect of HgCl$_2$ on the immune system. The disease may also be induced giving HgCl$_2$ by different routes (orally, respiratory) or by using other inorganic compounds (HgCl$_2$, HgNH$_2$Cl). By contrast, HgCl$_2$ does not induce autoimmunity in Lewis (LEW) rats. It induces in that strain an increase in the number of CD8+ T cells responsible for a non-specific immunosuppression (inhibition of the proliferative response of T cells to mitogens and of the local graft-versus-host response). More interestingly, $HgCl_2$ attenuates or even abrogates organ-specific autoimmunity (experimental allergic encephalomyelitis or Heymann's nephritis) when LEW rats receive $HgCl_2$ three to four weeks before immunization with the autoantigen. That CD8+ T cells are responsible for the immunosuppression observed has been confirmed since treatment with an anti-CD8+ T cell monoclonal antibody (OX8) completely abrogates the protection. Finally, susceptibility to autoimmunity is genetically controlled and 4 genes localized within and outside the major histocompatibility complex are involved. In conclusion: inorganic mercury may induce either autoimmunity or immunosuppression depending upon genetic factors.

INTRODUCTION

Mercurials have been known to be associated with autoimmune manifestations in humans for a long time. Since mercurials are nowadays rarely used as pharmaceutical agents, the frequency of such manifestations has apparently decreased. It is still possible that small amounts of mercury in the environment or in the work place may play a role in apparently idiopathic immunologically-mediated disorders. It is interesting in that respect that, while 6 to 9% of cases of membranous glomerulopathy (MG), a classical form of autoimmune glomerulonephritis (AIGN) appear as a consequence of drug or toxic exposure, including Hg exposure, 76 to 79% are still of unknown origin (Adu and Cameron, 1989). Mechanisms of Hg-induced autoimmunity in humans are unknown.

During the past twenty years, several experimental models have been developed which allowed to better understand possible mechanisms at play. In addition, these models showed that susceptibility was genetically-controlled and that, depending upon the species or the strain tested, either autoimmunity or immunosuppression could be achieved.

In this review, we will first consider mercury-induced immunologically-mediated manifestations in humans, and then the available experimental models.

**Mercury-induced autoimmunity or hypersensitivity reactions in humans**

Autoimmune glomerulonephritis

Besides tubular lesions, which are the consequence of a direct toxic effect of mercury, various mercurials may induce an immunologically-mediated glomerulonephritis of the immune complex type (reviewed in Fillastre et al., 1988). Organo-mercurials were used in the past as diuretics and the nephrotic syndrome has been reported to be associated with such treatment. Inorganic mercurials may also induce similar glomerular lesions. Ammoniated mercury and mercury-containing creams were used for the treatment of psoriasis and as skin-lightening creams respectively. Although ammoniated mercury is no longer used nowadays, skin-lightening creams are still used in Africa and have been reported to be responsible for several cases of immunologically-mediated glomerulonephritis (Barr et al., 1972). Interestingly these creams are also able to induce quite similar glomerular lesions in the rabbit (Lindqvist et al., 1974). Mercury-containing laxatives are still in use in some countries and are also responsible for MG. Finally, similar glomerular lesions have been reported following occupational or environmental exposure. Taken together, about 100 cases of presumable or histologicaly proven AIGN in Hg-exposed individuals have been published suggesting that such an association is not fortuitous (reviewed in Fillastre et al., 1988).

Systemic angeitis

Kawasaki described in 1967 an acute febrile mucocutaneous syndrome in children that is due to a necrotizing vasculitis. Various organs (heart, kidney, joints) may also be affected. More recently immunological abnormalities including an increase in total serum IgE concentration (Kusakawa and Heiner, 1976), a polyclonal activation of B cells and a decrease in the number of suppressor/cytotoxic T cells (Leung et al., 1982) have been reported suggesting that this disease was indeed immunologically-mediated. Leung et al. (1986) showed that these patients do produce cytotoxic antibodies to mediator-induced endothelial cell antigens that could be responsible for the vascular damage observed during the acute phase of the disease. It is interesting to note that several of these abnormalities have also been observed in susceptible strains of rats injected with mercury (see below) and that mercury exposure has been reported in some children with Kawasaki's disease (Adler et al., 1982).

Other skin reactions following mercury exposure which have received various names (baboon syndrome, pink disease, acrodynia) have also been reported (Kazantzis, 1978; Nakayama et al., 1983; Andersen et al., 1984). These manifestations are probably also the consequence of a vasculitis and/or of contact dermatitis.

Oral manifestations

Patients may develop oral lichen planus, cutaneous lichen planus or lichenoid reactions adjacent to amalgam fillings. It has been suggested that these manifestations could be of immune origin and, interestingly, oral lichen planus is preferentially observed in HLA DR3 positive patients (Jontell et al., 1987). The role of amalgam fillings in inducing immunologically-mediated diseases must be considered in the future.

Immunological abnormalities associated with mercury exposure

The mechanisms of the above-mentioned disorders are still unknown. It is likely that oral manifestations, associated with T cell infiltration, are a consequence of the binding of Hg to self-constituents. However this does not explain easily the increase in total serum IgE, nor the angeitis associated with polyclonal activation of B cells. Several immunological abnormalities have been reported in humans. $HgCl_2$, combined to the B cell mitogen pokeweed, increases IgE production *in vitro* by B cells from non-atopic patients (Kimata et al., 1983). $HgCl_2$ has also been reported to act as a mitogen on human lymphocytes (Caron et al., 1970). These data suggest that Hg has a profound effect on the immune system. We will see now that experimental models that have been developed do help understand better the possible mechanisms of action of $HgCl_2$.

Effect of inorganic mercury on lymphocyte functions in experimental animals

Inorganic mercurials may have either a suppressive or a potentiating effect on lymphocyte functions depending upon the species or the strain chosen and on the amount of Hg used. Numerous assays (mitogenic effect of $HgCl_2$ on T cells, effect of $HgCl_2$ on the mixed lymphocyte reaction, on T cell mitogen responses, on polyclonal B cell activation induced by lipopolysaccharide, or on specific antibody production), (Gaworski and Sharma, 1978; Lawrence, 1981; Dieter et al., 1983; Nakatsuru et al., 1985; Reardon and Lucas, 1987) have been used. Most studies showed that $HgCl_2$

has a suppressive effect that was considered to be the consequence of direct or indirect alteration or inhibition of enzymes such as pyruvate kinase (Dieter et al., 1983). However Nordlind (1985) and Reardon and Lucas (1987) showed that $HgCl_2$ is a mitogen for T cells in guinea pigs and in Balb/c mice. It was suggested that $HgCl_2$ could act by modifying "self" cell-surface structures such as class I or class II molecules leading to a type of mixed-lymphocyte reaction. However, the ability of $HgCl_2$ to induce autoimmunity, hypersensitivity reactions or immunosuppression with pathological consequences (suppression of autoimmune disorders, or increased susceptibility to infections or to cancer development) has not been considered in these studies.

## Induction of hypersensitivity reactions and autoimmunity in experimental animals

### Guinea-pig

Polak et al. (1968) showed that the ability of guinea-pigs to develop contact dermatitis was genetically-controlled. Strain II was found to be resistant while 80% of animals from strain XIII were susceptible. Genes involved have not been studied further. The ability of $HgCl_2$ to induce autoimmunity in these species has not been evaluated either.

### Rat

Extensive studies have been performed in the Brown-Norway (BN) rat, which appeared to be highly susceptible to $HgCl_2$-induced autoimmunity. In this model, $HgCl_2$ is subcutaneously injected three times a week for at least two months at a dose of 100 µg/100 g body weight.

#### Description of the disease

Rats injected with $HgCl_2$ exhibit after five to six days, the first autoimmune abnormalities that reach a peak from day 15 to day 21 and then decline (Bellon et al., 1982). These abnormalities include a proliferation of CD4+ T cells and B cells in the spleen and lymph nodes (Hirsch et al., 1982; Aten et al., 1988a; Pelletier et al.,

1988a), a hyperimmunoglobulinemia which affects mainly IgE (Prouvost-Danon et al., 1981; Pelletier et al., 1988a). Numerous antibodies to self (DNA, laminin, collagen II and IV, thyroglobulin) and non self (TNP) antigens are produced (Hirsch et al. 1982; Bellon et al., 1982; Fukatsu et al., 1987; Aten et al., 1988b; Pusey et al., 1990). Antibodies to glomerular basement membrane (GBM) components, mainly laminin, are responsible for an AIGN. This glomerulonephritis is initially characterized by linear IgG deposits along the GBM and then by the presence of granular IgG deposits typical of a MG (Sapin et al., 1977; Druet et al., 1978). The first phase is the consequence of the binding of circulating autoantibodies to the GBM; the pathogenesis of the second phase is still unclear. Heavy proteinuria and the nephrotic syndrome are usually observed at the acme of the disease and 30 to 50% of the animals may die at that time. The remaining animals recover and then become resistant (Bowman et al., 1984). The mechanisms responsible for the induction and for the regulation phases will be considered later.

### Effect of other inorganic mercurials, of the route of administration and of the dose

Several inorganic mercurials have been tested for their ability to induce autoimmunity. Interestingly $HgCl_2$ given orally as well as ammoniated mercury ($HgNH_2Cl$) applied on the skin induced quite similar manifestations as $HgCl_2$. Mercurobutol applied on the skin or given in the form of a pessary in female BN rats also induced the same manifestations (Druet al al., 1981). Finally, $HgCl_2$ -induced autoimmunity was observed whatever the route of introduction (Bernaudin et al., 1981) of the chemical (intravenously, orally or respiratory).

More recently the effect of the dose was evaluated in rats receiving 20, 10, 5, 2.5 or 1.25 μg of $HgCl_2$ per 100 g body weight three times a week for a five month period. The effect of Hg was followed on serum IgE concentration which is a highly sensitive marker. Quite interestingly a clear dose effect relationship was observed and it was possible to define a non effect dose. Indeed serum IgE concentration did not increase in rats receiving 1.25 μg of $HgCl_2$ while it increased in the other groups with a highly significant correlation ($r = 0.989$) between serum IgE concentration at the peak and the amount of $HgCl_2$ injected (unpublished).

### Genetic control of susceptibility

The ability of several strains of rats to develop autoimmunity has been studied. As mentioned above, BN rats with the RT-1[n] haplotype at the major histocompatibility complex (MHC) (equivalent of the human HLA or mouse H2) were found to be highly susceptible. Two other strains that are derived from the BN strain namely MAXX rats (Henry et al., 1988b) and DZB rats (Aten et al., 1988b) were also found to be susceptible. In contrast, all the strains that we tested which bore the RT-11 haplotype (LEW, Fischer 344, AS and BS) were completely resistant whatever the dose of HgCl$_2$ injected (Druet et al, 1982). Rats bearing the u haplotype (LOU and WAG) were also resistant but Wistar Furth rats developed a typical MG when given high doses of HgCl$_2$ (0.4 mg/100 g body weight). Finally several other strains with the c (PVG/c, AUG), a (AVN, DA), d (BDV), b (BUF), k (OKA) or f (AS2) haplotype developed a mild form of autoimmunity (Weening et al., 1978; Druet et al., 1982).

Susceptibility of segregants between BN and LEW rats was then tested in order to characterize the genes responsible. We found that susceptibility to AIGN depended upon about 3 genes (Druet et al., 1977; Sapin et al., 1982) and that the increase in serum IgE concentration depended upon 4 genes (Sapin et al., 1984). Congenic rats with the LEW background and the BN MHC (LEW.1N) as well as those with the BN background and the LEW MHC (BN.1L) were resistant while F1 obtained by crossing both strains were susceptible. This shows clearly that both MHC linked genes and genes outside the MHC locus are required for the autoimmunity to occur. These findings are interesting since drug-induced autoimmunity is also genetically determined in humans (Wooley et al., 1980).

Mechanisms of induction

All the findings reported above suggested that Hg was acting by disturbing the regulatory mechanisms of the immune system and induced a polyclonal activation of B cells. This was confirmed using *in vitro* and *in vivo* experiments. It was shown that T cells were required for this polyclonal activation to occur (Hirsch et al., 1982; Pelletier et al., 1987a). Further experiments were then performed in order to elucidate how T cells could act on B cells. It was found that T cells from HgCl$_2$-injected BN rats were able to proliferate when cultured in the presence of normal B cells (Pelletier et al., 1985, 1986). This proliferation was blocked when the stimulator B cells were cultured in the presence of anti-class II monoclonal antibodies. These autoreactive anti-class II T cells were highly frequent (1/5000) at the acme of the disease (Rossert et al., 1988).

The role of T cells from HgCl$_2$-injected BN rats was then further confirmed since such cells were able to transfer autoimmunity into normal BN rats provided the recipients were treated with an anti-CD8 monoclonal antibody (Pelletier et al., 1988b). Other studies are required to determine the exact specificity of these T cells. Several non exclusive hypotheses can be proposed. 1) Hg could modify class II molecules in such a way that the "modified" class II become similar to all class II molecules. 2) It is also possible that following the interaction with Hg, the binding of self-peptides on class II molecules is modified and that T cells then recognize class II molecules presenting self-peptides. 3) Finally Hg may also interact with the T cell receptor which would then interact with class II molecules. This latter hypothesis is more likely since T cells from $HgCl_2$-injected rats recognize normal B cells and are able to transfer the disease in normal rats. Whatever the hypothesis, the resulting situation is expected to be quite similar to that in the chronic graft-versus-host (GVH) reaction (Gleichmann et al., 1984) in which autoimmunity is the consequence of alloreactive T cells that stimulate polyclonally B cells bearing allo class II molecules. Recent observations strengthen the analogy between the GVH model and the mercury model. In the chronic GVH reaction obtained by transferring T cells from normal BN rats into (LEW x BN)F1 hybrids, autoimmune manifestations were observed which are quite similar to those described in the mercury model (Tournade et al., 1990). In addition anti-laminin antibodies produced in both circumstances share a similar idiotype showing that the same B cell clones are activated (Guery et al., 1990a).

Finally, there is now good evidence that autoimmunity appearing in the context of allogeneic reactions such as the GVH reaction are, at least in mice, mediated by TH2 cells (Abramowitz et al., 1990). CD4+ T cells have been divided into two different subsets in mice (Mosmann and Coffman, 1989). TH2 cells produce IL4 and IL5 which provide help for B cells to synthethize IgE, IgG1 and IgA antibodies. Until now it is difficult to measure directly IL4 in the rat. However one of the effects of this interleukin is to enhance class II expression at the B cell surface. We could recently demonstrate that, in BN rats injected with $HgCl_2$, class II expression increased quite significantly at the B cell surface from day 3 (Dubey et al., 1991). It is therefore likely that TH2-like cells are activated very early and are responsible for the polyclonal activation of B cells in this model.

## Regulation of $HgCl_2$-induced autoimmunity

We mentioned above that mercury-induced autoimmunity was autoregulated in BN rats. Several experiments have been performed to elucidate the mechanisms responsible for this regulation. Bowman et al. (1987) showed that the number of CD8+ T cells increases at the beginning of the convalescence phase which was compatible with a role for suppressor/cytotoxic T cells. The same group (Bowman et al., 1984) showed that T cells from convalescent rats when transferred into BN rats were able to attenuate Hg-induced autoimmunity. They also showed that CD8+ cells were responsible for this effect. Weening et al. (1981) also found a defect of suppressor function in PVG/c rats injected with $HgCl_2$. However treatment of BN rats with an anti-CD8 monoclonal antibody did not modify the course of the disease although the rats were indeed depleted in CD8+ T cells (Pelletier et al., 1990).

Chalopin and Lockwood (1984) provided evidence that anti-idiotypic antibodies were also involved in the regulation process. In agreement with this, we could recently demonstrate that BN rats injected with $HgCl_2$ produce autoanti-idiotypic antibodies (Guery and Druet, 1990b). Their role remains to be clearly established.

Other mechanisms may also be involved. It is well known that TH1 cells in mice produce gamma interferon which antagonizes the effects of IL4. It is therefore possible that, following activation of TH2 cells, activation of TH1 cells occurs and attenuates the effects of TH2 cells. Finally, Mirtcheva et al. (1989) demonstrated in mice that expression of H-2E (IE) encoded molecules was associated with resistance or low responsiveness to $HgCl_2$. It was therefore suggested that H-2A (IA) would act as an immune response gene while H-2E would act as a suppressive locus. It is possible that these loci act successively which would explain the first phase of activation and the second suppressive phase.

To conclude, the mechanisms at play are still unclear but are likely to depend upon several factors.

## Mouse

Three groups have more recently started working in mice which obviously offers a great advantage to study more precisely the genetic control of susceptibility. They showed that $HgCl_2$ induces in H-$2^S$ (B10.S, A.SW) mice an autoimmune disease characterized by the production of anti-nuclear and anti-nucleolar autoantibodies (Robinson et al., 1984; Hultman and Enestrom, 1987; Stiller-Winkler et al., 1988). These mice produce high levels of IgE (Pietsch et al., 1989). They also develop an autoimmune glomerulonephritis which seems to be due, at least in part, to the deposition of anti-nucleolar antibodies (Hultman and Enestrom, 1988). The disease is also T cell-dependent (Stiller-Winkler et al., 1988).

Two very important points could be demonstrated in mice. First, as mentioned above, susceptibility to anti-nuclear antibody production was shown to depend on class II genes (Robinson et al., 1986) and more precisely on the H-2A$^S$ locus while resistance was associated with H-2E (Mirtcheva et al., 1989). Second, the role of IL4 has been more clearly demonstrated than in the rat since treatment with a monoclonal anti-IL4 antibody abrogates the production of IgG1 anti-nuclear antibodies and the increase of total serum IgE level (Ochel et al., 1991). This finding clearly demonstrates the role played by TH2 cells.

## Rabbit

Interestingly, Roman-Franco et al. (1978) reported that $HgCl_2$-induced in New Zealand rabbits an autoimmune glomerulonephritis quite similar to that described in BN rats. Rabbits also produced anti-nuclear antibodies.

### Induction of immunosuppression in the rat

We mentioned earlier that LEW rats are resistant to Hg-induced autoimmunity even when injected with very high doses of $HgCl_2$. These rats remain apparently healthy when treated as BN rats with 0.1 mg of $HgCl_2$. However, several abnormalities can be detected when these rats are carefully examined. In contrast to what was observed in BN rats, the number of CD8+ cells is significantly increased in LEW rats injected with $HgCl_2$. This increase was found to be responsible for an inhibition of T cell functions: the response to PHA and the mixed lymphocyte reaction were profoundly depressed. These functions were restored after removal of CD8+ cells suggesting that the function of CD4+ T cells was not altered but that these cells were down-modulated by the CD8+ cells (Pelletier et al., 1987b). These findings prompted us to test the effect of $HgCl_2$ on the development of autoimmune diseases to which LEW rats are highly susceptible. We could thus demonstrate that pretreatment of LEW rats with $HgCl_2$ prevented the occurrence of Heyman's nephritis, an antibody-mediated autoimmune glomerulonephritis (Pelletier et al., 1987c). Similar treatment was also able to prevent the appearance of experimental allergic encephalomyelitis, a T cell mediated autoimmune disease considered to be the counterpart of multiple sclerosis in humans (Pelletier et al., 1988c). As expected the proliferative response to myelin basic protein was abolished and could be restored following removal of CD8+ cells. More recently, it was shown that treatment of these rats with an anti-CD8 monoclonal antibody inhibited the protection provided by $HgCl_2$ (Pelletier et al., 1991).

There are therefore a number of data showing that $HgCl_2$ induces CD8+ T cells which have a suppressor function in LEW rats. It will be interesting to test whether these rats are more prone to infections or to tumor development. The reason why these T cells are induced is presently unknown. It is possible that recognition of class II molecules is also involved, like in BN rats. It is tempting to speculate that the equivalent of IE region could play a pivotal role in that respect. It must be mentioned to that point that the Hg-induced CD8+ T cells in LEW rats are not responsible for the resistance of LEW rats to Hg-induced autoimmunity because treatment with the anti-CD8 monoclonal antibody did not allow expression of autoimmunity in that strain (Pelletier et al., 1991).

CONCLUSION

HgCl$_2$ given at non toxic doses affects differently the immune system depending upon the strain tested. Either autoimmunity or immunosuppression are observed. There is evidence that these effects are mediated by T cells reacting with class II molecules, at least in the former situation. The experimental models developed in rats and mice will certainly allow to understand better the mechanisms of autoimmunity and the role of chemicals in inducing either autoimmunity or immunosuppression.

REFERENCES

Abramowicz, D., Doutrelepont, J.M., Lambert, P., Van der Vorst, P., Bruyns, C., and Goldman, M., 1990, Increased expression of Ia antigens on B cells after neonatal induction of lymphoid chimerism in mice: role of interleukin 4, Eur. J. Immunol., 20:469-475.

Adler, R., Boxstein, D., Schaff, P., and Kelly, D., 1982, Metallic mercury vapor poisoning simulating mucocutaneous lymph node syndrome, J. Pediatrics, 101:967-968.

Adu, D., and Cameron, J.S., 1989, Aetiology of membranous nephropathy, Nephrol. Dial.Transplant, 4:757-758.

Andersen, K.E., Hjorth, N., and Menne, T., 1984, The baboon syndrome: systemically-induced allergic contact dermatitis, Contact Derm., 10:97-100.

Aten, J., Bosman, C.B., Rozing, J., Stijnen, T., Hoedemaeker, P.J., and Weening, J.J., 1988a, Mercuric chloride-induced autoimmunity in the Brown Norway rat, Am. J. Pathol., 133:127-138.

Aten, J., Bruijn, J.A., Veninga, A., de Heer, E., and Weening, J.A., 1988b, Anti-laminin autoantibodies and mercury-induced membranous glomerulopathy, Kidney Int., 33:309 (Abstract).

Barr, R.D., Rees, P.H., Cordy, P.E., Jungu, A., Woodger, B.A., and Cameron, H.M., 1972, Nephrotic syndrome in adult Africans in Nairobi, Brit. Med. J., 2:131-134.

Bellon, B., Capron, M., Druet, E., Verroust, P., Vial, M.C., Sapin, C., Girard, J.F., Foidart, J.M., Mahieu, P., and Druet, P., 1982, Mercuric chloride induced autoimmune disease in Brown-Norway rats: sequential search for anti-basement membrane antibodies and circulating immune complexes, Eur. J. Clin. Invest., 12:127-133.

Bernaudin, J.F., Druet, E., Druet, P., and Masse, R., 1981, Inhalation or ingestion of organic or inorganic mercurials produces auto-immune disease in rats, Clin. Immunol.Immunopathol., 20:129-135.

Bowman, C., Mason, D.W., Pusey, C.D., and Lockwood, C.M., 1984, Autoregulation of autoantibody synthesis in mercuric chloride nephritis in the Brown-Norway rat. I. A role for T suppressor cells, Eur. J. Immunol., 14:464-470.

Bowman, C., Green, C., Borysiewicz, L., and Lockwood, C.M., 1987, Circulating T-cell populations during mercuric chloride-induced nephritis in the Brown Norway rat, Immunology, 61:515-520.

Caron, G.A., Poutala, S., and Provost, T.T., 1970, Lymphocyte transformation induced by inorganic and organic mercury, Int. Arch. Allergy, 37:76-87.

Chalopin, J.M., and Lockwood, C.M., 1984, Autoregulation of autoantibody synthesis in mercuric chloride nephritis in the Brown Norway rat. II. Presence of antigen-augmentable plaque-forming cells in the association with humoral factors behaving as auto-anti-idiotypic antibodies, Eur. J. Immunol., 14:470-475.

Dieter, M.P., Luster, M.I., Boorman, G.A., Jameson, C.W., Dean, J.H., and Cox, J.W., 1983, Immunological and biochemical responses in mice treated with mercuric chloride, Toxicol.Appl. Pharmacol., 68:218-228.

Druet, E., Sapin, C., Gunther, E., Feingold, N., and Druet, P., 1977, Mercuric chloride induced anti-glomerular basement membrane antibodies in the rat, Genetic control, Eur. J. Immunol., 7:348-351.

Druet, P., Druet, E., Potdevin, F., and Sapin, C., 1978, Immune type glomerulonephritis induced by $HgCl_2$ in the Brown Norway rat, Ann. Immunol., Paris, 129C:777-792.

Druet P., Teychenne, P., Mandet, C., Bascou, C., and Druet, E., 1981, Immune-type glomerulonephritis induced in the Brown-Norway rat with mercury-containing pharmaceutical products, Nephron, 28:145-148.

Druet, E., Sapin, C., Fournie, G., Mandet, C., Gunther, E., and Druet, P., 1982, Genetic control of susceptibility to mercury-induced immune nephritis in various strains of rat, Clin. Immunol. Immunopathol., 25:203-212.

Dubey, C., Bellon, B., Hirsch, F., Kuhn, J., Vial, M.C., Goldman, M. an Druet, P., 1991. Increased expression of class II major histocompatibility complex molecules on B cells in rats susceptible or resistant to $HgCl_2$-induced autoimmunity. Clin. Exp. Immunol., in press.

Fillastre, J.P., Druet, P., and Mery, J.P., 1988, Proteinuric nephropathies associated with drugs and substances of abuse, in "The nephrotic syndrome", J.S. Cameron, and R.J. Glassock, eds., 697-744, Marcel Dekker, New York and Basel.

Fukatsu, A., Brentjens, J.R., Killen, P.D., Kleinman, H.K., Martin, G.R., and Andres, G.A.,1987, Studies on the formation of glomerular immune deposits in Brown Norway rats injected with mercuric chloride, Clin. Immunol. Immunopathol., 45:35-47.

Gaworski, C.L., and Sharma, R.P., 1978, The effects of heavy metals on [$^3$H] thymidine uptake in lymphocytes, Toxicol. Appl. Pharmacol., 46:305-313.

Gleichman, E., Pals, S.T., Rolink, A.G., Radaszkiewicz, T., and Gleichmann, H., 1984, Graft-versus-host reactions. Clues to the etiopathology of a spectrum of immunological diseases, Immunol. Today, 5:324-332.

Guery, J.C., Tournade, H., Pelletier, L., Druet, E., and Druet, P., 1990a, Rat anti-glomerular basement membrane antibodies in toxin-induced autoimmunity and in chronic graft-vs-host reaction share recurrent diotypes, Eur. J. Immunol., 20:101-105.

Guery, J.C., and Druet, P., 1990b, A spontaneous hybridoma producing autoanti-idiotypic antibodies that recognize a $V_k$ associated idiotope in mercury-induced autoimmunity, Eur. J. Immunol., 20:1027-1031.

Henry, G.A., Jarnot, B.M., Steinhoff, M.M., and Bigazzi, P.E., 1988, Mercury-induced renal autoimmunity in the MAXX rat, Clin. Immunol. Immunopathol., 49:187-203.

Hirsch, F., Couderc, J., Sapin, C., Fournie, G., and Druet, P., 1982, Polyclonal effect of $HgCl_2$ in the rat, its possible role in an experimental autoimmune disease, Eur. J.Immunol., 12:620-625.

Hultman, P., and Enestrom. S., 1987, The induction of immune complex deposits in mice by peroral and parenteral administration of mercuric chloride: strain dependent susceptibility, Clin. Exp. Immunol., 67:283-298.

Hultman, P. and Enestrom, S., 1988, Mercury induced antinuclear antibodies in mice: characterization and correlation with renal immune complex deposits, Clin. Exp. Immunol., 71:269-274.

Jontell, M., Stahlblad, P.A., Rosdahl, I., and Lindblom, B., 1987, HLS-DR3 antigens in erosive oral lichen planus, cutaneous lichen planus, and lichenoid reactions, Acta Odontol. Scand., 45:309-344.

Kazantzis, G., 1978, The role of hypersensitivity and the immune response in influencing susceptibility to metal toxicity, Environ. Health Persp., 25: 111-118.

Kawasaki, T., 1967, Acute febrile mucocutaneous syndrome with specific desquamation of the fingers and toes in children: clinical observations of 50 cases, Jap. J. Allergol., 16:178-185.

Kimata, H., Shinomiya, K., and Mikawa, H., 1983, Selective enhancement of human IgE production *in vitro* by synergy of pokeweed mitogen and mercuric chloride, Clin. Exp. Immunol., 53:183-191.

Kusakawa, S., and Heiner, D.C., 1976, Elevated levels of immunoglobulin E in the acute febrile mucocutaneous lymph node syndrome, Pediat. Res., 10:108-111.

Lawrence, D.A., 1981, Heavy metal modulation of lymphocyte activities. I. *In vitro* effects of heavy metals on primary humoral immune responses, Toxicol. Appl. Pharmacol., 57:439-451.

Leung, D.Y.M., Siegle, R.L., Grady, S., Krensky, A., Meade, R., Reinherz, E.L., and Geha, R.S., 1982, Immunoregulatory abnormalities in mucocutaneous lymph node syndrome, Clin. Immunol. Immunopathol., 23:100-112.

Leung, D.Y.M., Geha, R.S., Newburger, J.W., Burns, J.C., Fiers, W., Lapierre, L.A., and Pober, J.S., 1986, Two monokines, interleukin 1 and thmor necrosis factor, render cultured vascular endothelial cells susceptible to lysis by antibodies circulating during Kawasaki syndrome, J. Exp. Med., 164:1958-1972.

Lindqvist, K.J., Makene, W.J., Shaba, J.K., and Nantulya, V., 1974, Immunofluorescence and electron microscopic studies of kidney biopsies from patients with nephrotic syndrome, possibly induced by skin lightening creams containing mercury, East Afr. Med. J., 51:168-169.

Mirtcheva, J., Pfeiffer, C., Hallmann, B., and Gleichmann, E., 1989, Immunological alterations inducible by mercury compounds. III. H-2A acts as an immune response and H-2E as an immune "suppression" locus for $HgCl_2$-induced antinucleolar autoantibodies, Eur. J. Immunol., 19:2257-2261.

Mosmann, T.R., and Coffman, R.L., 1989, Heterogeneity of cytokine secretion patterns and functions of helper T cells, Adv. Immunol., 46:111-147.

Nakatsuru, S., Ooshashi, J., Nozaki, H., Nakada, S., and Imura, N., 1985, Effect of mercurials on lymphocyte functions in vitro, Toxicology, 36:297-305.

Nakayama, H., Niki, F., Shono, M., and Hada, S., 1983, Mercury exanthem. Contact Derm., 9:411-417.

Nordlind, K., 1985, Binding and uptake of mercuric chloride in human lymphoid cells, Int. Arch. Allergy, Appl. Immunol., 77:405-408.

Ochel, M., Nohr, H.W., Pfeiffer, C. and Gleichmann, E., 1991, IL-4 is required for the IgE and IgFl increase and IgFl autoantibody formation in mice treated with mercuric chloride. J. Immuno. 143:3006-3011.

Pelletier, L., Pasquier, R., Hirsch, F., Sapin, C., and Druet, P., 1985, *In vivo* self-reactivity of mononuclear cells to T cells and macrophages exposed to $HgCl_2$, Eur. J. Immunol., 15:460-465.

Pelletier, L., Pasquier, R., Hirsch, F., Sapin, C., and Druet, P., 1986, Autoreactive T cells in mercury-induced autoimmune disease: *in vitro* demonstration, J. Immunol., 137:2548-2554.

Pelletier, L., Pasquier, R., Vial, M.C., Mandet, C., Moutier, R., Salomon, J.C., and Druet, P., 1987a, Mercury-induced autoimmune glomerulonephritis: requirement for T-cells, Nephrol. Dial. Transplant, 1:211-218.

Pelletier, L., Pasquier, R., Rossert, J., and Druet, P., 1987b, $HgCl_2$ induces nonspecific immunosuppression in Lewis rat, Eur. J. Immunol., 17:49-54.

Pelletier, L., Galceran, M., Pasquier, R., Ronco, P., Verroust, P., Bariety, J., and Druet, P., 1987c, Down modulation of Heymann's nephritis by mercuric chloride, Kidney Int., 32:227-232.

Pelletier, L., Pasquier, R., Guettier, C., Vial, M.C., Mandet, C., Nochy, D., Bazin, H., and Druet, P., 1988a, $HgCl_2$ induces T and B cells to proliferate and differentiate in BN rats, Clin. Exp. Immunol., 71:336-342.

Pelletier, L., Pasquier, R., Rossert, J., Vial, M.C., Mandet, C., and Druet, P., 1988b, Autoreactive T cells in mercury-induced autoimmunity. Ability to induce the autoimmune disease, J. Immunol., 140:750-754.

Pelletier, L., Rossert, J., Pasquier, R., Villarroya, H., Belair, M.F., Vial, M.C., Oriol, R., and Druet, P., 1988c, Effect of $HgCl_2$ on experimental allergic encephalomyelitis in Lewis rat. $HgCl_2$-induced down-modulation of the disease, Eur. J. Immunol., 18:243-247.

Pelletier, L., Rossert, J., Pasquier, R., Vial, M.C., and Druet, P., 1990, Role of CD8+ T cells in mercury-induced autoimmunity or immunosuppression in the rat, Scand. J. Immunol., 31:65-74.

Pelletier, L., Rossert, J., Pasquier, R., Villarroya, H., Oriol, R. and Druet, P., 1991. HgCl2-induced perturbation of the T cell network in experimental allergic encephalomyelitis. Cell. Immunol., in press.

Pietsch, P., Vohr, H.-W., Degitz, K., and Gleichmann, E., 1989, Immunological alterations inducible by mercury compounds. II.$HgCl_2$ and gold sodium thiomalate enhance serum IgG concentrations in susceptible mouse strains, Int. Arch. Allergy Appl. Immunol., 90:47-53.

Polak, L., Barnes, J.M., and Turk, J.L., 1968, The genetic control of contact sensitization to inorganic metal compounds in guinea pigs, Immunology, 14:707-711.

Prouvost-Danon, A., Abadie, A., Sapin, C., Bazin, H., and Druet, P., 1981, Induction of IgE synthesis and potentiation of anti-ovalbumin IgE antibody response by HgCl$_2$ in the rat, J. Immunol., 126:699-702.

Pusey, C.D., Bowman, C., Morgan, A., Weetman, A.P., Hartley, B., and Lockwood, C.M., 1990, Kinetics and pathogeneticy of autoantibodies induced by mercuric chloride in the Brown Norway rat, Clin. Exp. Immunol., 81:76-82.

Reardon, C.L., and Lucas, D.O., 1987, Heavy-metal mitogenesis: Zn$^{++}$ and Hg$^{++}$ induce cellular cytotoxicity and interferon production in murine T lymphocytes, Immunobiol., 175:455-469.

Robinson, C.J.G., Abraham, A.A., and Balazs, T., 1984, Induction of antinuclear antibodies by mercuric chloride in mice, Clin. Exp. Immunol., 58:300-306.

Robinson, C.J.G., Balazs, T., and Egorov, I.K., 1986, Mercuric chloride, gold sodium thiomalate and D-penicillamine induced antinuclear antibodies in mice, Toxicol. Appl.Pharmacol., 86:159-169.

Roman-Franco, A., Turiello, M., Albini, B., Ossi, E., Milgrom, F., and Andres, G.A., 1978, Anti-basement membrane antibodies and antigen-antibody complexes in rabbits injected with mercuric chloride, Clin. Immunol. Immunopathol., 9:464-481.

Rossert, J., Pelletier, L., Pasquier, R., and Druet, P., 1988, Autoreactive T cells in mercury-induced autoimmunity. Demonstration by limiting dilution analysis, Eur. J. Immunol., 18:1761-1766.

Sapin, C., Druet, E., and Druet, P., 1977, Induction of anti-glomerular basement membrane antibodies in the Brown-Norway rat by mercuric chloride, Clin. Exp. Immunol., 28:173-178.

Sapin, C., Mandet, C., Druet, E., Gunther, E., and Druet, P., 1982, Immune complex type disease induced by HgCl$_2$ in Brown-Norway rats: genetic control of susceptibility, Clin Exp. Immunol., 48:700-704.

Sapin, C., Hirsch, F., Delaporte, J.P., Bazin, H., and Druet, P., 1984, Polyclonal IgE increase after HgCl$_2$ injections in BN and LEW rats: a genetic analysis, Immunogenetics, 20:227-236.

Stiller-Winkler, R., Radaszkiewicz, T., and Gleichmann, E., 1988, Immunopathological signs in lymph node assay of responder and nonresponder strains, Eur. J. Immunopharmacol., 10:475-484.

Tournade, H., Pelletier, L., Pasquier, R. Vial, M.C., Mandet, C., and Druet, P., 1990, Graft-versus-host reactions in the rat mimic toxin-induced autoimmunity, Clin. Exp. Immunol. 81:334-338.

Weening, J.J., Fleuren, G.J., and Hoedemaeker, Ph.J., 1978, Demonstration of antinuclear antibodies in mercuric chloride-induced glomerulopathy in the rat, Lab. Invest., 39:405-411.

Weening, J.J., Hoedemaeker, Ph.J., and Bakker, W.W., 1981, Immunoregulation and antinuclear antibodies in mercury-induced glomerulopathy in the rat, Clin. Exp. Immunol., 45:64-71.

Wooley, P.H., Griffin, J., Panayi, G.S., Batchelor, J.R., Welsh, K.I., and Gibson, T.J., 1980, HLA-DR antigens and toxic reactions to sodium aurothiomalate and D-penicillamine in patients with rheumatoid arthritis, N. Engl. J. Med., 303:300-302.

# EXPOSURE TO MERCURY IN THE POPULATION

Staffan Skerfving

Department of Occupational
  and Environmental Medicine
University Hospital
S-221 85 Lund
Sweden

## ABSTRACT

In the population, there is uptake of mercury (Hg) as inorganic (mainly elemental Hg vapor, Hg°),  and organic (mainly methyl-Hg, MeHg)Hg.  We have monitored biologically the uptake of inorganic Hg through measurements of the Hg level in urine (U-Hg) and of MeHg through the Hg level in blood (B-Hg).

Hg exposure may occur from occupational and non-occupational sources.  The most prevalent occupational exposure is to Hg° among dental personnel.  However, with use of adequate methods, the exposure is low.  Thus, in a recent study of 244 dentists and dental nurses, we found that the average U-Hg was only 3.3 µg/g creatinine (range up to 23), which was only slightly higher than the level found in a reference population, 2.0 µg/g crea (up to 10).  Considerably higher occupational exposure to Hg° occurs in workers in chloralkali plants and fluorescent-tube factories.  Such workers had U-Hg levels up to 78 µg/g crea.

In the general population, without occupational Hg exposure, the main source of exposure to Hg° is usually dental amalgam.  In a recent study of 81 subjects, we found that there was a pronounced, and exponential, increase of U-Hg with a rising number of amalgam fillings.  In subjects with no fillings, the level corresponded to less than 0.5 µg/g crea, in those with many fillings to 5 µg/g crea.  Also, removal of all fillings resulted in a dramatic decrease of U-Hg; after one year, the level was only about 25% of that before removal.

In Sweden, the main source of exposure of MeHg is <u>fish</u>. All fish contain MeHg. However, the levels are particularly high in fish from lakes, rivers, and coastal waters contaminated with Hg. We have recently studied the association between fish intake and Hg exposure in a population of 396 subjects. In subjects who never had fish, the average B-Hg was 1.8 ng/g, in those who had at least two fish meals per week 6.7 ng/g. The importance of fish was also clearly demonstrated by a close association between levels of marine n-3 polyunsaturated fatty acids in serum and B-Hgs. The situation will deteriorate, as acid rain increases the fish MeHg level.

## INTRODUCTION

In this paper, a summary will be given of the exposure to mercury in the population. In particular, studies performed in Sweden during the last years will be reviewed.

Man is exposed to mercury through ambient air, water, drugs, amalgam fillings, and foods, and in the occupational setting (Clarkson et al., 1988a; WHO, 1990 and 1991).

The exposure through ambient air and water is generally low. Exposure through drugs may sometimes - and in some areas of the world - be considerable (WHO, 1991). However, this problem will not be dealt with here. Instead, the focus will be upon exposure through foods and amalgam fillings, as well as on some aspects of occupational exposure.

The exposure through foods occurs mainly from fish, and is then in the form of methylmercury. The exposure from amalgam fillings is to elemental mercury, which is also the main exposure in the occupational setting.

### Biological Indices of Exposure

<u>Mercury levels in blood and urine</u>

We have assessed the exposure to mercury mainly through biological monitoring. In the blood, methylmercury accumulates mainly in the red cells (Birke et al., 1972; Skerfving, 1974; Schütz and Skerfving 1975; Clarkson et al., 1988a). Thus we have used the mercury levels in red cells as an index of exposure to methylmercury. Exposure to elemental mercury vapour causes an increase of mercury levels in plasma and urine (Einarsson et al., 1974; Roels et al., 1987; Clarkson et al., 1988a). Thus, we have used the concentrations in these media as indices of exposure.

Total mercury levels were determined by atomic absorption spectrometry, the most important analytical method (Schütz et al., in press), or - in some cases - neutron activation analysis. Methylmercury was analyzed by gas chromatography.

Mobilization of Mercury

At exposure to elemental mercury vapour, the kinetics of mercury in blood and urine is fairly rapid, while it seems that mercury levels in some organs may change far more slowly (Clarkson et al., 1988a). Thus, the mercury level in blood and urine may not reflect the accumulation in sensitive organs, such as the brain and kidney.

For metals other than mercury, chelating agents are used to mobilize metal from organs into the urine, to give an estimate of the body burden. For example, in the case of lead, EDTA and penicillamine are widely used, for iron and aluminum, deferrioxamine.

Several compounds have the ability to mobilize mercury into the urine. We have tested 2,3-dimercaptopropane-1-sulfonate (DMPS), to find out whether it would add information on the mercury accumulation status (Molin et al., in press a). An oral dose of 0.3 g DMPS was given to subjects with varying exposure to mercury: workers and dentists exposed to elemental mercury vapour and subjects with varying dental amalgam status. We collected urine and blood before and 24h after the dosing.

DMPS caused pronounced but brief increase of the urinary mercury level (Table 1). The urinary increase was not associated with the estimated time-integrated mercury exposure, but was simply proportional to the urinary mercury level before the DMPS dosing.

Thus, the mobilization test reflects the recent exposure, but did not add information on the body burden, possibly because a main part of this is present as a selenium complex (Kosta et al., 1975; Hargreaves et al., 1988; Nylander and Weiner 1989; Nylander, 1990), the mercury of which is not readily available for the agent. Further, the DMPS-mobilized amount of mercury corresponded to the etimated uptake during only a few days.

METHYLMERCURY EXPOSURE VIA FISH

In Sweden, mainly due to industrial contamination, fish from lakes, rivers, and coastal waters often have high methylmercury levels. In most of the country, the levels are above 0.5 mg/kg, in some areas above 5 mg/kg.

The Swedish population have a high intake of fish (Johsson et al., 1972). Most of this is caught in the ocean, where fish mercury levels generally are much lower than in fish from lakes. However, sport fishing is very popular, and therefore a major part of the catch is from lakes, rivers, and coastal waters.

A few years ago, we sampled blood from a sample of subjects living in a mercury-contaminated area in the south of Sweden (Svensson et al., 1987). There was a clearcut association between blood-mercury levels and fish intake (Figure 1). In each consumption category, subjects who only had fish bought in shops had lower levels than those who also caught fish themselves. The reason is, of course, that the fish caught by the subjects themselves (or family members, etc.) often originated from mercury-contaminated water areas.

The dominant role of fish as a determinant of the blood-mercury level is even clearer shown by the association between levels of the marine polyunsaturated n-3 fatty acids in blood serum (as determined by gas chromatography; Svensson et al., 1991) and the mercury concentrations in red cells, seen in a samples of subject with no moderate, or high intake of fish (Figure 2); Svensson et al., submitted for publication). In subjects with high fish intake, more than 90% of the mercury in blood cells is present as methylmercury (Brunmark et al., submitted for publication).

The highest blood-mercury level encountered in these studies (27 µg/l in whole blood) was moderate, much lower than was found in Swedes two decades ago, when ten times higher concentrations were encountered, let go in exceptional subjects (Birke et al., 1972; Skerfving, 1974; Skerfving et al., 1974). The reduction is mainly the result of blacklisting a great number of water areas, and warning from the health authorities to the public concerning consumption of contaminated fish.

However, there are still population strata at risk. Some subjects will not follow advice. Also, some groups, e.g. fishermen, have a particularly high fish intake. In such groups, the risk for fetuses have to be considered, as methylmercury passes the placental barrier, and accumulates in the fetus, where it may cause central nervous damage (WHO, 1990).

Thus, in fetuses of fishermen's wives, the exposure may be substantial (Skerfving, 1988). Also, because of their monotonous diet, and the fact that the mercury level in breast milk reflects the level in blood plasma of the woman (Figure 3; Skerfving, 1988), breast-fed infants may suffer considerable mercury intake. However, a remarkably small fraction (in average 16%) of the mercury in breast milk was methylmercury, and thus the main part of the mercury probably was present as inorganic mercury, which probably reflects the situation in the mother's blood plasma.

In the most heavily exposed fetuses and infants in the fishermen population, the exposure is comparable to that possibly associated with slight retardation of the central nervous function development (Kjellström et al., 1989), but much lower than that

Table 1. Mobilization of mercury into urine (means) by 0.3 g sodium 2,3-dimercaptopropane-1-sulfonate (DMPS) per os (Molin et al., in press).

| | Urinary mercury level (µg/g creatinine) | | | |
| | | | Referents | |
| | Workers (N=10) | Dentists (N=8) | Amalgam (N=18) | Without (N=5) |
|---|---|---|---|---|
| Before DMPS | 26 | 3.0 | 1.4 | 0.5 |
| 0-6 h after DMPS | 780 | 38 | 23 | 4.9 |
| Increase (times) | 30 | 13 | 16 | 10 |

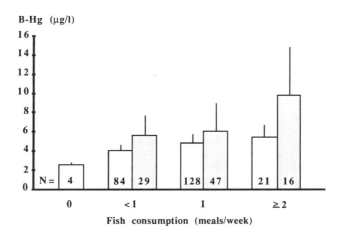

Fig. 1. Fish intake and levels of mercury in blood (B-Hg; means and upper ranges) from a sample of 329 subjects from the south of Sweden. Open bars denote subjects who only had fish bought in shops, closed bars those who also had fish caught by themselves (or family members etc.). Data from Svensson et al. (1987).

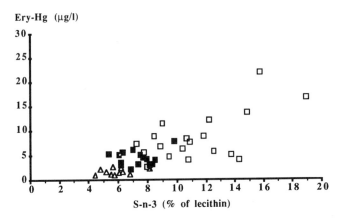

Fig. 2  Serum levels of marine polyunsaturated n-3 fatty acids (S-n-3) and mercury
concentrations in red cells (Ery-Hg) in samples of Swedes with no (N=11;
triangles), moderate (N=15; closed squares), or high (N=19; open squares) intake of
fish.  Data from Svensson et al., submitted for publication.

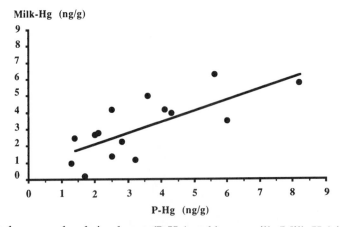

Fig. 3.  Total mercury levels in plasma (P-Hg) and breast milk (Milk-Hg) in 12 Swedish
fishermen's wives and three referent women.  A regression line (Y=0.67X+0.83;
r=0.74; p<0.01) is shown.  Data from Skerfving (1988).

associated with more advanced brain damage as in the outbreaks of prenatal poisoning in Minamata and Iraq (WHO, 1990).

The methylmercury situation in Swedish fish was, for a period considered to be under control, after substantial reductions of mercury emissions around 1970. However, recently, theoretical models of environmental mercury turnover have given a more pessimistic view. Mercury emissions must be reduced substantially if the levels in fish are to go down.

A particular problem is the fact that a low pH in water, through unknown routes, is associated with high methylmercury levels in the fish. As Sweden is a country severely affected by acid precipitation, and as large parts of the country has low soil buffer capacity, the acid rain is a major concern for the future.

To study whether the population is affected at present by acid precipitation, we sampled blood from subjects in two acid areas and one non-acid area, all of them in the south of Sweden (Svensson et al., 1987). We did not observe any significant difference between the groups. The median blood-mercury level for 195 subjects in the acid areas was 4.5 µg/l, and for 91 in the non-acid area 4.9 µg/l. However, the conclusion was obscured by the fact that a large fraction of the fish consumed in both areas was from the ocean, and because sport-fishing activities sometimes took place outside the individual's area of residence.

## EXPOSURE TO MERCURY THROUGH AMALGAM

It has been known for a long time that there is mercury exposure from dental amalgam (Stock, 1939). An exposure to elemental mercury vapour occurs when the dentist installs amalgam fillings. In connection with the operation, there is a rapid but brief rise of mercury levels in plasma (Molin et al., 1990). In urine, the mercury level also increases, but the peak is delayed a few days (Figure 4), probably because the absorbed mercury, before reflected in the excretion, has accumulated a temporarily somewhere in the body, most likely in the kidneys. When all amalgam fillings were removed in a subject, there was a gradual decrease of the urinary mercury levels during half a year, whereafter a new steady state was attained, at about a fourth of the original concentration. A corresponding increase has been observed after placement of amalgam (Molin et al., in press b).

This clearly demonstrates that there is a continuous release of mercury, probably as elemental mercury vapor (Olsson and Bergman, 1987; Clarkson et al., 1988b), from the fillings, between the amalgam installments. Thus, chewing of gum increases the mercury

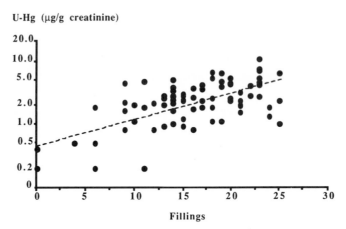

Fig. 4. Urinary-mercury levels (U-Hg; means) before and after removal of dental amalgam fillings (at time 0) and installment of gold inlays (mean 11 per person) in 10 subjects. Data from Molin et al. (1990).

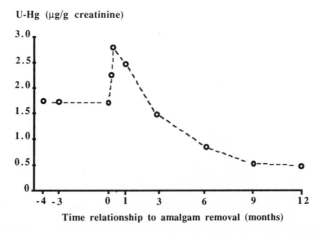

Fig. 5. Amalgam fillings and urinary mercury level (U-Hg; logarithmic) in a sample of 80 subjects from the south of Sweden. A regression line is shown (Log Y=0.042X-0.35; r-0.66; p<0.01). Data from Akesson et al. (1991).

release. However, it seems that foods may modify the release; chewing of hard-boiled egg decreased the oral mercury levels (Aronsson et al., 1989). Other foods have not yet been studied.

We studied the relationship between the number of amalgam fillings and mercury levels in urine and blood in a sample of the population in the south of Sweden (Figure 5; Akesson et al., 1991). In subjects without amalgam, the urinary levels were low (below 0.5 μg/g creatinine). With increasing number of amalgam fillings, there was a rise of the mercury level in urine. Interestingly, the increase was non-linear, the reason for this is not known, one possible explanation is the non-linear metabolism of inorganic mercury (cf. Nordberg and Skerfving, 1972). Gold fillings did not affect the relationship, but other metals together with amalgam caused a significant increase of urinary mercury. The mercury levels observed fit with the calculations made by Clarkson et al. (1988b) on the absorption of mercury released from amalgam fillings. Also, they are in agreement with other reports on associations between amalgam fillings and mercury levels in blood plasma and urine (Olstad et al., 1987; Langworth et al., 1988; Molin et al., 1989, 1990, and in press a and in press b), as well as in brain and kidney (Egglestone and Nylander, 1987; Nylander et al., 1987; Schiele, 1988; Nylander, 1990).

The levels caused by the uptake from amalgam are low, as compared to those associated with effects on the human brain and kidney (WHO, 1991). In a population study of women, there was no increase of symptoms with rising number of amalgam fillings (Ahlqwist et al., 1988). However, as to the kidney, animal experiments suggest the possibility of a substantial inter-individual variation in sensitivity (Hultman and Eneström, 1987 and 1988; Druet, in press). Such a variation is clear as to the skin; mercury exposure in connection with amalgam work is sufficient to induce reactions in rare subjects, with a type-IV sensitivity to metallic mercury.

## OCCUPATIONAL EXPOSURE TO ELEMENTAL MERCURY VAPOUR

Nowadays, occupational exposure to mercury is mainly to elemental mercury vapour. The largest exposed group of occupationally exposed subjects is dental personnel. However, at least in Sweden, the exposure is low. We studied 244 dentists and dental nurses (Akesson et al., in press). They had a urinary mercury level, which was only marginally higher than in the occupationally non-exposed referents. The average increase in the dental personnel corresponded to 19 amalgam surfaces in their own mouths, which fits well with other studies of mercury in biological media of dental personnel (Nilsson and Nilsson, 1986b; Molin et al., 1989; Skare et al., 1990), and with estimates of mercury exposure from amalgam (Clarkson et al., 1988b), and with the levels of mercury in air measured in dental surgeries (Nilsson and Nilsson, 1986a).

Among other occupationally mercury-exposed populations, workers in the chloralkali industry is an important group. They are, however, relatively few, only a couple of hundred, even in a country like Sweden, which is heavily dependent upon chlorine for the paper pulp-industry. Also, their exposure has decreased during the last decades. The highest urinary-mercury level recorded in several recent studies was 78 µg/g creatinine (Barregärd et al., ; 1988; Erfurth et al., 1990).

Other occupations may occasionally cause heavy exposure. Recently, two cases of classical mercury poisoning (with central nervous system symptoms and signs) were recorded among workers distilling mercury from amalgam residues (Skerfving et al., to be published). Their urinary-mercury levels amounted to 160 and 340 µg/g creatine, respectively.


EFFECTS OF EXPOSURE TO ELEMENTAL MERCURY VAPOUR

A few comments will be given here on recent studies on effects of exposure to elemental mercury vapour in man. As mentioned above, elemental mercury may induce skin allergy. Besides that, of main interest are the effects on the central nervous system, the kidney, and the fetus. Not much is known about possible prenatal effects (WHO, 1991).

## Central nervous system

Classical poisoning by elemental mercury vapour mainly involves symptoms and signs from the central nervous system (CNS: WHO, in press).

Slight effects on the CNS may occur in groups of workers with average urinary mercury levels of about 50 µg/g creatinine (Piikivi et al., 1984; Roels et al., 1985), and possible even lower (Piikivi and Hänninen, 1989; Piikivi and Tolonen, 1989).

## Kidneys

Inorganic mercury may induce severe kidney damage with a nephrotic syndrome. In this context should be mentioned the experimental evidence of an immune-complex disorder, with involvement of the kidney, in sensitive strains of rodents at very low doses of inorganic mercury (Hultman and Eneström, 1987 and 1988; cf. Druet, 1991).

Table 2. Response to thyrotropin- and gonadotropin-releasing hormones in subjects with varying exposure to elemental mercury vapour. All figures are means. Data from Erfurth et al. (1990).

| | Mercury workers (N=11) | Referents (N=10) | Dentists (N=9) | Referents (N=11) |
|---|---|---|---|---|
| U-Hg (μg/g creatinine) | 37*** | 1.0 | 2.3** | 0.7 |
| Hormones (IA/Basal) | | | | |
| S-PRL | 660 | 630 | 715 | 540 |
| S-TSH | 320 | 330 | 270 | 270 |
| S-LH | 170 | 170 | 110 | 120 |
| S-FSH | 55 | 50 | 40 | 30 |

U-Hg = Urinary mercury; IA = Incremental area; Basal =Basal level; S-PRL = Serum prolactin level; S-TSH = Thyroid stimulating hormone; S-LH = Leuteinizing hormone; S-FSH = Follicle stimulating hormone
*** = $p<0.001$; ** = $p<0.01$

## Endocrine organs

The pituitary seems to accumulate mercury at exposure to elemental mercury vapour (Kosta et al., 1975; Nylander et al., 1989), possibly partly due to the fact that its anterior lobe is situated outside the blood-brain barrier. A similar accumulation occurs in the thyroid (Kosta et al., 1975; Nylander et al., 1989).

We thus found it worthwhile to study, in some detail, the pituitary and thyroid functions in subjects with varying exposure to elemental mercury vapour (Table 2; Erfurth et al., 1990). We assessed mercury workers and dentists and matched referent groups. However, there was no association between mercury exposure and basal serum levels of pituitary and other hormones. Neither was there any effect on the hormone response after intravenous infusion of thyrotropin- and gonadotropin-hormones releasing factors.

Similar results have been reported in another recent study of dentists (Langworth et al., 1990). Thus, the anterior pituitary and the thyroid seemed not to be affected by the exposure. This may be because the glands are not sensitive to exposure to elemental mercury vapour, or because the exposure was too low, or because the mercury in the pituitary is in a chemical form, which is not biologically active. As to the latter point, the selenium content of the glands are high, and the mercury may be present as an inactive selenium complex; there are indications of the existence of such (Kosta et al., 1975; Hargreaves et al., 1988; Nylander and Weiner, 1989; Nylander, 1990).

## ACKNOWLEDGEMENTS

Some of the studies reviewed here were supported by grants from the National Swedish Environmental Protection Board and the Swedish Work Environment Fund. Björn Akesson, Ingrid Akesson, Lars Barregard, Per Brunmark, Eva Marie Erfurth, Margareta Molin, Anita Nilsson, Andrejs Schutz, Gunnar Skarping, and BG Svensson have participated in different studies.

## REFERENCES

Ahlguist, M., Bengtsson, G., Furuness, B., Hollender, L. and Lapidus, L., 1988. Number of amalgam fillings in relation to objectively experienced symptoms in a study of Swedish women. Comm. Dent. Oral Epidemiol. 16:227-231.

Akesson, I., Schütz, A., Attewell, R., Skerfving, S., and Glantz, P.O., 1991 Mercury and selenium status in dental personnel - impact of amalgam work and fillings, Arch. Environ. Health, in press.

Aronsson, A.M., Lind, B., Nylander, M. and Nordberg, M., 1989. Dental amalgam and mercury. Biol. Metals, 2:25-30.

Barregard, L., Thomasen, Y., Schütz, A. and Marklund, S., 1988. Enzymuria in workers exposed to inorganic mercury. Int. Arch. Occup. Environ. Health, 61:65-69.

Birke, G., Hagman, D., Johnels, A.G., Plantin, L.O., Sjöstrand, B., Skerfving, S., Westermark, T., and Österdahl, B., 1972. Studies on humans exposed to methylmercury through fish consumption. Arch. Environ. Health, 25:77-91.

Brunmark, P., Skarping, G. and Schütz, A. Submitted for publication. Determination of methylmercury in biological samples using capillary gas chromatography and selected ion monitoring.

Clarkson, T.W., Hursh, J.B., Sager, P.R. and Syversen, T.L.M., 1988a. Mercury, in "Biological Monitoring of Biological Metals", T.W. Clarkson, L. Friberg, G.F. Nordberg and P.R. Sager, eds., Plenum Press, New York, pp. 199-246.

Clarkson, T.W., Friberg, L., Hursh, J.B. and Nylander, M. 1988b. The prediction of intake of mercury vapor from amalgams, in "Biological Monitoring of Biological Metals", T.W. Clarkson, L. Friberg, G.F. Nordberg and P.R. Sager, eds., Plenum Press, New York, pp. 247-264.

Druet, P., Effect of inorganic mercury on the immune system, in "Advances in Mercury Toxicology", T.Suzuki, N. Imura, and T.W. Clarkson, eds., Plenum Press, New York, 1991.

Egglestone, D.W. and Nylander, M., 1987. Correlation of dental amalgam with mercury in brain tissue. J. Prosthet. Dent. 58:704-707.

Einarsson, K., Hellström, K., Schütz, A., Skerfving, S., 1974. Intrabronchial aspiration of metallic mercury. Acta Med. Scand. 195:527-531.

Erfurth, E.M., Schütz, A., Nilsson, A. and Skerfving, S., 1990. Normal pituitary hormone response to thyrotropin and gonadotropin releasing hormones in subjects exposed to elemental mercury vapour. Brit. J. Ind. Med. 47:639-644.

Hargreaves, R.J., Evans, J.G., Janota, I., Magos, L. and Cavanagh, J.B., 1988. Persistent mercury in nerve cells 16 years after metallic mercury poisoning. Neuropathol. Appl. Neurobiol. 14:443-452.

Hultman, P. and Eneström, S., 1987. The induction of immune complex deposits in mice by peroral and parenteral administration of mercuric chloride: Strain dependent susceptibility. Clin. Exp. Immunol. 67:283-292.

Hultman, P. and Eneström, S., 1988. Mercury induced antinuclear antibodies in mice: characterization and correlation with renal immune complex deposits. Clin. Exp. Immunol. 71:269-274.

Jonsson, E., Nilsson, T., Skerfving, S. and Svensson, P., 1972. Consumption of fish and exposure to methylmercury through fish in Sweden, Var Föda 24:59-69.

Kjellström, T., Kennedy, P., Wallis, S., Stewart, A., Friberg, L., Lind, B., Wutherspoon, P. and Mantell, C., 1989. Physical and mental development of children with prenatal exposure to mercury from fish. Stage 2. Interviews and psychological tests at age 6. National Swedish Environmental Protection Board, Solna, 112 p.

Kosta, L., Byrne, A.R. and Zelenko, V., 1975. Correlation between selenium and mercury in man following exposure to inorganic mercury. Nature, 254:238-239.

Langworth, S., Elinder, C.G. and Akesson, A., 1988. Mercury exposure from dental fillings. I. Mercury concentrations in blood and urine. Swed. Dent. J. 12:69-70.

Langworth, S., Röjdmark, S. and Akesson, A., 1990. Normal pituitary response to thyrotropin releasing hormone in dental personnel exposed to mercury. Swed. Dent. J. 14:101-103.

Molin, M., Bergman, B., Marklund, S.L., Schütz, A. and Skerfving, S., 1990. Mercury, selenium and glutathione peroxidase before and after amalgam removal in man. Acta Odont. Scand. 48:189-202.

Molin, M., Bergman, B., Marklund, S.L., Schütz, A. and Skerfving, S. The influence of dental amalgam placement on mercury, selenium, and glutathione peroxidase in man. Acta Odont. Scand., in press b.

Molin, M., Marklund, S.L., Burgman, B. and Nilsson, B., 1989. Mercury, selenium, and glutathione peroxidase in dental personnel. Acta Odont. Scand. 47:383-390.

Molin, M., Schütz, A., Skerfving, S. and Sallsten, G. Mobilized mercury in subjects with varying exposure to elemental mercury vapour. Int. Arch. Occup. Environ. Health, in press a.

Nilsson, B. and Nilsson, B., 1986a. Mercury in dental practice. I. The working environment of dental personnel and their exposure to mercury vapor. Swed. Dent. J. 10:1-14.

Nilsson, B. and Nilsson, B., 1986b. Mercury in dental practice. II. Urinary mercury excretion in dental personnel. Swed. Dent. J. 10:221-232.

Nordberg, G. and Skerfving, S., 1972. Metabolism, in "Mercury in the Environment. A Toxicological and Epidemological Appraisal." L. Friberg and J. Vostal, eds., Chemical Rubber Co., Cleveland, Ohio, pp. 29-91.

Nylander, M. and Weiner, J., 1989. Relaiton between mercury and selenium in pituitary glands of dental staff. Brit. J. Ind. Med. 46:751-752.

Nylander, M., 1990. Accumulation and biotransformation of mercury and its relationship to selenium after exposure to inorganic mercury and methylmercury. A study on individuals with amalgam fillings, dental personnel, and monkeys. Department of Environmental Hygiene, Karolinska Institute, Stockholm, 1990, ISBN 91-628-0092-2. Thesis.

Nylander, M., Friberg, L., Eggleston, D.W., and Björkman, L., 1989. Mercury accumulation in tissues from dental staff and controls in relation to exposure. Swed. Dent. J. 13:235-243.

Nylander, M., Friberg, L. and Lind, B., 1987. Mercury concentrations in the human brain and kidneys in relation to exposure from amalgam fillings. Swed. Dent. J. 11:179-187.

Olsson, S. and Bergman, M., 1987. Letter to the editor. Scand. J. Dent. Res. 66:1288-1289.

Olstad, M.L. Holland, R.I., Wandel, N. and Pettersen, A.H., 1987. Correlation between amalgam restorations and mercury concentrations in urine. J. Dent. Res. 66:1179-1182.

Piikivi, L. and Hänninen, H., 1989. Subjective symptoms and psychological performance of chlor-alkali workers. Scand. J. Work Environ. Health. 15:69-74.

Piikivii, L. Hänninen, H., Martelin, T. and Mantere, P., 1984. Psychological performance in long-term exposure to mercury vapors. Scand. J. Work Environ. Health 10:35-41.

Piikivii, L. and Tolonen, U., 1989. EEG findings in chlor-alkali workers subjected to low long-term exposure to mercury vapors. Brit. J. Ind. Med. 46:370-375.

Roels, H., Abdeladim, S., Ceulemans, E., Lauwerys, R., 1987. Relationshs between the concentrations of mercury in air and in blood and urine in workers exposed to mercury vapor. Ann. Occup. Hyg. 31:135-145.

Roels, H., Gennart, J.P., Lauwerys, R., Buchet, J.P., Malchaire, J. and Bernard, A., 1985. Surveillance of workers exposed to mercury vapour: Validation of previously proposed biological threshold limit value for mercury concentration in urine. Am. J. Ind. Med. 7:45-71.

Schiele, R., 1988. Quecksilberabgabe aus Amalgam und Quecksilberablagerung in Organismus undToxikologishe Bewertung, in "Amalgam - Pro und Contra". G. Knolle, ed., Deutsche Arzte-Verlag, Köin, pp. 123-131.

Schütz, A., Skarping, G. and Skerfving, S., in press. Mercury, in "Trace element analysis in biological specimens". M. Stoeppler and R.F.M. Herber, eds. Year Book Medical Publishers, Chicago, Ill, USA.

Schütz, A. and Skerfving, S., 1975. Blood cell $\partial$-aminolevulinic acid dehydratase activity in humans exposed to methylmercury. Scand. J. Work Environ. Health 1:54-59.

Skare, I., Bergström, T., Engqvist, A. and Weiner, J.A., 1990. Mercury exposure of different origins among dentists and dental nurses. Scand. J. Work Environ. Health 16:340-347.

Skerfving, S., 1974. Methylmercury exposure, mercury levels in blood and hair, and health status in Swedes consuming contaminated fish. Toxicology 2:3-23.

Skerfving, S., 1988. Mercury in women exposed to methylmercury through fish consumption, and in their newborn babies and breast milk. Bull. Environ. Contam. Toxicol. 41:475-482.

Skerfving, S., Hansson, K., Mangs, C., Lindsten, I. and Ryman, N., 1974. Methylmercury induced chromosome damage in man. Environ. Res. 7:83-98.

Stock, A., 1939. Die chronische Quecksilber- und Amalgamvergiftung. Zahnärtzl. Rundsch. 48:403-407.

Svensson, B.G., Björnham, A., Schütz, A., Lettewall, U., Nilsson, A. and Skerfving, S., 1987. Acidic deposition and human exposure to toxic metals. Sci. Tot. Environ. 67:101-115.

Svensson, B.G., Schutz, A., Nilsson, A., Akesson, I., Akesson, B. and Skerfving, S. Fish as a source of exposure to mercury and selenium, Submitted for publication.

Svensson, B.G., Nilsson, A., Rappe, C., Hansson, M., Akesson, B. and Skerfving, S., 1991. Exposure to dioxins and dibenzofurans through fish consumption. N. Engl. J. Med.

WHO, 1990. Environmental Health Criteria 101, Methylmercury, World Health Organization, Geneva, 144 pp.

WHO, 1991. Environmental Health Criteria, Inorganic mercury, World Health Organization, Geneva, 168 pp.

# BLOOD AND URINARY MERCURY LEVELS AS INDICATORS OF EXPOSURE
# TO MERCURY VAPOR

Yukio Yamamura, Minoru Yoshida, and Shigehiko Yamamura

Department of Public Health
St. Marianna University School of Medicine
2-16-1, Sugao, Miyamae-ku
Kawasaki 213, Japan

## ABSTRACT

It has been established that there is no correlation between the signs of mercury poisoning and the concentrations of mercury in blood and urine. Measurements of mercury concentration in urine have nevertheless been used for evaluation of exposure to vapor of elemental mercury. This paper reports some interesting observations we made in the mercury thermometer industry, in which workers are exposed to vapor of elemental mercury. The industry consists of small-scale plants scattered throughout Japan, but mainly in the district of Tokyo.

## INTRODUCTION

Today, mercury poisoning due to exposure to mercury vapor is extremely rare. However, in Japan, workers in small-sized or household factories manufacturing, for example, thermometers, are still being exposed to high concentrations of mercury vapor due to working environments which are slow to improve. We investigated the actual status of exposure to mercury vapor at working sites in such plants, and found several victims of mercury poisoning and a number of workers exposed to high concentrations of mercury vapor. Based on the results of this investigation, we are going to discuss the value of measurement of blood and urinary levels of inorganic mercury or elemental mercury as indicators of exposure to mercury

*Advances in Mercury Toxicology*, Edited by T. Suzuki *et al.*
Plenum Press, New York, 1991

vapor. In addition, we measured the blood and urinary levels of inorganic mercury in workers at chloralkali plants in Japan, and report the levels of exposure to mercury vapor at these plants.

ANALYTICAL METHODS

For total mercury determination, blood and urine samples were placed in a decomposition vessel (Uniseal Ltd., Israel). After the addition of nitric acid, the samples were digested in a heating oven or heating block. Mercury was detected by the reductive vaporization-atomic absorption method (Pan et al. 1980). Inorganic mercury (ionic) in blood and urine was determined by Magos's method (1971). Elemental mercury was determined in urine by the method of Henderson et al (1974).

RESULTS

**Elimination Kinetics**

Elimination kinetics was studied in a single subject, a Japanese male who developed chronic mercury poisoning in the retail sale of metallic mercury. At the time of hospitalization, the subject was 54 years old, had a history of 26 years of exposure to mercury, and had a concentration of 57.2 µg/dl of inorganic mercury (60.2 µg/dl of total mercury) in whole blood. After hospitalization, inorganic mercury concentration in plasma and red cells started to decline. The elimination was triphasic, with half-lives for plasma of 10, 28, and 105 days and for red cells of 7, 27, and 70 days (Figure 1; Yamamura, 1983).

During exposure, inorganic mercury concentration in urine of this subject was 2,000 µg/L or more. On days 5 and 6, the mercury concentration decreased to 500 and 200 µg/L, respectively, and remained elevated for another 100 days. It appears that urinary excretion of inorganic mercury reflects the second and third phase of mercury disappearance from red blood cells.

**Mercury Concentration in Blood and Urine of Workers Exposed to Mercury Vapor**

Inorganic mercury concentrations in blood and urine of 110 employees in 28 small mercury thermometer plants are indicated in Figure 2. Inorganic mercury concentrations in urine of 36 workers exceeded 100 µg/L. Five workers from this group had tremors, however there was no definite dose-response relationship between

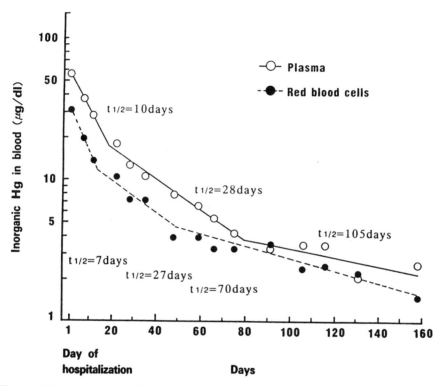

Fig. 1. Elimination rate of inorganic mercury from red blood cells and plasma of the presented patient. Before hospitalization, mercury concentrations in whole blood were 57.2 μg/dl (inorganic mercury) and 60.2 μg/dl (total mercury).

the mercury concentrations in blood or urine and the severity of the tremor. Two workers suffered from nephropathy. In one case, the diabetic nephropathy was associated with tremor. In this case, the inorganic mercury concentration in the blood was 28 µg/dl, while his mercury excretion in urine was low, 117 µg/1. In the other case with chronic nephritis, the inorganic mercury concentration in the blood was high (64 µg/dl) but with a moderate mercury concentration in urine (300 µg/L) (Yamamura, 1985).

Inorganic mercury was measured in the urine and blood of 190 mercury cell workers from 13 chloralkali plants. The mercury cell room at one of the plants was a closed type, but the rooms of the other 12 plants were open. The inorganic mercury concentrations in urine was 100 µg/L or less in 95% of the workers, and in blood was 5 µg/dl or less in 95% of them. None of the workers had any symptoms of mercury poisoning (Yoshida et al., 1979).

## Biological Monitoring of Exposure

In the residential section of one of the small mercury thermometer makers, air mercury concentration was in the range of 100-200 µg/m$^3$. In this family, the inorganic mercury concentration in urine was 446 µg/L for the older daughter, and 543 µg/L for the younger daughter; there is a large difference in the inorganic mercury concentrations in urine between them. On the other hand, the inorganic mercury concentration in blood is 11.0 µg/dl for the older daughter, and 9.1 for the younger daughter; the difference was small. This finding suggests that the inorganic mercury concentration in the blood may be a better indicator of mercury vapor than inorganic mercury concentration in urine. It is clear in these cases, the air mercury concentrations of 100-200 µg/m$^3$ corresponds to the inorganic mercury concentration of about 10 µg/dl in blood (Table 1).

Area monitoring of mercury was done for 4 days in two plants (A and B). In Plant A, where the time-weighted average (TWA) concentration of mercury was 80 µg/m$^3$, three workers had been employed for 10 to 30 years. In Plant B, where TWA concentration of mercury was 20 µg/m$^3$, nine workers had been employed for 2 to 3 years. The average concentration ratio of air to blood (inorganic mercury) in Plant A was 1.4; in Plant B the ratio was 1.8 (Figure 3). This finding indicates that the concentration ratios were similar in both groups, despite the fact that the degrees of the exposure were very different. On the other hand, the concentration ratios of air to urine (inorganic mercury) were very different: ratio was equal to 9.4 in Plant A, and equal to

Table 1. The inorganic mercury concentrations in blood and urine of a household mercury thermometer plant

| Occupation | Age | Inorganic Hg | Inorganic Hg in blood μg/dl | in urine μg/dl |
|------------|-----|--------------|------------------------------|----------------|
| Patient    | Filler     | 45 | 24.5 | 795 |
| Wife       | Assistant  | 42 | 19.7 | 215 |
| Daughter*  | Bank Clerk | 19 | 11.0 | 446 |
| Daughter*  | Student    | 16 | 9.1  | 543 |

*Non-occupational

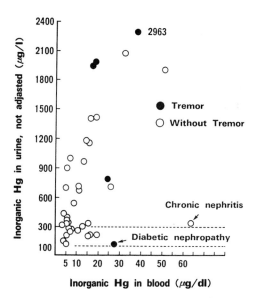

Fig. 2. Relationship between inorganic mercury concentrations in blood and urine of workers exposed to mercury. Urine specimens were collected from 36 workers of 28 mercury thermometer plants; open circle - without tremor, close circle - with tremor. Urine concentration is given as measured without any adjustment.

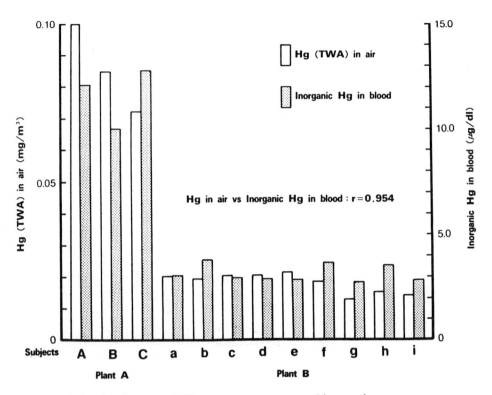

Fig. 3. Relationship between TWA mercury exposure and inorganic mercury
concentration in blood. TWA mercury exposure represents average
concentration measured over 4 days. Subjects A-C are mercury fillers in a small
thermometer plant with a long history of occupational exposure to mercury
vapor (Plant A). Subjects e-i are workers in a large thermometer plant with a
relative short history of occupational exposure to mercury vapor (Plant B).

4.4 in Plant B (Figure 4). This means that mercury retention was higher at high
exposure (A group) than at low exposure (B Group).

In another study, 22 workers from 3 plants were monitored in order to obtain
information on the relationship between the degree of exposure and inorganic mercury
concentration in blood. Personal monitors were used to evaluate inhalation exposure.
There was a linear correlation between TWA concentration in the air of the workplace
and the inorganic mercury concentration in the blood of the workers (r = 0.862). The
regression line indicates that occupational exposure to a TWA of 100 µg/m³
corresponds to an inorganic mercury concentration in blood of 7-8 µg/dl (Yamamura,
1978).

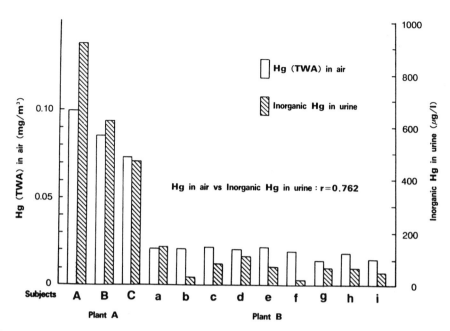

Fig. 4  Relationship between TWA mercury exposure and inorganic mercury concentration in urine. Urine samples were collected over 8 hour and mercury concentrations were adjusted to 1.024 of specific gravity. The TWA mercury exposure represents average concentration measured over 4 days. Subjects A-C are mercury filler in a small thermometer plant with a long history of occupational exposure to mercury vapor (Plant A). Subjects e-i are workers of a large thermometer plant with a relative short history of occupational exposure to mercury vapor (Plant B).

## Relationship Between Degree of Exposure and Elemental and Inorganic Mercury in Urine

Elemental and inorganic mercury concentrations were measured in the urine of 10 workers from 2 plants (7 workers from Plant B and 3 workers from Plant C). Spot samples of urine were collected prior to the work shift and at the end of the shift. The measured concentration of mercury in air (TWA), and of inorganic mercury and elemental mercury concentrations in urine of employees are depicted in Figure 5 (Yoshida, unpublished). The correlation coefficient between the TWA and the concentration of elemental mercury in urine collected at the end of shift was 0.831 (Figure 6) it was obviously higher than that between the TWA concentration in air and the concentration of inorganic mercury in urine (r = 0.348) (Figure 7) (Yoshida, 1985).

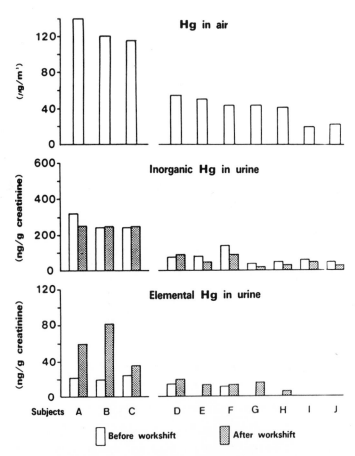

Fig. 5. Effect of sampling time on the relationship TWA mercury exposure and mercury concentrations in urine. Inorganic and elemental mercury concentrations were measured in spot urine samples collected from ten workers in two thermometer plants prior to starting the shift and at the end of the shift.

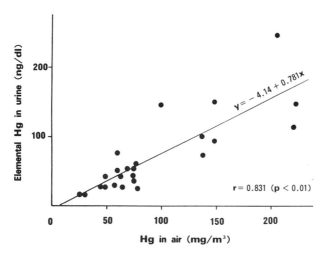

Fig. 6. Correlation of elemental mercury concentration in urine collected at end of the workshift from ten workers in two thermometer plants with TWA mercury exposure (by personal sampling).

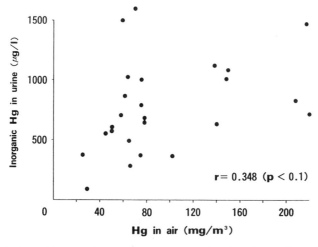

Fig. 7. Correlation of inorganic mercury concentration in urine collected at the end of the work shift from ten workers with TWA mercury exposure (by personal sampling).

## DISCUSSION

Smith et al. (1970) estimated that the mercury exposure to TWA of 0.1 mg/m$^3$ by area monitoring corresponds to a mercury concentration in blood of 6 µg/dl, and 230 µg/L in urine. On the basis of personal monitoring, Bells et al. (1973) estimated a value of 100 µg/L for total mercury in urine. Our findings in the present study reveal that inorganic mercury concentration in blood corresponding to mercury exposure to TWA of 0.1 mg/m$^3$ was in the range of 10 to 12 µg/dl, and urinary concentrations in the range of 300~500 µg/L (based on area monitoring) or 150 ~ 200 µg/L (based on personal monitoring). This discrepancy of Bells et al. (1973) and our experience has probably resulted from the fact that Bells collected urine in the morning (16 hours composite post-exposure experiments), whereas we collected urine in the afternoon, at the end of shift.

Gompertz (1982) considered total mercury concentration in urine as an indicator of a long-term integrated exposure, and total mercury concentration in blood as an indicator of a recent exposure. Our experience with exposure monitoring is consistent with Gompertz's description. In our practice, we have been using the criteria recommended by the Department of Health of the State of California. We consider inorganic mercury concentrations in urine above 50 µg/L as a sign of increased absorption, and consider concentrations above 100 µg/L as a warning level. However, these reference limits can be applied only after 3 months of exposure, when steady state has been reached.

Henderson et al. (1974) described that at some rate of absorption of inhaled elemental mercury vapor, the mercury is rapidly oxidized, organically bound, and excreted in bound form in urine. At a higher rate of absorption, some of the mercury circulates to the kidney in the elemental form and is excreted into urine in this form as well as ionic (inorganic) and organically bound mercury. In this study the correlation coefficients between the TWA mercury exposure and the concentration of elemental mercury in urine collected at the end of the shift was high. These findings suggest that the determination of elemental mercury in urine may serve as a useful indicator for assessing levels of recent exposure to mercury vapor.

## CONCLUSION

The monitoring of inorganic mercury in blood is a useful tool in the evaluation of recent exposure to vapor of elemental mercury. Measurement of elemental mercury in urine specimens collected at the end of the shift is a useful indicator of health hazards.

REFERENCES

Bells, Z.G., Lovejoy, H.B., Vizena, T.R., 1973. Mercury exposure relations and their correlation with urine mercury excretion, III. J. Occup. Med. 15: 501-508.

Henderson, R., Shotwell, H. P., Krause, I., 1974. Analyses for total, inorganic and elemental mercury in urine as basis for a biological standard. Am. Ind. Hyg. Assoc. J. 35: 576-580.

Gompertz, D., 1982. Biological monitoring of workers exposed to mercury vapor. J. Soc. Occup. Med. 32: 141-145.

Magos, L., 1971.Selective atomic absorption determination of inorganic mercury and methylmercury in undigested biological samples. Analyst. 96: 847-853.

Pan, S.K., Imura, N., Yamamura, Y., Yoshida, M., Suzuki, T., 1980. Urinary methylmercury excretion in persons exposed to elemental mercury vapor. Tohoku J. Exp. Med. 130: 91-45.

Smith, R.G., Vorwald, A.J., Patel, L.S. et al., 1970. Effects of exposure of mercury in the manufacture of chlorine. Am. Ind. Hyg. Assoc. J. 31: 687-700.

Yamamura, S., 1978. Relationship of time-weighted mercury levels in the urine and blood inorganic mercury levels in thermometer workers (in Japanese). Jap. J. Ind. Health. 2: 110-111.

Yamamura, Y., 1983. Relationship between the inorganic mercury concentration in the blood and expired air in mercury poisoning. Fifth International Conference "Medichem", San Francisco, California (September 5-10, 1983).

Yamamura, Y., 1985. Inorganic mercury in the blood, urine and expired air in workers exposed to elemental mercury vapor (in Japanese). Assoc. Ind. Hyg. Jap. 24: 46-51.

Yoshida, M., Yamamuchi, H., Hirayama, F. et al., 1976. Mercury contents in hair, blood cells and plasma from workers exposed to metallic mercury, inorganic mercury and ethylmercury (in Japanese). The St. Marianna Med. J. 4: 41-54.

Yoshida, M., 1985. Relation of mercury exposure to elementary mercury levels in the urine and blood. Scand. J. Work. Eviron. Health 11: 33-37.

Yoshida, M., and Yamamura, Y. Unpublished data.

# EPIDEMIOLOGICAL AND CLINICAL FEATURES

## OF MINAMATA DISEASE

Akihiro Igata

Kagoshima University
Kagoshima, Japan

## ABSTRACT

Minamata disease is the methylmercury intoxication, through contaminated fish by a chemical factory in Minamata city. Based on the results of our regional survey, cardinal clinical features of it were clarified, by a multivariant analysis of all symptoms in inhabitants in the polluted area. The clinical features were found to be essentially the same as those of Hunter Russell syndrome, however, some other additional symptoms were also found. Those are influenced by many factors, such as degree of exposure, duration of pollution etc. The disposition of each inhabitant also plays a role in clinical manifestation. This analysis contributes to a correct individual diagnosis and to the correct estimation of patients in polluted areas. It was also found by long term studies that a few inhabitants began to claim some neurological symptoms, after ceasing of pollution, which were suggested mainly due to aging.

As many inhabitants with mild neurological complaints were not easily diagnosed, a questionable borderline group should be postulated, for social settlement of Minamata disease. Compared with those in other areas in the world, the characteristics of Minamata disease will be discussed.

## INTRODUCTION

Since 1956, methylmercury pollution has been found near Minamata city in Kumamoto prefecture, which was induced by methyl mercury, discharged from a chemical factory in Minamata city. The pollution was so widespread that the number of victims increased to more than 2,000 in Kumamoto and Kagoshima prefectures. Since 1974 much effort has been undertaken to elucidate the pathomechanism of the disease and to develop a new treatment under the sponsorship of the Environment Agency of the Japanese government (Environmental Agency 1974-89, 1975).

Table 1. Patients with Minamata Disease (June 1990)

|  | Diagnosed | Rejected | Pending | Not examined |
|---|---|---|---|---|
| Kumamoto Pref. | 1,760 | 8,234 | 844 | 1,859 |
| Kagoshima Pref. | 474 | 2,780 | 50 | 320 |

Quantitative Diagnostic Procedure

The diagnosis of Minamata disease is usually not difficult in typical and severe cases, but it is not always easy to diagnose in mild cases since damage to the health of inhabitants ranges from severe to healthy persons. Because of this difficulty, it is important to decide where the line should be drawn in recognizing Minamata disease. This problem should be settled only on a medical basis. In actual situations, the objective, not subjective, diagnosis should be made, so that everyone can agree with the conclusion based on reasonable evidence (Igata, 1974). To diagnose Minamata disease objectively, a quantitative method was proposed by us, using multivariant analysis based on all information obtained in the population survey performed in 1973 and 1974 in the polluted area of Kagoshima prefecture, neighboring Minamata city, with a population of 80,000. Independently of individual diagnosis, by which many cases with Minamata disease were newly found, all data for these inhabitants were analyzed by multivariant analysis. As the first step, in a population survey, patients with any neurological symptoms included among those indicating Minamata disease were discovered from the answers to inquiries, the recovery rate being higher than 97%. After a simple neurological examination as the second step, the third precise neurological examination was undertaken in patients showing any indication of neurological symptoms and signs, and this took 2 hours for each patient. Thereafter all information on this third extracted group was analyzed by principal-factor and discriminant analysis. As a result, a group with Minamata disease, was found as shown in Figure 1, differentiated from non-affected inhabitants. From this diagram, a reasonable dividing point between Minamata disease and the non-affected group would be 9.13.

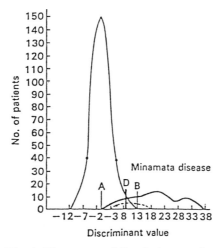

Fig. 1 Histogram of discriminant values

If A is made the dividing point, all Minamata disease patients will be included with a lot of erroneously diagnosed normal inhabitants. If B is made the dividing point, all patients are definite Minamata cases but some on the left side of this point are erroneously discarded. Therefore point D (9.13) was settled on as the most reasonable dividing point, to minimize the misdiagnosis. Using this point D to separate Minamata disease from healthy persons, the ratio of erroneous diagnosis is kept to be less than 2.5% of individual diagnoses. Using a two-dimensioned diagram, each case can be checked, as to whether the diagnosis is correct or not. Seen from diagrams of inhabitants in each area by principal analysis as in Figure 2, the number suspected of having Minamata disease increases in parallel to the distance from Minamata city.

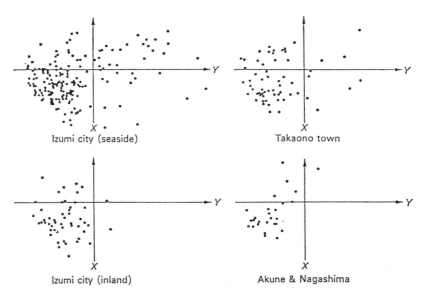

Fig. 2  Distribution of patients in various area.
X: factor I, Y: factor II (principal analysis)
In parallel to the distance from Minamata city, the groups toward the right disappear.

Table 2. Misdiagnosis of Minamata disease

1) Based on 52 items of information

|  | Minamata disease | Non-Minamata disease | Not decided |
|---|---|---|---|
| Minamata disease | 37 | 0 | 5 |
| Non-Minamata disease | 6 | 228 | 2 |

Misdiagnosis 6/236  2.5%

2) Based on 24 items of information

|  | Minamata disease | Non-Minamata disease | Not decided |
|---|---|---|---|
| Minamata disease | 36 | 4 | 6 |
| Non-Minamata disease | 4 | 225 | 1 |

Misdiagnosis 8/230    3.5%

3) Based on 12 items of information

|  | Minamata disease | Non-Minamata disease | Not decided |
|---|---|---|---|
| Minamata disease | 34 | 7 | 8 |
| Non-Minamata disease | 7 | 223 | 2 |

Misdiagnosis 14/232  6.0%

In addition, it became possible to check whether any symptom should be included among those of Minamata disease or not. The results made clear, e.g., that symptoms of cervical spondylosis are not included among those of Minamata disease, and that arteriosclerosis of the retina or systemic hypertension are not related to those of Minamata disease, on the other hand, polyneuritis (Figure 3), visual field constriction, cerebellar ataxia and others have a close severity-response relation to Minamata disease. Thus some new symptoms in addition to those of Hunter-Russell syndrome, for example, anosmia, hyperrefexia etc. were found.

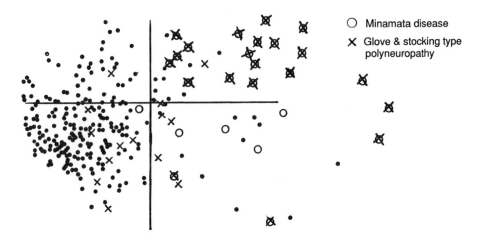

Fig. 3 Sensory polyneuritis and Minamata disease

Thus, this diagnostic method is the most reliable one, if symptoms are correctly identified. Still there are some minor disadvantages in this method, i.e., epidemiological conditions, are not input, which are not easy to be quantitatively expressed. In diagnosing Minamata disease, there are possibly some errors, originating form the erroneous confirmation of each symptom, because the symptoms sometime fluctuate from day to day or from morning to night, even within a day to a slight extent. In addition,t here are psychological factors in patients, especially who are eager to be compensated being kept waiting for a long time to be diagnosed. Needless to say, these errors are not caused by this method, but are inevitable in any. Concerning the importance of each symptom contributing to the correct diagnosis of Minamata disease, the quantitative scores for each symptom (discriminant values) were calculated on the basis of our own statistical analysis, which is shown in Table 2. These scores can be regarded as the quantitative criteria for Minamata disease, valid for methylmercury poisoning under one mode of pollution. It is, therefore, not always applicable to organic mercury intoxication in other areas, where the mode of pollution is different. Through our experience, it was proposed to postulate a border line inhabitants, who can not be diagnosed as definitely having Minamata disease, but still have complaints partially compatible with those of Minamata disease. This proposal is now being adopted as the administrative countermeasure. In fact, through this step, social settlement is being achieved, except for some who are now on trials after rejection. This countermeasure will be suitable for evaluation of any health impairment by pollution.

## Epidemiological Information

Concerning epidemiological information contributing to the correct diagnosis, the amount of fish intake and mercury concentration in blood, hair, and urine are useful in diagnosing acute intoxication. In chronic Minamata disease, after the end of pollution, new procedures to indicate contamination in the past, were devised, e.g. the measurement $\beta_2$-microglobulin in urine was found to be parallel to the mercury content in the hair during the period of pollution (Figure 4), and the grade of contamination in the past can be estimated from it. However, in the patients now, in whom contamination ceased 20 years ago, such procedures are not useful.

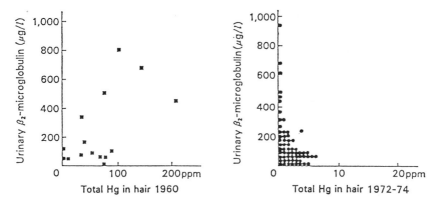

Fig. 4   Relationship between total Hg in hair and urinary $\beta_2$-microglobulin.

The excretion of mercury after the administration of a chelating agent is also helpful in estimating past contamination (Figure 5). The history of residency in polluted areas, career as fisherman and food qualities suggest exposure to a high degree of pollution. In congenital Minamata disease, the content of methylmercury in the umbilical cord, which is usually preserved by Japanese custom, helps to indicate the severity of pollution at the time of delivery.

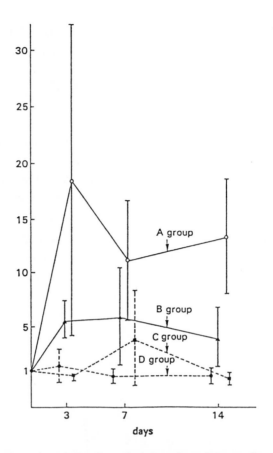

Fig. 5   Excretion of total Hg after administration of tiopronin. A) Hair content
of Hg was higher than 15 ppm in the past. B) Hair content of Hg was
lower than 15 ppm. C) placebo and D) without any drug.

## Identification of Each Symptom

There is no available pathognomonic test indicating Minamata disease such as the
TPHA test for syphilis. The diagnosis should be based on a characteristic combination
of symptoms. Therefore, it is very important to identify each symptom, in other
words, it is very important to confirm each symptom objectively and correctly.

### A. Peripheral Neuropathy

Peripheral neuropathy is one of the cardinal symptoms of Minamata disease.
Sensory impairment of the extremities is of the glove-stocking type, sometimes with

perioral dysesthesia (Hamada, 1981). There are practically no cases of Minamata disease without peripheral neuropathy, though some have reported that there are rare exceptions. Evidence of peripheral neuropathy is manifested by morphological changes in the peripheral nerves, both on autopsy and in biopsy. The characteristics of pathological involvement of peripheral nerves in Minamata disease are the general loss of myelinated fibers with relative increase in small sized myelinated fibers which can be regarded as the result of regeneration (Figure 6).

A typical histogram of myelinated fiber diameters in a patient with Minamata disease is shown in Figure 7. The disturbance of peripheral autonomic nerves can also be confirmed by loss of unmyelinated fibers and autonomic dysfunctions such as low skin temperature and low sweating especially in distal extremities. In thermography (Kamitsuchihashi, 1985) the general temperatures are low and also the recovery from low temperature is slower than in normal age-matched controls (Figure 8). Deep tendon reflexes are usually diminished or absent, although exaggerated ones can be confirmed in 40% of the patients. The motor and sensory conduction velocities are usually delayed in accordance with the severity of the disease, the sensory one being more severely affected, although there are none in whom conduction velocities are normal even when there is numbness (Figure 9).

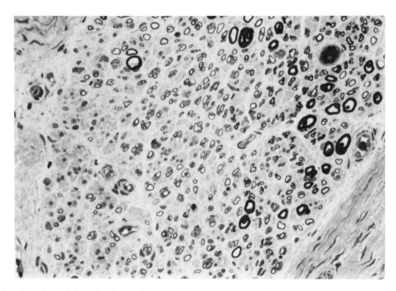

Fig. 6    Typical biopsied sural nerve in a patient: 35 year old male with Minamata disease

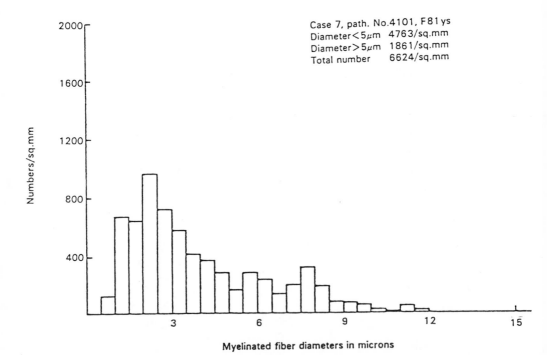

Case 7, path. No.4101, F81ys
Diameter$<5\mu$m   4763/sq.mm
Diameter$>5\mu$m   1861/sq.mm
Total number      6624/sq.mm

Fig. 7   Histogram of the biopsied sural nerve fibers in a patient with Minamata disease.

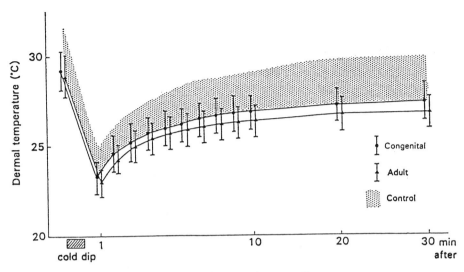

Fig. 8  Cold dip test in Minamata disease.

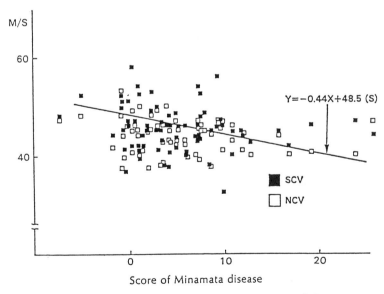

Fig. 9  Index of Minamata disease and NCV.

Sensory impairment does not always mean diminished sensation but sometimes the subjective complaints of numbness, a tingling sensation or hyperethesia. Recently somatosensory evoked potentials have come to be used to check lesions in the brain (Arimura, 1985). This method assumes that lesions in the central nervous system might be mainly responsible for the sensory impairment seen in Minamata disease.

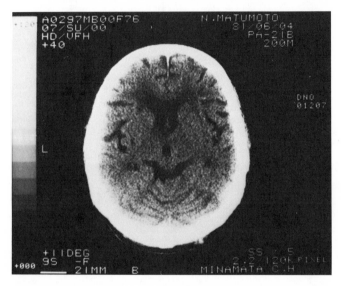

Fig. 10 Cerebella atrophy in a patient with Minamata disease (64 year old male).

### B. Cerebellar ataxia

Although ataxia can be identified by the usual neurological examination, it is sometimes not easy, especially in borderline cases, to identify the ataxia objectively and quantitatively. For the objective evaluation of cerebellar ataxia, many procedures are now being applied, including use of the gravidimeter, objective registration of finger-nose test, analysis of voice, writing, gait, etc. (Hamada, 1980) In the results, clumsy movement is not directly related to cerebellar ataxia but to concomitant rigidity, tremor, weakness, spasms, etc. A CT scan is also useful in evaluating the morphological changes in the central nervous system, including the occipital lobe, cerebellum and cerebrum (Figure 10).

Comparing the age matched controls, some quantitative scores of Minamata disease based on CT findings are proposed, although the cerebellar atrophy due to other diseases cannot be excluded. Using many parameters for the cerebellum and also the cerebrum and brain stem in the CT scan, the differential diagnosis of Minamata disease can be made with an accuracy higher than 80% Akinesia-like slowness of movement can also be observed.

### C. Tremor

Tremor can also be objectively evaluated by computer analysis. The frequency and amplitude of tremor in this disease is rather characteristics, compared to tremors of other diseases such as Parkinson's disease or hyperthyroidism.

### D. Gait

Gait disturbance is also rather common in Minamata disease and can be analyzed with a special precise gait-analyzer using a computer. The characteristic gait in Minamata disease, which includes elements of weakness, ataxia, rigidity and/or apasticity, partially influenced by numbness, cannot be by itself differentiated from other diseases with certainty.

### E. Mental deterioration

In severe cases of Minamata disease, mental deterioration is inevitable and mild mental symptoms such as erroneous calculation, disturbed memories and character changes can be found in mild cases. However, severe dementia like Alzheimer disease could not be found in the affected patients.

### F. Ophthalmological and otological symptoms

Even in a routine neurological examination, visual field constriction and disturbed ocular movement etc. are characteristic signs of Minamata disease. Abnormal visual evoked potential can be an objective parameter of this disease. From the otological point of view, hearing loss of impairment of hearing, discriminative understanding of voice and hearing fatiguability are specific. Disturbance of equilibrium is also characteristic of Minamata disease. These aspects will be discussed in the corresponding chapters for each. Abnormal auditory brain stem response is also suggestive of Minamata disease.

Clinical Course

In any intoxication, the symptoms usually improve gradually after contamination ceases. However, there are some patients with Minamata disease, in whom the symptoms worsen after contamination ceases or in whom the symptoms appear a few years later (Shirakawa, 1979; Igata 1978; 1975). This is called Minamata disease of late onset. Generally speaking, late onset is difficult to understand. Nevertheless, such cases do exist and some of them are verified on autopsy. This late onset Minamata disease might be accounted for in the following way: 1) effect of aging on latent Minamata disease, 2) the long lasting but slight damage due to a minimal amount of organic mercury remaining in the brain, although this is difficult to believe in the light of many data on the accumulation and metabolism of ingested mercury, 3) the psychological condition of people who are eager to be compensated. For each possibility, we have no definite evidence for any of these, but it is an important problem to be solved on a medical and social basis. Roughly speaking, such late onset and late progression are limited to some patients and the peak was reached before 1975, incidence now being rare.

Other neurological signs and symptoms

Some other neurological signs and symptoms in Minamata disease have been clarified by a study group sponsored by the Japanese government for the past years (Igata, 1986). These signs include anosmia, loss of taste, abnormal latency in evoked potentials, tonic painful spasm, paroxysmal fainting and bladder disturbance, etc. Among the inhabitants of the polluted areas, there are some patients with a total loss of superficial sensation (Igata, 1981) and/or total ophthalmoplegia with reflectoric eye jerks (Arimura, 1980). This peculiar symptom might be explained as a psychological reaction. In addition, no abnormal evoked potentials are confirmed in these cases. No relation to the severity or to the high incidence of typical symptoms was found. However, it might be a symptom due to a peculiar effect on the brain due to organic mercury, since there are some in whom no other psychological or hysterical reactions are found.

Diagnostic criteria of Minamata disease

The committee, sponsored by the Japanese government, had proposed official guidelines for the diagnosis of Minamata disease, compatible with our quantitative diagnostic criteria. Cases with subjective numbness in the extremities only, without any objective signs, such as areflexia, delayed sensory conduction velocity or abnormal findings in biopsied sural nerves are excluded from Minamata disease by these criteria.

Involvement of other organs

Although the target organ in organic mercury poisoning is nervous tissue, including the peripheral nerves, other organs such as kidney, liver, etc. are sometimes also involved, especially in severely affected patients (Igata, 1973). According to our data, the methylmercury has some effect on platelet aggregation *in vitro*. This result might explain the diffuse ischemic change in the acute stage of very severe cases, which are confirmed on autopsy, although general convulsions can also be the cause of them (Shaw, 1979). In the chronic cases now in question, we have no evidence that contamination in the past can be a causative factor in the arteriosclerosis or cerebral vascular diseases, although many cases, complicated with cerebrovascular diseases, have been reported probably due to the aging in the polluted areas (Nagashima, 1985). To check the real causal relationship between cerebrovascular diseases and mercury, the severity-response relationship should be confirmed, i.e. the more severe the case is, the higher the incidence should be. No data suggesting the possibility of arteriosclerosis directly induced by methylmercury have been reported. According to the neuropathological studies, arteriosclerosis-like vascular changes can be frequently found in severely damaged and atrophied brain, regardless of its causes. In fact, our experience revealed no relationship between the severity of Minamata disease and the grade of arteriosclerosis or incidence of Strokes. The situation is the same in the complication of hypertension or diabetes mellitus. Among the patients now being considered, there are some cases complicated by kidney and liver damage. In the autopsied cases, the concentration of total mercury is slightly higher in kidney and liver than in other organs. Since the content of methylmercury is not always high in the organs, it is possible that such damage is due to transformed inorganic mercury, or at least partly. There is a higher incidence of proteinuria, high $B_2$-microglobulin or n-acetylhexosaminidase than in non-affected inhabitants (Igata, 1985; Ohkatsu and Igata, 1978).

Congenital Minamata Disease

In the period of severe pollution, the incidence of cerebral palsy was found to be much higher than in other areas (Harada, 1979; Moriyama, 1974). This was the first hint of the existence of congenital Minamata disease born from affected mothers. Until now, about 50 cases have been found altogether in Kumamoto and Kagoshima prefectures. These cases are sometimes quite difficult to differentiate from a child with MInamata disease who became ill in childhood after a normal birth. To diagnose congenital Minamata disease, such epidemiological information as the mother's history of residence in the polluted areas during the severe pollution prior to 1967 is important. This end of pollution was suggested from the data indicating that the methylmercury content in preserved umbilical cords had normalized by this year. The methylmercury content in their mothers' hair is also contributory. No late onset or late progression have been seen in congenital cases. Usually the symptoms improve gradually in parallel with dvelopment of the child. The symptoms of this congenital MInamata disease are essentially similar to those of adult Minamata disease. However, peripheral neuropathy is not manifest as a cardinal symptom, but mental retardation with symmetrical motor disturbance is rather characteristic. If any characteristic symptoms such as cerebellar ataxia, visual congenital constriction, peripheral neuropathy etc. are confirmed, a diagnosis of Minamata disease is likely to be mad.e it is also very important that their mother's symptoms are rather contributory to the diagnosis of congenital Minamata disease, although it is a general rule that the disease is less severe in mothers than in affected children. Brain damage in congenital Minamata disease is diffuse and symmetrical as seen in the CT scan, so diseases with focal lesions can be excluded as non-Minamata diseases. The evoked potentials including auditory brain stem response on both sides are sometimes abnormal. These findings without any focal lesions sometimes indicate congenital Minamata disease.

DISCUSSION AND CONCLUSION

Paralleling the progress of society, health hazards due to some industrial products and byproducts have come to be reported, so that the public pollution and hazards have become a very important medical and social problem of our ear, of which Minamata disease is a typical example. The outbreak of Minamata disease should serve as a warning to the developed world. This problem should therefore should be properly dealt with to avoid its resurgence in any part of the world. In Minamata disease, the pollution was stopped many years ago, but, to our regret, many problems remain unsolved, although a considerable effort has been maintained for many years. As the results of our studies show, the pathomechanism, clinical characteristics and procedure for prevention and treatment became clear for typical Minamata disease. however, the diagnosis of mild cases cannot be made in the sense of "all or none." To rationalize

Table 3.  Discriminant values of Minamata disease

| | |
|---|---|
| Clumsy finer nose test | 5.02 |
| Clumsy movement of tongue | 4.13 |
| Perioral dysesthesia | 2.68 |
| No coarse nystagmus | 2.66 |
| Abnormal ocular movement | 2.53 |
| Weakness | 2.25 |
| Visual field constriction | 2.02 |
| Dysesthesia of glove-stocking type | 0.61 |
| Anosmia | 0.30 |
| Hearing loss | 0.30 |

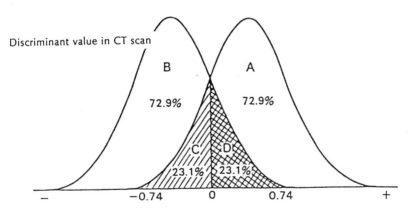

Fig. 11  Diagnosis of Minamata disease using 13 parameters of CT scan. A, Minamata disease. B, Normal controls. C, Minamata disese, which was diagnosed as non-Minamata disease. D, Non-Minamata disease, which was diagnosed as Minamata disease.

diagnosis of mild cases of Minamata disease, quantitative evidence is necessary, so that anyone can agree with the conclusion. In 1971-72 an outbreak of organic mercury pollution was reported in Iraq (Bakir et al., 1973). An effort was made to compare the clinical aspects of this outbreak with those of Minamata disease (Rustam et al., 1974). Although the symptoms in both outbreaks are quite similar to each other, there are some differences between the two. In Iraq, the pollution was not of long duration, compared with that in Minamata. The loss of visual acuity was reported there but no residual or chronic symptoms have been reported. In addition, in other parts of the world, a lot of new outbreaks of mercury pollution have been reported, including those of natural origin (Clarkson, 1976). Nevertheless, the results of studies on Minamata pollution should be a textbook of new mercury pollution or health hazards or possible ones in the future. In this sense, the clinical studies should be continued to obtain a thorough understanding of mercury poisonings.

REFERENCES

Annual reports of the study group on Minamata disease (in Japanese), 1974-1989 Environment Agency, Tokyo.
Arimura, K., 1985. Somatosensory evoked potential in Minamata disease. Sixth workshop on Minamata Disease. Environmental Agency, Nov., Kumamoto, Japan.
Arimura, Y., 1980. Studies on ocular motor apraxia (in Japanese). Saishin Igaku (Modern Medicine) 35:2-5.
Bakir, F. et al., 1974. Methylmercury poisoning in Iraq. Science 181:230-241, 1974.
Clarkson, T.W., 1976. Exposure to methylmercury grossy narrowed white dog rivers. Report to Canadian government.
Hamada, R., 1981. Studies on methylmercury poisoning (in Japanese). Kagoshima Med. J. 23:107-117.
Hamada, R., 1980. Objective evaluation of cerebellar speech (in Japanese). Saishin Igaku (Modern Medicine) 35:11-14.
Harada, M., 1979. Congenital Minamata disease (in Japanese). In: Minamata disease. S. Arima, ed. Seirinsha, pp. 345-370.
Igata, A., 1986. Clinical aspects of Minamata disease studies edited by T. Tsubaki, and H. Takahashi. Kodansha, Tokyo.
Igata, A., 1985. Urinary b-D N-acetylglucosamines in the patients with Minamata disease. Annual report of the study group on Inamata disease, pp. 56-57.
Igata, A., 1982. objective diagnosis of Minamata disease by CT scan. Annual report of the study group on Minamata disease, pp. 54-55.
Igata, A., 1981. Studies on the total anesthesia in patients with Minamata disease (in Japanese). Annual report of the study group on Minamata disease, pp. 52-53.
Igata, A., 1978. Minamata disese of late onset (in Japanese). Igakuno Ayumi (Progress of Medicine) 96:890-894.

Igata, A., 1975. The late onset of organic mercury intoxication after exposure. Studies of the effects of alkylmercury intoxication. Environmental Agency, Tokyo, pp. 178-179.

Igata, A., 1974. A quantitative analysis of diagnosis of Minamata disease (in Japanese). Shinkei Kenkyu no Shinpo (Progress of Neurological Research) 18:890-900.

Igata, A., 1973. Minamata disease in Kagoshima prefecture (in Japanese),. Nihon Ijishinpo 2578:23-28.

Kamitsuchihashi, H., 1985. Vasodilator response in patients with Minamata disease. Annual report of the study group on Minamata disease, pp. 99-100.

Moriyama, H., 1974. Congenital methylmercury intoxication. 18:901-911.

Nagashima, K., 1985. Arteriosclerosis in underdeveloped children (in Japanese). Annual report of the study group on Minamata disease, pp. 90-98, 1985.

Niina, K. Clinical and experimental methyl mercurial encephalopath studies on organic mercury intoxication (in Japanese). Kagoshima Med. J. 35:203-220.

Ohkatsu, Y. and Igata, A., 1978. Urinary B2 microglobulin in patients with Minamata disease (in Japanese). Shinkeinaika (Neurological Medicine) 8:271-273.

Rustum, H. et al., 1974. Methylmercury poisoning in Iraq. Brain 97:499-507.

Shaw, C.M., 1975. Cerebrovascular lesions in experimental and methyl mercurial encephalopathy. Neurotoxicol. 1:57-74.

Shirakawa, K., 1979. Minamata disease of late onset. In: Minamata disease. S. Arima, ed., Seirinsha, pp. 331-337.

Studies on the health effects of alkyl mercury in Japan, 1975. Environment Agency.

Takizawa, Y., 1972. Studies on the distribution of mercury in several body organs. Acta Med. et Biol. 19:193-197.

Tsubaki, T., 1968. Organic mercury intoxication along Agano-river (in Japanese). Rinshoshinkeigaku (Clinical Neurology) 8:511-520.

Uchino, M. , 1985. Studies on CT scan findings of Minamata disease (in Japanese). Rinshoshinkeigaku (Clinical Neurology) 25:1024-1029.

# MERCURY IN HUMAN ECOLOGY

Tsuguyoshi Suzuki

Department of Human Ecology, School of Health
Sciences, Faculty of Medicine, The University of
Tokyo, Hongo 7-3-1, Bunkyo-ku,
Tokyo 113, Japan

## ABSTRACT

Mercury has existed in the human environment at every stage of human evolution. Therefore, how was, and is, human adaptation to the existing level of mercury to be clarified to approach the limitation in our adaptability. For this, it is necessary to evaluate the environmental characteristics of each local human ecosystem.

1. Mercury in a tropical human ecosystem with traditional ways of living.

In an ecosystem in the Papua lowlands, mercury, mostly in the form of methylmercury, is found in fishes and reptiles at elevated levels. Mercury levels are correlated with the stable isotope ratio of nitrogen ($^{15}N/^{14}N$) in animal food which indicates a biomagnification of mercury accumulation through the food chain. In animal foods, mercury to selenium ratio is getting close to unity on the molar basis with increasing mercury levels. Dietary mercury intake and hair mercury levels in the fish-eating sector are comparable with those in developed countries. Balance between Se and Hg in the dietary intake is inclined to relative excess of Se to Hg.

2. Mercury in an urbanized ecosystem (Tokyo).

In an urbanized ecosystem in an industrialized country, the major source of mercury is via fish consumption. In organs obtained from forensic autopsy cases in Tokyo, methylmercury levels are uniform through all the organs (30 to 50 ng/g on

*Advances in Mercury Toxicology*, Edited by T. Suzuki *et al.*
Plenum Press, New York, 1991

average) except the liver (113 ng/g), but inorganic mercury levels are high in the liver and kidney (266 to 456 ng/g) and low in the brain, heart and spleen (4 to 9 ng/g). Selenium levels are significantly correlated with mercury levels in some organs, particularly in the kidney, where % inorganic to total mercury is as high as 85%.

From these results and additionally the results obtained on harbor seals caught in the Okhotsk sea, the role of selenium in adapting environmental mercury will be hypothetically discussed.

## INTRODUCTION

Mercury is considered to be a long-lasting environmental contaminant from various pieces of evidence; thus, it has existed in the human environment at every stage of human evolution, even though the existing level might be variable. Human beings, as well as animals, should have acquired ability to deal with mercury by keeping it in a non-toxic state in the process of absorption, metabolism, accumulation and excretion. There is certainly a limitation in this ability, as has been proven by repeated epidemics of methymercury poisoning, and the limitation is closely related to not only the biological characteristics of the organism but also the relevant environmental characteristics, in particular, to coexisting chemical agents which may modify the mercury toxicity. Bearing this notion in mind, I would like to introduce our mercury studies in field research of human ecology in the lowlands of Papua, New Guinea and in the cases of forensic autopsy in Tokyo. In the studies of human ecology, or the ecological approach to mankind, the principal goal is to understand human adaptive mechanisms in relation to the environmental setting (Ohtsuka and Suzuki, 1990). Mercury has been an active agent in this setting with its natural abundance in company with artificial enrichment. Hence, the adaptive mechanism to environmental mercury must not be missed in human ecology studies.

## I. Mercury in a Tropical Human Ecosystem with Traditional Way of Living in Lowland Papua, New Guinea

The subjects for this study were the Gidra-speaking people inhabiting the eastern part of the deltaic lowland between the Fly River in the north and the Torres Stait in the south in the Western Province of Papua, New Guinea (Figure 1). The details of the people and the land were already reported by us (Ohtsuka, 1983; Ohtsuka and Suzuki, 1990), but briefly, they live depending on Metroxylon sago starch, tubers such as taro and yam, hunting and/or fishing, and imported foods such as rice, wheat flour and canned fish/meat. Their foods vary with the village locality in the Gidraland, i.e., the villagers living inland eat negligible amount of fish, ˙ ⎧s those in a coastal village consume less amounts of local land animals such ⎧ ɔwary, wallabies, deer and pig, and more amounts of imported foods.  In 1971 ⎧ hair samples collected from villagers on an inland village called Wonie, contair ⎧ cury at very low levels of 1.8 µg/g for men and 1.4 µg/g for women (Ohtsuka ⎧ ⎧uki, 1978).

In our field research in 1980 and 1981, fo⎧ ⎧ges, in different localities with different subsisting activities, were selected for the intensive survey.  Those were Rual in the northern Gidraland close to the Bituri River, Wonie in the inland, Ume on the bank of the Binaturi River and Dorogori on the coast (Figure 1).  Foods, mainly local and several imported, and villagers' hair were collected and analysed for the content of various elements including mercury and selenium and the stable isotopic ratios of the carbon and the nitrogen.

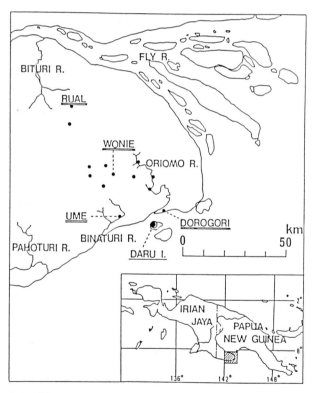

Fig. 1. Map of the Gidraland showing the locaiton of the four study villages: northern village (Rual), in land village (Wonie), riverine village (Ume), and coastal village (Dorogori).

### I.-1. Mercury Levels and Ratio of Inorganic to Total Mercury in the Gidra Foods

Among the Gidra foods, plant foods contained a small amount of mercury, usually below 0.01 µg/g; the highest concentration was 0.026 µg/g in galip seeds. Among animal foods, fish showed the highest level of total mercury ranging from 0.038 to 1.86 µg/g, and reptiles had the next highest total mercury levels. As is shown in Figure 2, the ratio of inorganic to total mercury becomes small (less than 10% in most of the cases) with elevation of total mercury levels.

From these mercury values in foods and the record of the household food consumption survey, the dietary intake of total and organic mercury was estimated on the village average basis; the highest estimate, 81.3 µg/day/adult male for total mercury and 70.4 µg/day/adult male for organic mercury, was found in the coastal village, Dorogori, and the lowest estimate, 8 µg/day/adult male for total mercury and 6 µg/day/adult male for organic mercury, was found in the inland village, Wonie, where fish was not consumed during the survey period. In the northern and riverine villages, it was 16.4 and 21.6 µg/day/adult male for total mercury, respectively. This level in Dorogori is quite high in comparison with the reported values in various communities, even in those with heavy fish eating habit, in the world and exceeds the tolerable weekly intake recommended by WHO (1972) (Hongo et al., 1989). A shark with a high mercury content in muscle as 1.9 µg/g was caught during the food consumption survey period and contributed to elevate the dietary intake of mercury there.

### I.-2. Characterization of the Gidra Food with Carbon and Nitrogen Stable Isotope Ratios

Ecological characteristics of each plant and animal, i.e., the photosynthetic character of plants and the position of animals in food-web of an ecosystem, can, to a considerable degree, be revealed by measuring the stable isotope ratio of carbon and nitrogen (Park and Epstein, 1961; O'Leary, 1981; Minagawa and Wada, 1984). Plants with $C_3$, $C_4$ and crassulacean (CAM) photosynthetic pathways show different discrimination against $^{13}C$ during photosynthesis; thus, isotopically light $C_3$ plants (wheat, rice, barley, oats, rye, potato, sugar beet, cassava, sweet potato, soybean, and grape), heavy $C_4$ plants (sugar cane, corn, sorghum and millet) and intermediate CAM plants (pineapple) can be distinguished by measuring the ratio of $^{13}C$ to $^{12}C$. The stable isotopic composition of carbon in an animal reflects the ratio of its diet (DeNiro and Epstein, 1978). This is true for human beings. Accordingly, the carbon isotope ratio of flesh of herbivorous animals differs depending upon the isotopic composition of the plant eaten. Similarly, in the case of nitrogen, the isotopic composition of nitrogen, $^{15}$N to $^{14}$N, in an animal reflects the nitrogen isotopic composition of its diet, but stepwise enrichment of $^{15}$N along food chains is greater than that of $^{13}$C (DeNiro and Epstein, 1981). This makes a difference in significance of the nitrogen isotope ratio from the carbon isotope ratio as a biological marker in analysing an ecological setup; the greater the $^{15}$N ratio, the higher the trophic level to which the animal concerned belongs.

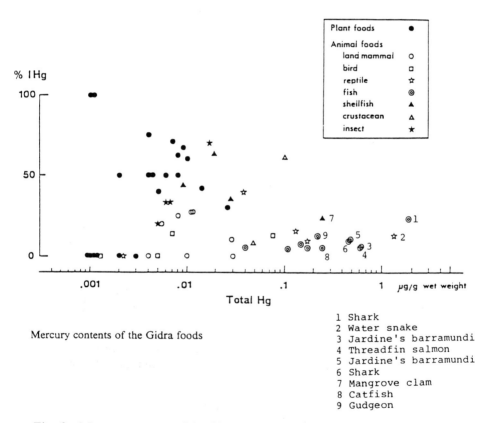

Mercury contents of the Gidra foods

1 Shark
2 Water snake
3 Jardine's barramundi
4 Threadfin salmon
5 Jardine's barramundi
6 Shark
7 Mangrove clam
8 Catfish
9 Gudgeon

Fig. 2. Mercury contents of the Gidra foods; Total mercury and % inorganic to total.

Figure 3 indicates the distribution of the isotopic composition of carbon and nitrogen in foods consumed by the Gidra. Simply saying, major foods of the Gidra were isotopically divided into 5 groups, i.e., $C_3$ plant, $C_3$ plant feeder (most of terrestrial animals and birds), $C_4$ plant feeder (grass wallaby and some deer), freshwater animals, and marine animals. Marine fishes have the highest isotopic ratios for both carbon and nitrogen. The next highest for nitrogen is found in freshwater and brackish water fishes and reptiles, of which the isotopic composition of carbon is the lowest. Terrestrial animals and birds are lower than fishes in the nitrogen composition. In the tropical savanna, $C_4$ grasses which have high drought- and fire-tolerance, dominate (Harris, 1980), and animals fed these grasses have higher carbon isotopic compositions. In the figure, preliminary results on the hair of the Gidra are shown. Only adult male samples were analysed and shown broken down by villages.

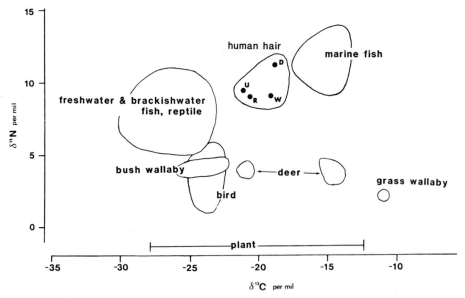

Fig. 3. The stable isotope ratio of nitrogen and carbon in the Gibra foods. Source: Yoshinaga et al., in press. The results are expressed in the d notation

$$d*X = \left[ \frac{(*X/X)\text{sample}}{(*X/X)\text{standard}} - 1 \right] .1000 \text{ per mil}$$

where *X and X are the heavier and the lighter stable isotopes of the element. For nitrogen, the standard is atmospheric nitrogen; for carbon, the standard is the Peedee belemnite (PDB) carbonate.

Reflecting the different dietary composition, the values from the coastal village, Dorogori (marked with D), were located close to marine fishes, those from the northern and riverine villages, Rual (R) and Ume (U), were closer to freshwater fishes than to marine fishes, and those from the inland village, Wonie (W), were in the middle of marine and freshwater fishes.

The stable isotope ratio of nitrogen is significantly positively correlated with the mercury content, but not with the lead content, in animal foods of the Gidra (Figure 4) (Yoshinaga et al., in press). The isotope ratio of carbon does not show any significant correlation with mercury or lead levels. If we suppose that the increase in the $^{15}N/^{14}N$ as $d^{15}N$ per mil is 3.5 per one trophic level (Minagawa and Wada, 1984), mercury concentrations will increase by a factor of 4-5 per one trophic level in the Gidra's ecosystem. It is not yet convincing how valid this estimate is, because the factor is far smaller than the values obtained in experiments as the bioconcentration factor (WHO IPCS 1989). However, our data may be supportive evidence for the notion that mercury undergoes bioconcentration along with food chains.

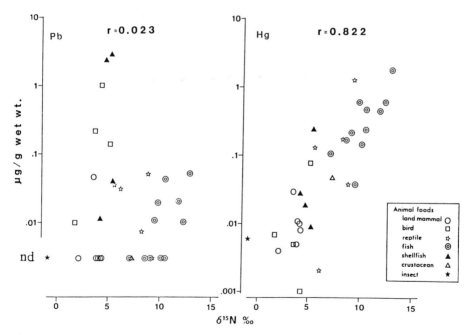

Fig. 4. Correlation of the stable isotope ratio of nitrogen with mercury or lead levels in the Gidra animal foods.
Source: Yoshinaga et al., in press.

## I-3. Mercury and Coexisting Elements in the Gidra Foods

Elemental composition of the Gidra foods was reported in detail in previous papers (Hongo et al., 1989). Among the elements measured, five elements (Na, Mg, P, K, and Ca) were found at relatively high levels in the range of 0.1 to 10 mg/g wet weight (Ca in some animal foods and fruits was below 0.1 mg/g), six elements (Al, Mn, Fe, Cu, Zn, and Sr) were usually contained at concentrations up to 1 mg/g wet weight, and four elements (Cr, Se, Hg and Pb) were usually contained at low levels, below 1 μg/g wet weight.

Table 1 shows correlations between mercury content and other element contents of foods. In animal foods, a significant positive correlation was found only for the Hg/Se pair (Yoshinga et al, in press), and reflecting the fact that terrestrial animal flesh contained Zn, Fe and Cu more and Hg less than aquatic animal flesh, significant negative correlation was found for the Hg/Fe, Hg/Cu and Hg/Zn pairs. In plant foods, the elements, of which content showed significant positive correlation with Hg content were Ca, Mn, Cu and Sr. Among these coexisting elements, selenium is one of the most noticeable, since the equimolar, or close to equimolar, coexistence with mercury has been repeatedly recorded in marine mammals and humans exposed to mercury (e.g., Koeman et al., 1973; Kosta et al., 1975). In the present animal samples, which were an edible part of the body and composed mainly of muscles, selenium is in a state of relative excess toward mercury in many samples, but the molar ratio of Se/Hg is approaching to 1 with an elevation of mercury concentration in animal tissues (Figure 5) (Yoshinaga et al., in press). If we suppose that the tissue selenium is saturated with mercury at the molar ratio, 1:1, the corresponding mercury concentration would be several nmol/g (about 1 μg Hg/g) and over. The identical pattern of Hg/Se accumulation, i.e., a relative excess of selenium at low level mercury accumulation and approaching to the unity of the molar ratio with an elevation of mercury concentration, was observed in the liver of marine mammals such as seals and porpoises (Figure 6) (Koeman et al., 1975; Himeno et al., 1989) and a very suggestive thing is the fact that most of the animals with high mercury concentrations were found dead (Koeman et al., 1975). The tissue mercury concentration at the molar ratio of 1:1 was several μg Hg/g in this case. Dietary mercury intake was compared with dietary selenium intake in the Gidra, with additional information on Japanese (Hongo et al., 1989) and a population in Tuscany, Italy (Rossi et al., 1976), in Figure 7. The molar ratio of Se to Hg is quite variable among populations, i.e., 25:1 in Japanese, 1.7:1 in an Italian population, 4:1 to 15:1 in the Gidra villagers. Studies of this sort have been quite few up to now, but very large variation in Se to Hg ratio in dietary intake must be kept in mind for comparing various populations.

Table 1. Correlation of mercury content with content of other elements of the Gidra Foods

| | Animal Foods n=39 (n=36 for Se) | Plant Foods n=20 (19 for Se) | All Foods n=59 (55 for Se) |
|---|---|---|---|
| Na | .162 | .375 | .557*** |
| Mg | -.141 | .385 | -.037 |
| P | -.013 | .354 | .546*** |
| K | .141 | .186 | .209 |
| Ca | .208 | .668** | .147 |
| Mn | -.245 | .590** | -.323* |
| Fe | -.517** | .430 | -.214 |
| Cu | -.613*** | .623** | -.397** |
| Zn | -.652*** | .166 | .107 |
| Sr | .106 | .497* | -.061 |
| Al | -.299 | -.192 | .046 |
| Cr | -.014 | .281 | .179 |
| Pb | -.070 | .085 | .210 |
| Se | .655*** | .451 | .754*** |

Figures in the table are correlation coefficient. To calculate the coefficient, logarithmic values of concentrations were used. *, **, ***: $p<0.05$, $p<0.01$, $p<0.001$. Source: Suzuki et al. (1988); Hongo et al. (1989); Yoshinga et al., in press.

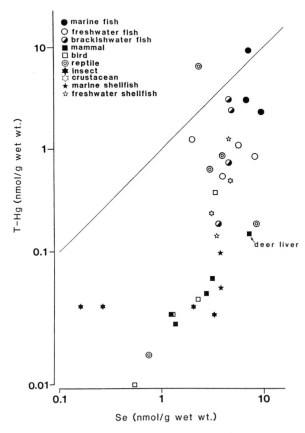

Fig. 5.  Correlation of total mercury and selenium contents of the
Gidra animal foods.
Source: Yoshinaga et al., in press.

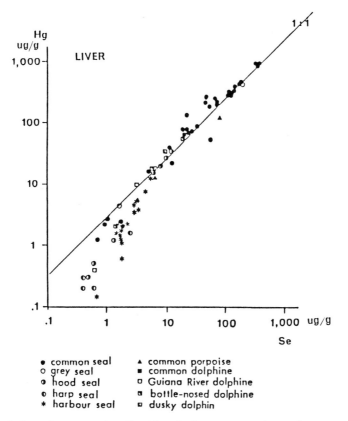

Fig. 6. Correlation of mercury (total) with selenium contents in marine mammal's liver.
Source: Koeman et al. (1975) and Himeno et al. (1989).

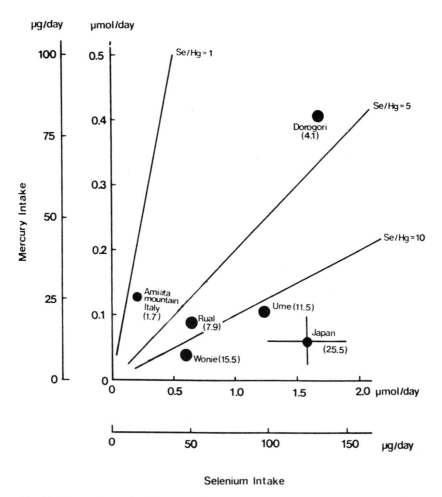

Fig. 7. The molar ratio of mercury to selenium in human dietary intake. Source: Hongo et al. (1989); Rossi et al. (1976) and Yoshinaga et al., in press.

## I.-4. Hair Mercury Concentrations in the Gidra

As already mentioned in the introductory description regarding this study, the Gidra inhabiting an inland village and eating a negligible amount of fish had very low hair mercury concentrations in 1971. Again in 1981, very low levels of hair mercury concentrations were found in the same village (Wonie). The remaining three villages showed higher levels, in particular, the levels found in the northern village were as

high as 6-7 µg/g on the average (Table 2). The average hair mercury level by villages
was not consistent with the estimate of dietary mercury intake by the food consumption
survey mentioned earlier. In Dorogori, the coastal village, where the estimated
mercury intake was the highest, the hair mercury levels were lower than those in Rual,
the northern village. Several reasons might be postulated for this inconsistency. For
instance, firstly, we missed the Rual villagers' fish consumption during their stay in
fishing camps, and secondly, in Dorogori, a rare happening of catching shark with
high mercury content elevated the estimate of dietary intake of mercury (Suzuki et al.,
1988). The hair mercury value can indicate the quantity of each individual's fish
consumption and the stable isotope ratio of the nitrogen can show the quality of it;
hence, the mercury level was compared to the stable isotope ratio of the nitrogen in hair
of selected villagers (Figure 8) (Yoshinaga et al., in press). In the figure, the village is
characterized with its respectively unique distribution; lower $d^{15}N$ and higher mercury
in Rual, also lower $d^{15}N$ and lower mercury in Wonie, higher $d^{15}N$ and intermediate
mercury in Dorogori, and intermediate $d^{15}N$ and lower mercury in Ume. These results
indicate that heavy consumption of freshwater fish is most likely to have occurred in
the northern village, Rual, at the time the hair sampled was sprouting, and marine fish
consumption was not so heavy as was expected from the location of the village in the
coastal village, Dorogori.

Table 2    Hair Mercury Levels in the Gidra in 1981

| Village | Sex | N | Hg level, µg/g |
|---------|-----|---|----------------|
| Northern | m | 28 | 7.1 |
| (Rual) | f | 35 | 6.4 |
| Inland | m | 31 | 1.5 |
| (Wonie) | f | 34 | 1.0 |
| Riverine | m | 50 | 3.8 |
| (Ume) | f | 40 | 3.4 |
| Coastal | m | 37 | 4.1 |
| (Dorogori) | f | 34 | 4.4 |

Figures in the table are arithemetic means of total mercury concentration. Hair samples
of ten adult males from each village were measured for inorganic mercury conent, and
the percent inorganic to total mercury was approximately 10% without the intervillage
difference. Source: Suzuki et al. (1988).

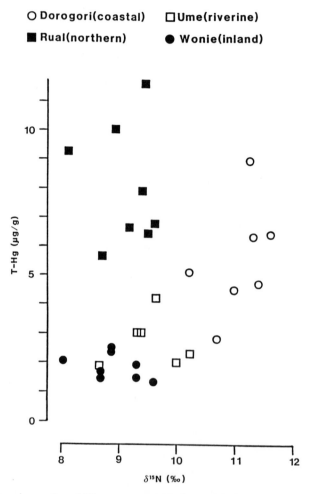

Fig. 8. Comparison of total Hg conetnet with the stable isotope ratio of the nitrogen of the Gibra's hair. Source: Yoshinaga et al., in press.

## II. Mercury Concentration in Organs of Contemporary Japanese and Its Correlation with Selenium Concentrations

The next study to be introduced was conducted for the autopsy cases in Tokyo. The human tissues analyzed were obtained during November 1986 to July 1987 at the Department of Forensic Medicine, Faculty of Medicine, University of Tokyo. The subjects (32 males and 14 females, aged 4 month to 82 years) were all residents of the Tokyo metropolitan area, and they had no known exposure to heavy metals including mercury. The causes of death (with numbers of subjects appearing within parenthesis) were as follows: suffocation (11), blood loss for various reasons (9), traumatic injuries and shock (14), subarachnoidal or intracerebral hemorrhage (3), and other (9). Six different types of tissue (cerebrum, cerebellum, kidney, liver, heart, and spleen) were removed from the subjects, but not all six types were removed from every subject. Additionally from 36 subjects, the scalp hair samples from the distal parts were collected. Mercury with speciation of total mercury, methylmercury and inorganic mercury, selenium, and other elements were analyzed with various methods such as flameless atomic absorption spectrometry (AES), ECD gas chromatography, fluorometry and ICP-AES.

### II-1. Mercury Concentration in Organs

As is shown in Figure 9, methylmercury (MeHg) concentrations in tissues showed uniform levels ranging from 30 ng/g in the spleen to 52 ng/g in the kidney cortex. The liver was an exception, containing higher levels of MeHg (average of 113 ng/g). While, inorganic mercury (IHg) concentration showed a marked contrast among the types of tissue; i.e., substantial accumulation of IHg was observed in the kidney (averages: 267 in medulla and 456 ng/g in cortex) and liver (316 ng/g), and very small amounts of IHg (several ng/g on average) were detected in the cerebrum, cerebellum, heart, and spleen. No age- and sex-related differences were recognized in MeHg concentrations, whereas IHg concentrations in the cerebrum and heart were correlated with age. Comparing the total mercury (THg) value to the MeHg and IHg values, the mercury in tissues of present subjects was judged to be composed of these two chemical forms. The distribution pattern of MeHg has suggested the continuous uptake of MeHg in the subjects (Matsuo et al., 1989). From both the hair mercury level (geometric average of 4.7 μ/g) and the proportion of organic to total Hg (87-99%) in the hair, the uptake of MeHg was considered to be at an average level for Japanese (Matsuo et al. unpublished). The IHg was accumulated in the liver and kidney, and sources of it should be multiple as follows: (1) intestinal absorption of IHg that was contained in foods, secreted in bile, or transformed by gut flora from MeHg in foods or in biliary secretion, (2) respiratory absorption of elemental mercury vapor in atmospheric pollution or from dental amalgam fillings, (3) conversion of MeHg to IHg in the tissue, and (4) transfer of IHg from other organs where the MeHg was demethylated. Since we did not have any information that made the estimation of the amount of uptake of IHg possible, it could not be evaluated which mechanisms had contributed most for accumulation of IHg in various tissues.

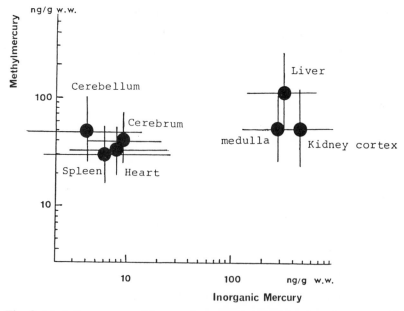

Fig. 9 Methylmercury and inorganic mercury concentrations in organs of contemporary Japanese. Source: Matsuo et al. (1989).
Dot and bar show the geometric mean and standard deviation.

## II-2. Selenium-Mercury Relationships in Organs

Levels of selenium (Se) in tissues and the relationships of Se to other elements, in particular mercury, in each organ are a matter of particular interest from a viewpoint of adaptive implications of metal-metal interactions to environmental pollution. As already mentioned, a pioneering work by Koeman et al. (1973) revealed the correlation between the concentrations of Se and Hg in the liver of marine mammals, which ingest large amounts of MeHg via fish consumption, with a molar ratio of Hg to Se of 1:1.

This finding was later confirmed repeatedly in other marine mammals, and coaccumulation of Se and Hg (mostly in the inorganic form) in the liver has been interpreted as a protective effect of Se against Hg. With regard to human beings, high levels of Se and Hg were detected in the organs of Minamata disease victims (Ujioka, 1960) and the coaccumulation of Se and Hg was reported in the organs of Idrija mercury miners (Kosta et al., 1975). The co-existence of Se with Hg in the cellular component of the brain of Minamata disease victims (Shirabe et al, 1979) and in the kidney tissue of patients with IHg intoxication (Aoi et al., 1985) were also reported. Nevertheless, detailed evaluation of the relation between Se and Hg has not yet been carried out in humans without any particular exposure to mercurials except for our recent report (Yoshinaga et al., 1990). If selenium plays a crucial role in adaptation of human beings to environmental mercury, the relation between Se and Hg in tissues of subjects, who live without receiving harmful effects of mercury, will be one of the most important pieces of information.

In the organs of contemporary Japanese, significant correlation between THg and Se was observed in the liver, kidney, and cerebellum, whereas highly significant correlation between IHg and Se was found only in the kidney (Figure 10), and that between MeHg and Se was observed in the heart, spleen, liver and kidney cortex, though the coefficients were not so high, if significant (Figure 11). In human kidney tissues, the pattern of coaccumulation of Se and Hg is identical to that observed in the liver of marine mammals, i.e., a relative excess of Se at low level accumulation of Hg (THg and IHg) and approaching unity of the molar ratio with an elevation of Hg concentration, albeit most of the subjects had not reached the unity of the ratio. In harbor seals significant correlation between Se and Hg (mostly in the inorganic form) was observed in the liver, but not in the kidney. In the kidney of the seals, mercury concentration was correlated with metallothionein (MT) concentration (Tohyama et al., 1986). In the present autopsy cases, MT could not be determined, but the tissue level of zinc (Zn), copper (Cu) and cadmium (Cd) is usually correlated with MT (Dunn et al., 1987). Hence, multiple regression analyses were applied to clarify the interrelationships between the metal concentrations. Table 3 summarizes the results: significant association of Se with IHg in the kidney was again confirmed; significant association of Se with MeHg in the spleen has attracted attention; among three heavy metals which may be related to MT, not the Zn or Cd, but the Cu, was noteworthy, since in the multiple regression equations, the Cu was a significant factor in accumulation of Se in the kidney medulla and heart, in accumulation of MeHg in the kidney, and in accumulation of IHg in the liver and kidney medulla, although the extent of association was not so strong as in the case of Se/Hg. It is not clear whether the Cu is related to the Hg by modifying the MT metabolism, or not. Although the metabolism of Cu-MT is known to be different from that of Zn-MT, and Cd-MT in the cell (Dunn et al., 1987), the meaning of the tissue Cu level in terms of MT metabolism has not been studied in humans. In animal experiments using rats, complicated patterns of interaction between Hg, Se, Zn and Cu were revealed; (1) coadministration of Hg, Se and Zn, or Hg, Se and Cu decreased Hg accumulation in the liver compared with that of Hg and Se, or administration of Hg alone (Komsta-Szumska and Chmielnicka, 1981), (2) coadministration of Hg and Zn, or Zn and Se, compared with administration of Zn alone, increased the level of endogenous Cu in the kidney (Chmielnicka et al., 1983), and (3) urinary excretion and the tissue level of endogenous Cu were increased by injection of $HgCl_2$, but the elevation was alleviated by a simultaneous injection of Se (selenite) (Chmielnicka et al., 1986). All these results indicate that the role of Cu in Hg metabolism should be carefully studied in future work.

## IV. A Speculative Model Regarding the Role of Selenium in Adaptability to Mercury

In animals eating fish (and reptiles), the uptake and accumulation of MeHg are inevitably an every day occurrence. Methylmercury ingested is subject to be demethylated by gut flora and by some unidentified mechanisms in the tissue. Inorganic mercury yielded by demethylation is detoxified with formation of Hg-Se complex or Hg-MT (Figure 12). Thus, the selenium status of the animals plays, at least in part, a key role in dealing with Hg to be inert (Suzuki, 1989). The amount of Se ingested is usually at a state of relative excess toward Hg; thus a considerable portion of Se is available to deal with toxic heavy metals including Hg, though the bioavailability of Se is variable with chemical forms of Se (Magos et al., 1987; Suzuki, 1988). If the amount of heavy metals taken into the body increases, the Se will be consumed with handling of metals and the status of Se deficiency will be manifest, even in the condition of apparently adequate dietary intake (van Fleet et al., 1981). Finally, the state of overflow of heavy metals will result in the manifested toxicity of metals (Figure 13).

What I should like to put emphasis on is summarized as follows: (1) In general populations, the balance between Se and Hg is inclined to relative excess of Se in terms of dietary intake and the level in the tissues, (2) mercury, in particular, the inorganic form of mercury, is handled by coaccumulating with selenium in the tissues, and

Table 3  Summarized results of multiple regression analyses

I. Dependent Variables = Se.

| Organ | IHG | MeHg | Zn | Cu | Cd | $R^2$ |
|---|---|---|---|---|---|---|
| Liver | NS | * | NS | NS | * | 0.451** |
| Kidney cortex | *** | NS | * | NS | NS | 0.712*** |
| Kidney medulla | *** | NS | NS | * | * | 0.860*** |
| Heart | NS | NS | NS | ** | -a | 0.534** |
| Spleen | NS | *** | *** | NS | - | 0.633*** |

II. Dependent Variable = MeHg.

| Organ | IHG | Se | Zn | Cu | Cd | $R^2$ |
|---|---|---|---|---|---|---|
| Liver | NS | NS | NS | NS | * | NS 0.455** |
| Kidney cortex | NS | NS | NS | * | NS | 0.472** |
| Kidney medulla | * | NS | NS | ** | * | 0.454* |
| Spleen | NS | *** | * | NS | - | 0.356** |

III. Dependent Variable = IHG.

| Organ | IHG | Se | Zn | Cu | Cd | $R^2$ |
|---|---|---|---|---|---|---|
| Liver | NS | NS | NS | * | NS | 0.356* |
| Kidney cortex | NS | *** | * | NS | NS | 0.653*** |
| Kidney medulla | * | *** | NS | -*b | -* | 0.798*** |

Only the significant multiple regression is described in the tables. NS, not significant; *, **, ***, significant at $p<0.05$, $0.01$, $0.001$. a, Cd was not included as a variable since it was not detected in cerebrum, cerebellum, heart and spleen. b, the mark prefixed with - shows that the regression coefficient is negative. Source: Yoshinaga et al. (1990).

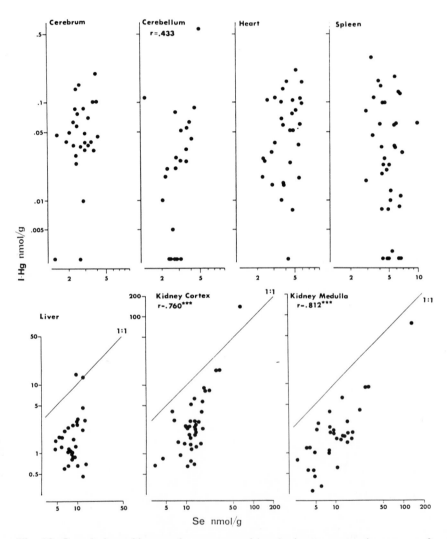

Fig. 10. Correlation of inorganic mercury with selenium contents in organs of contemporary Japanese. Source: Yoshinaga et al. (1990).

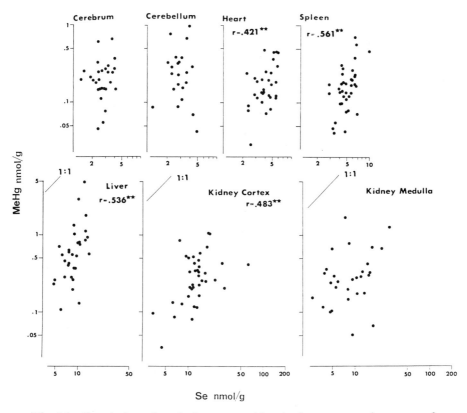

Fig. 11. Correlation of methylmercury with selenium contents in organs of contemporary Japanese. Source: Yoshinaga et al. (1990)

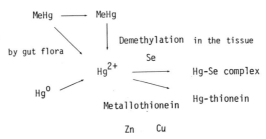

Fig. 12. Process of detoxification of mercurial compounds. Source: Suzuki (1989).

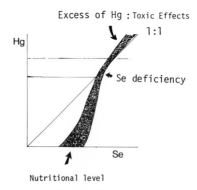

Fig. 13. A speculative model on the interaction of mercury and selenium.

adequate reserve of selenium is usually recognized in general populations, and (3) increasing retention of Hg would result in Se deficiency, which would precede manifestation of mercury toxicity.

ACKNOWLEDGEMENT

The works on which this review is written are supported by Grants-in-aid from the Japanese Ministry of Education, Science and Culture. The author would like to express his heartfelt thanks to collaborators, the names of whom are not mentioned individually, the Gidra people who accepted us, and Prof. Ishiyama of Forensic Medicine, Faculty of Medicine, University of Tokyo, who kindly provided specimens.

REFERENCES

Aoi, T., Higuchi, T., Kidokoro, R., Fukumura, R., Yagi, A., Ohguchi, S., Sasa, A., Hayashi, H., Sakamoto, N., and Hanaichi, T., 1985, An association of mercury with selenium in inorganic mercury intoxication, Hum. Toxicol., 4:637.

Chmiielnicka, J., Komsta-Szuumska, E., and Zareba, G., 1983, Effect of interaction between 65Zn, mercury and selenium in rats (retention, metallothionein, endogenous copper), Arch.Toxicol., 53:165.

Chmielnicka, J., Brzeznicka, E., Sniady, A., 1986, Kidney concentrations and urinary excretion of mercury, zinc and copper following the administration of mercuric chloride and sodium selenite to rats, Arch. Toxicol., 59:16.

DeNiro, M.J., and Epstein, S., 1978, Influence of diet on the distribution of carbon isotopes in animals, Geochim. Cosmochim. Acta, 42:495.

DeNiro, M.J., and Epstein, S., 1981, Influence of diet on the distribution of nitrogen isotopes in animals, Geochim. Cosmochim. Acta, 45:341.

Dunn, M.A., Blalock, T.L., and Cousins, R.J., 1987, Metallothionein, Proc. Soc. Experim. Biol. Med., 185:107.

Harris, D.R. (ed.) 1980, "Human Ecology in Savanna Environment", Academic Press, London.

Himeno, S., Watanabe, C., Hongo, T., Suzuki, T., Naganuma, A., Imura, N., Morita, M., 1989, Body size and organ accumulation of mercury and selenium in young harbor seals, (Phoca vitulina), Bull. Environ. Contam. Toxicol., 42:503.

Hongo, T., Suzuki, T., Ohtsuka, R., Kawabe, T. Inaoka, T., and Akimichi, T., 1989, Element intake of the Gidra in lowland Papua: Inter-village variation and the comparison with contemporary levels in developed countries, Ecol. Food Nutr., 23:293.

Keoman, J.H., Peeters, W.H.M., Koudstaal-Hol, C.H.M., Tjioe, P.S., and de Goeij, J.J.M., 1973, Mercury-selenium correlations in marine mammals, Nature, 245:385.

Koeman, J.H., van de Ven, W.S.M., de Goeij, J.J.M., Tjioe, P.S., and van Haaften, J.L., 1975, Mercury and selenium in marine mammals and birds, Sci. Total Environ., 3:279.

Komsta-Szumska, E., and Chmielnicka, J., 1981, Organ and subcellular distribution of mercury in rats in the presence of cadmium, zinc, copper, and sodium selenite, Clin. Toxicol., 18:1327.

Kosta, L., Byrne, A.R., and Zelenko, V., 1975, Correlation between selenium and mercury in man following exposure to inorganic mercury, Nature, 254:238.

Magos, L., Clarkson, T.W., Sparrow, S., and Hudson, A.R., 1987, Comparison of the protection given by selenite, selenomethionine and biological selenium against the renotoxicity of mercury, Arch. Toxicol., 60:422.

Matsuo, N., Suzuki, T., and Akagi, H., 1989, Mercury concentration in organs of contemporary Japanese, Arch. Environ. Health, 44:298.

Matsuo,. N., Suzuki, T., Yoshinaga, J., Hongo, T., and Akagi, H., Hair vs.organs: comparison of mercury concentrations in contemporary Japanese unpublished.

Minagawa, M., and Wada, E., 1984, Stepwise enrichment of $^{15}$N along food chains: further evidence and the relation between $^{15}$N and animal age, Geochim. Cosmochim. Acta, 48:1135.

Ohtsuka, R., 1983, "Oriomo Papuans: Ecology of Sago-Eaters in Lowland Papua", University of Tokyo Press, Tokyo.

Ohtsuka, R. and Suzuki, T., 1978, Zinc, copper and mercury in Oriomo Papuan's Hair, Ecol. Food Nutr. 6:243.

Ohtsuka, R., and Suzuki, T. (eds.), 1990, "Population Ecology of Human Survival, Bioecological Studies of the Gidra in Papua, New Guinea", University of Tokyo Press, Tokyo.

O'Leary, M.H., 1981, Carbon isotope fractionation in plants, Phytochem., 20:553.

Park, R., and Epstein, S., 1961, Metabolic fractionation of $^{13}$C and $^{13}$C in plants, Plant Physiol., 36:133.

Rossi, L.C., Clemente, G.F., and Santaroni, G., 1976, Mercury and selenium distribution in a defined area and its population, Arch. Environ. Health, 36:160.

Shirabe, T., Eto, K., and Takeuchi, T., 1979, Identification of mercury in the brain of Minamata disease victims by electron microscopic X-ray microanalysis, Neurotoxicology, 1:349.

Suzuki, T., Watanabe, S., Hongo, T., Kawabe, T., Inaoka, T., Ohtsuka, R., and Akimichi, T., 1988, Mercury in scalp hair of Papuans in the Fly estuary, Papua, New Guinea, Asia-Pacific J. Publ. Health, 2:39.

Suzuki, T., 1988, Selenium: Its Role in Metal-Metal Interaction, in: "Environmental and Occupational Chemical Hazards, Proceedings Asia-Pacific Symposium on Environmental and Occupational Toxicology," K. Sumino, S. Iwai, H.P. Lee, C.N. Ong, and K. Saijoh, eds., 21-30, International Center for Medical Research, Kobe University School of Medicine, Kobe.

Suzuki, T., 1989, Human adaptability to environmental pollutants, in particular methylmercury, Sangyo-Igaku Rebyu, 2:25, (in Japanese).

Tohyama, C., Himeno, S., Watanabe, C., Suzuki T., and Morita, M., 1986, The relationship of the increased level of metallothionein with heavy metal levels in the tissue of the harbor seal (*Phoca vitulina*), Ecotoxicol. Environ. Safety, 12:85.

Ujioka, T., 1960, Analytical studies on methylmercury in animal organs and foodstuff, J. Kumamoto Med. Assoc., 34 (Suppl. 1):383. (in Japanese)

van Fleet, J.F., Boon, G.D., and Ferrans, V.J., 1981, Induction of lesions of selenium-vitamin E deficiency in weanling swine fed silver, cobalt, tellurium, zinc, cadmium and vanadium, Am. J. Vet. Res., 42:789.

World Health Organization, 1972. Evaluation of certian food additives and contaminants, mercury, lead and cadmium. WHO Tech. Rep. Ser. No. 505.

World Health Organization, IPCS International Programme on Chemical Safety, 1989, Environmental Health Criteria 86, Mercury-Environmental Aspects, WHO, Geneva.

Yoshinaga, J., Matsuo, N., Imai, H., Nakazawa, M., Suzuki, T., Morita, M., and Akagi, H., 1990, Interrelationship between the concentrations of some elements in the   organs of Japanese with special reference to selenium-heavy metal relations, Sci. Total  Environ., 91:127.

Yoshinaga, J., Minagawa, M., Suzuki, T., Ohtsuka, R., Kawabe, T., Hongo, T. Inaoka, T., and Akimichi, T. Carbon and nitrogen isotopic characterization for New Guinea   Foods, Ecol. Food Nutr., in press.

Yoshinaga, J., Suzuki, T., Hongo, T., Minagawa, M., Ohtsuka, R., Kawabe, T., Inaoka, T., and Akimichi, T. Mercury concentration correlates with nitrogen stable isotope ratio in animal food of Papuans. Ecotoxicol. Environ. Safety, in press.

Yoshinaga, J., Suzuki, T., Ohtsuka, R., Kawabe, T., Hongo, T., Imai, H., Inaoka, T., and Akimichi, T.  Dietary selenium intake of the Gidra, Papua, New Guinea. Ecol. Food. Nutr., in press.

APPENDIX I

## PROGRAM

### *WEDNESDAY AUGUST 1*

A.M.   Registration

### Opening Lecture
(Chairperson: T. Suzuki)

T. W. Clarkson          A history of mercury

*Break*

### Session 1:   Fate in the Environment

(Chairperson: N. Imura)

R. Hecky          Increased methylmercury contaminations in fish in
                  newly formed freshwater reservoirs
H. Akagi          Speciation of mercury in the environment

*Lunch*

### Session 2:   Disposition in the Body
(Chairperson: S. Halbach)

R. Doi            Individual difference of methylmercury metabolism in
                  animals and its significance in methylmercury toxicity
P. Kostyniak      Mechanisms of urinary excretion of methylmercury

*Break*

(Chairperson: P. Kostyniak)

A. Naganuma       Role of glutathione in mercury disposition
K. Hirayama       Mechanism of renal handling of methylmercury
B. Kargacin       Methods for decreasing $^{203}$Hg retention in relation to age
                  and route of exposure.
C. Tohyama        Immunohistochemical localization of metallothionein in organs
                  of rats treated with either cadmium, organic or inorganic
                  mercurials

*Break*

485

### Session 3: Biotransformation
(Chairperson: L. Magos)

S. Halbach — Intracellular utilization of hydrogenperoxide and oxidation of mercury vapor

H. Takahashi — Differential determination of ionizable and unionizable forms of inorganic mercury in tissues

*THURSDAY, AUGUST 2*

### Session 4: Molecular Mechanisms of Toxicity
(Chairperson: M. Costa)

T. Narahashi — Role of neuronal ion channels in mercury intoxication

M.A. Verity — Role of oxidative injury in the pathogenesis of methylmercury neurotoxicity

*Break*

(Chairperson: M.A. Verity)

S. Omata — Alterations in gene expression due to methylmercury in central and peripheral nervous tissues of the rat

K. Miura — Microtubules, a sensitive target of methylmercury cytotoxicity

M. Costa — DNA damage by mercury compounds

*Lunch*

### Session 5: Selenium as a Modifying Factor of Mercury Toxicity
(Chairperson: H. Takahashi)

N. Imura — Possible mechanism of detoxifying effect of selenium on the toxicity of mercury compounds

L. Magos — Overview on the protection given by selenium against mercurials

*Break*

### Session 6: Characteristic Toxicities
(Chairperson: T. Narahashi)

H. Endou — Effect of mercuric chloride on angiotensin II-induced $Ca^{++}$ transient in the proximal tubule of rats

B. H. Choi — Effects of methylmercury on the developing brain

M. Inouye — Experimental approaches to developmental toxicity of methyl mercury

*FRIDAY AUGUST 3*

**Session 6: (continued)**
(Chairperson: B.H. Choi)

T. Sato        Neuropathology of methylmercury intoxication
H. Satoh       Behavioral toxicology of mercury compounds

*Break*

H. Arito        Effect of methylmercury on sleep patterns in the rat
P. Druet       Effect of inorganic mercury on the immune system

*Lunch*

**Session 7: Clinical and Epidemiological Aspect**
(Chairperson: T.W. Clarkson)

S. Skerfving     Exposure to mercury compounds in the population
Y. Yamamura   Blood and urinary mercury levels as indicators in biological
                   monitoring of exposure to mercury vapor
A. Igata       Clinical and epidemiological features of Minamata Disease

*Break*

**Special Lecture and Closing Remarks**
(Chairperson: A. Igata)

T. Suzuki      Mercury in human ecology

# APPENDIX II

## PUBLICATIONS FROM ROCHESTER INTERNATIONAL CONFERENCES IN ENVIRONMENTAL TOXICITY

1968 Conference:    Chemical Fallout: Current Research on Persistent Pesticides,
(eds. Miller and Berg),
Charles C. Thomas Publishers, Inc., 1969.

1969 Conference:    Effects of Metals on Cells, Subcellular Elements, and Macromolecules,
(eds. Maniloff, Coleman and Miller),
Charles C. Thomas Publishers, Inc., 1970.

1970 Conference:    Assessment of Airborne Particles,
(eds. Mercer, Morrow and Stoeber),
Charles C. Thomas Publishers, Inc., 1971.

1971 Conference:    Mercury, Mercurials and Mercaptans,
(eds. Miller and Clarkson),
Charles C. Thomas Publishers, Inc., 1973.

1972 Conference:    Behavioral Toxicology,
(eds. Weiss and Laties),
Plenum Press, 1975.

1973 Conference:    Molecular and Environmental Aspects of Mutagenesis,
(eds. Prakash, Sherman, Miller, Lawrence and Taber),
Charles C. Thomas Publishers, Inc., 1973.

1974 Conference:    Fundamental and Applied Aspects of Non-Ionizing Radiation,
(eds. Michaelson, Miller, Magin and Carstensen),
Plenum Press, 1975.

1975 Conference:    Environmental Toxicity of Aquatic Radionuclides: Models and Mechanisms,
(eds. Miller and Stannard),
Ann Arbor Science Publications, 1976.

1976 Conference:    Membrane Toxicity,
(eds. Miller and Shamoo),
Plenum Press, 1977.

1977 Conference:    Environmental Pollutants: Detection and Measurement,
(eds. Toribara, Coleman, Dahneke and Feldman),
Plenum Press, 1978.

1978 Conference:    Neurotoxicity of the Visual System,
(eds. Merigan and Weiss),
Raven Press, 1980.

1979 Conference:    Polluted Rain,
(eds. Toribara, Miller and Morrow),
Plenum Press, 1980.

1980 Conference:    Measurement of Risks,
(eds. Berg and Maillie),
Plenum Press, 1981.

1981 Conference:    Induced Mutagenesis: Molecular Mechanisms and their Implications
for Environmental Protection,
(ed. Lawrence),
Plenum Press, 1983.

1982 Conference:    Reproductive and Developmental Toxicity of Metals,
(eds. Clarkson, Nordberg and Sager),
Plenum Press, 1983.

1984 Conference:    The Cytoskeleton: A Target for Toxic Agents,
(eds. Clarkson, Sager and Syversen),
Plenum Press, 1986.

1986 Conference:    Biological Monitoring of Toxic Metals,
(eds. Clarkson, Friberg, Nordberg and Sager),
Plenum Publishing Corp., New York, 1988.